AF819233

Methods in Molecular Biology

Series Editor
John M. Walker
School of Life and Medical Sciences
University of Hertfordshire
Hatfield, Hertfordshire, UK

For further volumes:
http://www.springer.com/series/7651

For over 35 years, biological scientists have come to rely on the research protocols and methodologies in the critically acclaimed *Methods in Molecular Biology* series. The series was the first to introduce the step-by-step protocols approach that has become the standard in all biomedical protocol publishing. Each protocol is provided in readily-reproducible step-by-step fashion, opening with an introductory overview, a list of the materials and reagents needed to complete the experiment, and followed by a detailed procedure that is supported with a helpful notes section offering tips and tricks of the trade as well as troubleshooting advice. These hallmark features were introduced by series editor Dr. John Walker and constitute the key ingredient in each and every volume of the *Methods in Molecular Biology* series. Tested and trusted, comprehensive and reliable, all protocols from the series are indexed in PubMed.

Mass Spectrometry Imaging of Small Molecules

Methods and Protocols

Edited by

Young-Jin Lee

Department of Chemistry, Iowa State University, Ames, IA, USA

Humana Press

Editor
Young-Jin Lee
Department of Chemistry
Iowa State University
Ames, IA, USA

ISSN 1064-3745 ISSN 1940-6029 (electronic)
Methods in Molecular Biology
ISBN 978-1-0716-2029-8 ISBN 978-1-0716-2030-4 (eBook)
https://doi.org/10.1007/978-1-0716-2030-4

© The Editor(s) (if applicable) and The Author(s), under exclusive license to Springer Science+Business Media, LLC, part of Springer Nature 2022
This work is subject to copyright. All rights are solely and exclusively licensed by the Publisher, whether the whole or part of the material is concerned, specifically the rights of translation, reprinting, reuse of illustrations, recitation, broadcasting, reproduction on microfilms or in any other physical way, and transmission or information storage and retrieval, electronic adaptation, computer software, or by similar or dissimilar methodology now known or hereafter developed.
The use of general descriptive names, registered names, trademarks, service marks, etc. in this publication does not imply, even in the absence of a specific statement, that such names are exempt from the relevant protective laws and regulations and therefore free for general use.
The publisher, the authors and the editors are safe to assume that the advice and information in this book are believed to be true and accurate at the date of publication. Neither the publisher nor the authors or the editors give a warranty, expressed or implied, with respect to the material contained herein or for any errors or omissions that may have been made. The publisher remains neutral with regard to jurisdictional claims in published maps and institutional affiliations.

This Humana imprint is published by the registered company Springer Science+Business Media, LLC, part of Springer Nature.
The registered company address is: 1 New York Plaza, New York, NY 10004, U.S.A.

Preface

The last decade has seen the evolution of mass spectrometry imaging (MSI) from an emerging technology to a matured and enabling technology that is commonly adopted for various biological and biomedical applications. Visualizing the localization of proteins was the early focus of this technology; however, the most popular application is now on small molecules including metabolites, lipids, and drugs that were once regarded as challenging in the early days of the technique. A method book in this subject was published in the Methods in Molecular Biology series six years ago in 2015, edited by Lin He. In the last few years, there has been a tremendous amount of new development in the field This new book, with the same title, is intended to bridge the gap by addressing many of the recent and exciting advancements.

This book is organized into five themes that represent recent trends in MSI. Part I is comprised of innovative instrumental methods that allow for technological breakthrough. Localization of structural isomers can be determined by MS/MS imaging; however, some positive ions do not produce useful structural information in MS/MS. Chapter 1 describes how to convert positive ions to negative charges through charge inversion reactions, allowing for structural characterization of these challenging ions. MALDI-2, laser-induced post-ionization in matrix-assisted laser desorption ionization (MALDI), is a new technique that can dramatically enhance ion signals for many classes of small molecules. Chapter 2 illustrates how to perform MALDI-2 MSI in both top-illumination and transmission mode. MSI of live cells is an inherently challenging task. Chapter 3 tackles this issue with an atmospheric pressure MSI method using laser ablation and atmospheric plasma, as well as secondary ion mass spectrometry (SIMS) imaging of in-solution capped cells with single-layer graphene. Single cell metabolomics has become possible with high-spatial and high-mass resolution MSI. For this purpose, Chapter 4 combines fiber-based laser ablation electrospray ionization (f-LAESI) with 21 Tesla Fourier Transform Ion Cyclotron Resonance (FTICR)-MS. Time-of-flight SIMS has the highest spatial resolution among MSI techniques but has had limitations in MS/MS capability until recently. Chapter 5 explains how to perform MS/MS of unknown ions by combining ToF-SIMS and hybrid SIMS for consecutive tissue sections.

MALDI is the most popular ionization method in MSI; however, the use of an organic matrix significantly limits its application to small molecules due to mass interference or ion suppression. Part II addresses a collection of alternative matrices, sometimes dubbed by various acronyms, that have become popular in the last decade to avoid the issues inherent to organic matrices. Chapters 6–9 describe how to synthesize or use alternative matrices for MSI including silicon nanopost arrays (NAPA), metal-assisted LDI, nanostructure initiator mass spectrometry (NIMS) with fluorinated gold nanoparticles, and nanoparticle-assisted laser desorption ionization (Nano-PALDI), respectively.

There are many challenges that hinder quantification with MS imaging. Part III encompasses new methods to address this issue of quantitative MSI. Chapter 10 details how to perform nanosecond photochemical reaction (nsPCR) on tissue during the MALDI process, which showed dramatic improvement for the ion signals of amine-containing compounds and their quantifications. Chapter 11 overcomes common issues in MALDI quantification using a microstructured PDMS stamp to precisely control the matrix deposition and the analyte extraction area. Extraction of small molecules followed by liquid chromatography (LC)-MS is becoming popular for quantitative MSI as described in Chapters 12 and 13 using dropletProbe MS and liquid microjunction surface sampling (LMJSS), respectively.

Part IV is composed of new types of applications in MSI. Chapter 14 adopts three different on-tissue chemical derivatization reactions to enhance metabolite coverage in MALDI-MSI. Chapter 15 extended the application of MSI for glycogen imaging using on-tissue enzyme digestion. Chapter 16 describes the multimodal imaging of free-floating brain sections using MSI and immunohistochemistry. In Chapter 17, silver ions are doped into nanospray desorption electrospray ionization (nano-DESI) to increase the specificity and sensitivity of double bond containing analytes.

Data analysis is becoming more and more important in MSI, particularly to understand the dataset beyond the visualization of a few targeted metabolites. As a sole chapter in Part V, Chapter 18 introduces multiple statistical methods for MSI including Multivariate Curve Resolution Alternating Square (MCR-ALS), Cluster Large Applications (CLARA) and Density-Based Spatial Clustering of Applications with Noise (DBSCAN) as unsupervised algorithms, Random Forest as a supervised machine learning, and Image Fusion to artificially enhance the spatial resolution of MSI by utilizing higher resolution optical images.

This book does not cover all the recent developments in this subject, which is simply impossible with new approaches and methods being developed and reported every day, but I am proud that the work assembled here represents many of the recent key developments. I hope this book enables newcomers to grasp the overall scope of the technical trend and the practical knowhow to adopt some of the methodology. This book will promote new ideas and encourage current practitioners to expand their research into new directions.

Lastly, I would like to thank all the contributors of this book. My appreciation is not just pertaining to this publication but also their dedication to the community. I am grateful for this opportunity to work with many excellent scientists in the field and for being able to accomplish my role as a coordinator.

Ames, IA, USA ***Young-Jin Lee***

Contents

Contributors

CHRISTOPHER R. ANDERTON • *Environmental Molecular Sciences Laboratory and Biological Sciences Division, Pacific Northwest National Laboratory, Richland, WA, USA*
TANJA BIEN • *Institute of Hygiene, University of Münster, Münster, Germany*
ALAIN BRUNELLE • *Sorbonne Université, CNRS, Laboratoire d'Archéologie Moléculaire et Structurale, LAMS, Paris, France*
MARK T. CANCILLA • *Department of Preclinical Development, Merck & Co., Inc., Kenilworth, NJ, USA*
PIERRE CHAURAND • *Department of Chemistry, Université de Montréal, Montreal, QC, Canada*
BINGMING CHEN • *Department of Preclinical Development, Merck & Co., Inc., Kenilworth, NJ, USA*
JOSEF CVAČKA • *Institute of Organic Chemistry and Biochemistry of the Czech Academy of Sciences, Prague, Czech Republic*
KLAUS DREISEWERD • *Institute of Hygiene, University of Münster, Münster, Germany*
MARIA EMILIA DUEÑAS • *Department of Chemistry, Iowa State University, Ames, IA, USA*
TREVOR T. FORSMAN • *Department of Chemistry, Iowa State University, Ames, IA, USA*
FRÉDÉRIC FOURNELLE • *Department of Chemistry, Université de Montréal, Montreal, QC, Canada*
TARA R. HAWKINSON • *Department of Neuroscience, University of Kentucky College of Medicine, Lexington, KY, USA*
VILMOS KERTESZ • *Bioanalytical Mass Spectrometry Group, Biosciences Division, Oak Ridge National Laboratory, Oak Ridge, TN, USA*
JAE YOUNG KIM • *School of Electronic and Electrical Engineering, College of IT Engineering, Kyungpook National University, Daegu, Republic of Korea*
INGELA LANEKOFF • *Department of Chemistry – BMC, Uppsala University, Uppsala, Sweden*
EVAN LARSON • *Department of Chemistry, Iowa State University, Ames, IA, USA*
CHANG YOUNG LEE • *School of Energy and Chemical Engineering, Ulsan National Institute of Science and Technology (UNIST), Ulsan, South Korea*
YOUNG-JIN LEE • *Department of Chemistry, Iowa State University, Ames, IA, USA*
GONGYU LI • *School of Pharmacy, University of Wisconsin-Madison, Madison, WI, USA; Research Center for Analytical Science and Tianjin Key Laboratory of Biosensing and Molecular Recognition, College of Chemistry, Nankai University, Tianjin, China*
LINGJUN LI • *School of Pharmacy, University of Wisconsin-Madison, Madison, WI, USA*
JOHAN LILLJA • *Department of Chemistry – BMC, Uppsala University, Uppsala, Sweden*
HEEJIN LIM • *Department of New Biology, Daegu Gyeongbuk Institute of Science and Technology (DGIST), Daegu, Republic of Korea*
YUAN LIU • *School of Pharmacy, University of Wisconsin-Madison, Madison, WI, USA*
HONGMEI LU • *College of Chemistry and Chemical Engineering, Central South University, Changsha, Hunan, P. R. China*
LENKA MALETÍNSKÁ • *Institute of Organic Chemistry and Biochemistry of the Czech Academy of Sciences, Prague, Czech Republic*

Dae Won Moon • *Department of New Biology, Daegu Gyeongbuk Institute of Science and Technology (DGIST), Daegu, Republic of Korea; School of Undergraduate Studies, Daegu Gyeongbuk Institute of Science and Technology (DGIST), Daegu, Republic of Korea*
Kelly C. O'Neill • *Department of Chemistry, Iowa State University, Ames, IA, USA*
Amelia Palermo • *Department of Molecular and Medical Pharmacology, University of California, Los Angeles, CA, USA*
Sanghwan Park • *School of Energy and Chemical Engineering, Ulsan National Institute of Science and Technology (UNIST), Ulsan, South Korea*
Boone M. Prentice • *Department of Chemistry, University of Florida, Gainesville, FL, USA*
Laith Z. Samarah • *Department of Chemistry, George Washington University, Washington, DC, USA*
Katsuhiro Shiono • *Department of Bioscience and Biotechnology, Fukui Prefectural University, Eiheiji, Japan*
Jens Soltwisch • *Institute of Hygiene, University of Münster, Münster, Germany*
Štěpán Strnad • *Institute of Organic Chemistry and Biochemistry of the Czech Academy of Sciences, Prague, Czech Republic*
Veronika Strnadová • *Institute of Organic Chemistry and Biochemistry of the Czech Academy of Sciences, Prague, Czech Republic*
Ramon C. Sun • *Department of Neuroscience, University of Kentucky College of Medicine, Lexington, KY, USA; Markey Cancer Center, Lexington, KY, USA*
David Sýkora • *University of Chemistry and Technology Prague, Prague, Czech Republic*
Shu Taira • *Faculty of Food and Agricultural Sciences, Fukushima University, Fukushima, Japan*
Xiang Tian • *Department of Chemistry and Biochemistry, University of Oklahoma, Norman, OK, USA; Dynamic Omics, Center of Genomics Research (CGR), R&D, AstraZeneca, Gaithersburg, MD, USA*
Sebastiaan Van Nuffel • *Faculty of Science and Engineering, Maastricht University, Maastricht, The Netherlands; Maastricht MultiModal Molecular Imaging Institute (M4i), Maastricht University, Maastricht, The Netherlands*
Marissa Vavrek • *Department of Preclinical Development, Merck & Co., Inc., Kenilworth, NJ, USA*
Akos Vertes • *Department of Chemistry, George Washington University, Washington, DC, USA*
Vladimír Vrkoslav • *Institute of Organic Chemistry and Biochemistry of the Czech Academy of Sciences, Prague, Czech Republic*
Qian Wu • *College of Chemistry and Chemical Engineering, Central South University, Changsha, Hunan, P. R. China*
Zhibo Yang • *Department of Chemistry and Biochemistry, University of Oklahoma, Norman, OK, USA*
Zhu Zou • *Department of Chemistry and Biochemistry, University of Oklahoma, Norman, OK, USA*

Part I

Innovations in Instrumental Methods for Mass Spectrometry Imaging

Chapter 1

Gas-Phase Ion–Ion Reactions for Lipid Identification in Biological Tissue Sections

Boone M. Prentice

Abstract

The unambiguous identification of isobaric (i.e., same nominal mass) and isomeric (i.e., same exact mass) lipids remains a challenging yet vital aspect of imaging mass spectrometry (IMS) workflows. This chapter presents a methodology for the preparation of biological tissue samples and the use of a hybrid mass spectrometer to perform gas-phase charge inversion ion/ion reactions for improved lipid identification. This gas-phase ion/ion reaction method provides lipid structural information beyond what can be obtained via conventional tandem mass spectrometry (MS/MS) experiments. While this procedure is described here for the identification of phosphatidylcholine (PC) analyte cations using 1,4-phenylenedipropionic acid reagent dianions, it can readily be generalized to perform a diverse array of ion/ion reaction chemistries.

Key words Lipid identification, Gas-phase, Ion/ion reactions, Imaging mass spectrometry

1 Introduction

Imaging mass spectrometry (IMS) has emerged as a powerful technology for visualizing biochemical processes directly in tissue specimens by combining the molecular specificity of mass spectrometry with the spatial fidelity of a microscopic imaging approach. Despite the high molecular specificity of the mass spectrometer, the complex chemical milieu present in biological tissues sections results in the presence of many compounds of similar mass, at both nominal (i.e., isobaric) and exact (i.e., isomeric) m/z ratios [1]. The high level of structural homology of lipids and metabolites especially complicates the analysis of these compounds by mass spectrometry [2, 3]. The failure to differentiate these compounds from one another in a mass spectrum results in the spatial distribution of a single peak no longer representing the spatial distribution of a single compound in an imaging mass spectrometry experiment, but instead representing the confluence of several compounds that

Young-Jin Lee (ed.), *Mass Spectrometry Imaging of Small Molecules: Methods and Protocols*,
Methods in Molecular Biology, vol. 2437, https://doi.org/10.1007/978-1-0716-2030-4_1,
© The Author(s), under exclusive license to Springer Science+Business Media, LLC, part of Springer Nature 2022

have overlapping *m/z* values. Each unresolved ion could have a unique spatial distribution, so these types of spectral interferences result in inaccurate depictions of molecular distributions.

Chromatographic separations such as liquid chromatography (LC) and gas chromatography (GC) coupled to tandem mass spectrometry (MS/MS or MS^n) are commonly employed to separate and identify components in a complex mixture. However, these technologies are often incompatible with the rapid spectral acquisition rates and direct surface sampling methods employed in imaging mass spectrometry [4–6]. Additionally, conventional tandem mass spectrometry ion dissociation techniques, such as collision induced dissociation (CID), are often insufficient to unambiguously identify biomolecules [7]. In the interest of broadening the range of ion dissociation chemistries accessible during tandem mass spectrometry experiments, we have developed gas-phase ion/ion reaction instrumentation and methods on a hybrid quadrupole–hexapole collision cell-Fourier transform ion cyclotron resonance (QhFTICR) mass spectrometer (Fig. 1) [8]. This approach decouples the ionization method from the nature of the ion to be interrogated and enables the ion type to be derivatized to any desirable form. This is achieved in the gas-phase following the ionization event, allowing subsequent MS/MS experiments to provide structural information independent of the initial ion type generated from the tissue surface. These gas-phase reactions offer a number of benefits that make them ideally suited for use in imaging mass spectrometry experiments, including their rapid reaction rates (i.e., reaction times of less than 1 s are common) and the ability to readily compare underivatized and derivatized sample types, since the initial sample is never altered [9, 10]. Unlike analogous condensed-phase tissue derivatization procedures, which can lead to complex mixtures due to side reactions and incomplete reactions, a gas-phase approach eliminates physical manipulation of the tissue. This minimizes the possibility of tissue deformation and analyte delocalization that can occur during sample preparation protocols [11].

Herein, we present (1) a sample preparation workflow to perform positive ion mode lipid imaging mass spectrometry on biological tissue sections, (2) instrumental methodology for performing charge inversion ion/ion reactions to determine the fatty acyl chain identities of phosphatidylcholine (PC) ions sampled directly from biological tissues, and (3) an experimental design to determine the relative levels of *sn*-positional isomers detected in tissue. While this method has been developed for the reaction of PC analyte cations with 1,4-phenylenedipropionic acid (PDPA) reagent dianions, the procedure can be adapted for alternative reaction chemistries using any desired analyte and reagent. This reaction is termed a "charge inversion" experiment because it involves the reaction of singly charged lipid cations produced via

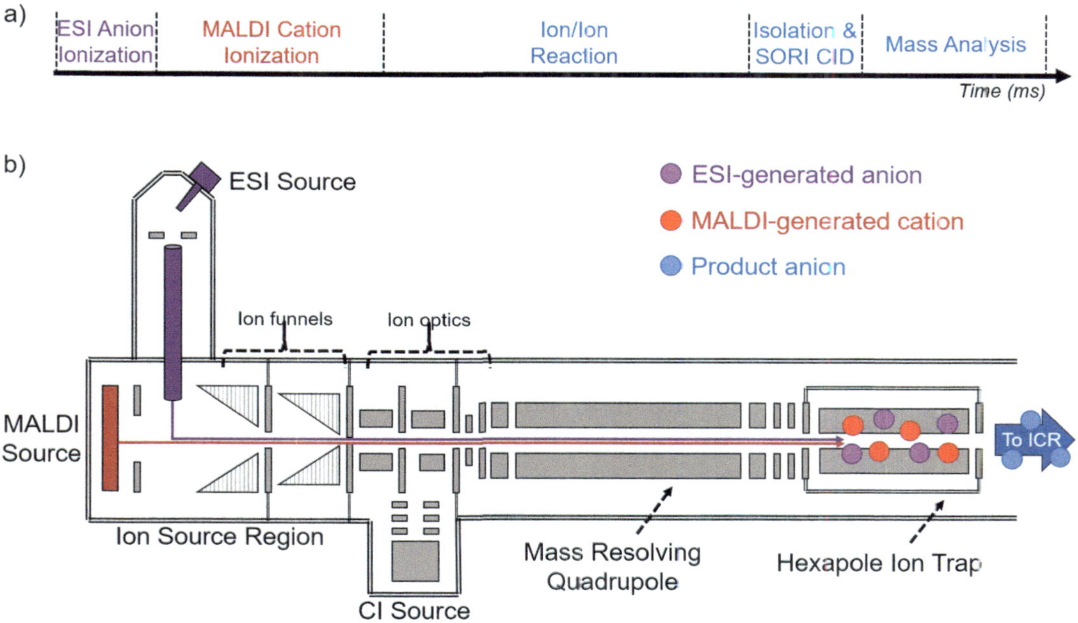

Fig. 1 (**a**) Instrument timing diagram showing reagent and analyte ion injection (50 ms and 500 ms, respectively), ion/ion reaction (1.5 s), product ion isolation and SORI CID (~10 ms), and mass analysis (~0.75 s) in the ICR cell. The entire ionization and reaction process takes ~2 to 3 s. (**b**) A simplified instrument diagram depicts the ion/ion reaction operation on the solariX instrument platform. This chemistry is enabled by sequentially injecting ESI-generated anions followed by MALDI-generated cations through a mass resolving quadrupole into a hexapole ion trap. Following a defined mutual storage reaction period, the product ions are transferred to the ICR for isolation, SORI CID, and mass analysis. The transfer optics from the hexapole ion trap to the ICR cell and the ICR electrodes have been omitted from the instrument diagram for simplicity. Note that the timing and instrument diagrams are not drawn to scale. (Reprinted with permission Specker JT, Orden SLV, Ridgeway ME, Prentice BM (2020) Identification of phosphatidylcholine isomers in imaging mass spectrometry using gas-phase charge inversion ion/ion reactions. Anal Chem 92:13192–13201 [8]. Copyright 2020 American Chemical Society)

MALDI with doubly charged reagent anions produced via electrospray ionization (ESI) to produce singly charged, charge-inverted analyte anions [12, 13]:

$$[PC + H]^{+} + [PDPA - 2H]^{2-} \rightarrow [PC + PDPA - H]^{-} \quad (1)$$

This product ion readily dissociates to give a demethylated PC anion:

$$[PC + PDPA - H]^{-} \rightarrow [PC - CH_3]^{-} + PDPA - CH_3 \text{ (neutral loss)} \quad (2)$$

which can be subjected to sustained off resonance irradiation (SORI) CID to produce fragment ions that are diagnostic for the fatty acid tail identities and positions:

$$[PC - CH_3]^- \rightarrow [R_1CH_2COO - H]^- \text{ or } [R_2CH_2COO - H]^- \quad (3)$$

The following protocol describes the experimental design and workflow for performing lipid identification directly from tissue using ion/ion reactions. A more detailed description of the instrumentation hardware can be found in a recent research article [8].

2 Materials

2.1 Tissue Cryosectioning

1. Cryomicrotome and accessories (cryomicrotome blade, sample specimen disk, antiroll plate).
2. Tissue samples.
3. Optimal cutting temperature (OCT) polymer compound.
4. Sample handling tools (forceps and small artist's paint brushes).
5. Teflon-coated glass slide.
6. Sample substrate (indium tin oxide-coated glass slides or common glass microscope slides).
7. Microscope slide box.
8. Styrofoam cooler with dry ice.
9. Ethanol.
10. Kimwipes.
11. Desiccator.

2.2 Matrix Application by Sublimation

1. Sublimation assembly (consisting of a rotary vane mechanical pump, valves, tubing, a vacuum trap, a vacuum measurement gauge, glass sublimation apparatus with a flat bottom, a hot plate, a baking dish, sand, a thermometer, and support stands and clamps for stabilization during operation).
2. 1,5-diaminonapthalene (DAN) (or other appropriate MALDI matrix).
3. Thermally conductive adhesive tape.
4. Ice.
5. Timer.
6. Reagent grade ethanol.
7. Kimwipes.

2.3 Solution Preparation

1. 1,4-phenylenedipropionic acid (PDPA) reagent (1 mg/mL in 49.5:49.5:1 water–acetonitrile–ammonium hydroxide by volume).
2. 1,5-diaminonapthalene (DAN) MALDI matrix (20 mg/mL in 70:30:0.1 acetonitrile–water–trifluoroacetic acid by volume).

3. Synthetic standards of lipid isomers (10 mM in methanol).
4. HPLC-grade water.
5. HPLC-grade acetonitrile.
6. Trifluoroacetic acid.
7. Ammonium hydroxide.
8. Hematoxylin and eosin solutions (H&E).

2.4 Instrumentation and Software

1. 7 T solariX XR FT-ICR mass spectrometer equipped with an Apollo II dual MALDI/ESI source and ETD (Bruker Daltonics, Billerica, MA, USA).
2. flexImaging 5.0 (Bruker Daltonics, Billerica, MA, USA).
3. DataAnalysis (Bruker Daltonics, Billerica, MA, USA).
4. Microsoft Excel or equivalent.
5. MTP Slide Adapter.
6. MTP AnchorChip MALDI.
7. Flatbed scanner.
8. Optical microscope (Zeiss Axioscan Z1 brightfield slide scanner [Carl Zeiss Microscopy LLC, White Plains, NY, USA] or equivalent).

3 Methods

3.1 Tissue Sectioning

Sample preparation is a critical aspect of any imaging mass spectrometry workflow. The quality of the resulting ion images depends on many steps, including tissue procurement and storage, tissue sectioning, and matrix application. Ensure that proper personal protective equipment is used (e.g., lab coat, lab goggles/glasses, gloves, cryostat sleeves).

1. Transport the tissues from cold storage to the cryochamber using a Styrofoam cooler filled with dry ice and allow roughly 15 min for the tissue to thermally equilibrate (*see* **Note 1**).
2. Mount the tissue to the specimen disk using a dime-sized area of OCT (use more or less OCT as required for larger or smaller tissues, respectively) (*see* **Note 2**).
3. Install the cryomicrotome blade, rinsing with ethanol and water prior to installation to remove any residual cartridge oil.
4. Trim the tissue to the desired organ depth, ensuring a flat and even cutting plane. Once the desired depth is reached, set the sectioning thickness (typically 8–20 μm).
5. Section the tissue at the desired thickness. Use artist's paintbrushes to gently position the tissue onto the Teflon-coated slide (*see* **Note 3**). Compress the MALDI microscope slide on top of the tissue as the tissue is resting on the Teflon-coated

slide (*see* **Note 4**). The tissue will adhere to the MALDI slide when the slide is removed, after which it can be thaw-mounted (*see* **Notes 5** and **6**). Place at least two serial sections on each slide via this process.

6. Store the samples in a −80 °C freezer until use. Warm the samples in a desiccator for 30 min prior to matrix application to prevent condensation buildup on the tissue surface.
7. Clean the cryomicrotome with ethanol or other appropriate disinfectant after use, taking care to properly handle and dispose of sharps and biohazardous materials.

3.2 Matrix Application

MALDI matrix selection is guided by several experimental parameters, including the analytes of interest, instrument ion polarity, and application method [14–17]. A standard matrix application method is described here using a custom-built sublimation apparatus for subsequent positive ion mode lipid imaging mass spectrometry [18]. A more reproducible matrix application procedure can be achieved using a robotic spraying system. Ensure that proper personal protective equipment is used (e.g., lab coat, lab goggles/ glasses, gloves) and that matrix application is performed in a chemical fume hood.

1. Prepare the sublimation assembly by turning on the vacuum pump, closing all valves, and adding crushed dry ice and ethanol to the bottom of the dewar containing the cold trap. Adjust the sand bath temperature to the desired temperature (*see* **Note 7**).
2. Prepare the matrix by adding ~200 mg of DAN matrix to the condenser sleeve, enough to evenly coat the bottom of the device [19].
3. Prepare the microscope slide by mounting it onto the bottom of the condenser using conductive adhesive tape to ensure thermal equilibration during the sublimation process. The side with the tissue section should be inverted and facing downward over the matrix on the bottom of the chamber when the condenser is placed in the sleeve. Assemble the apparatus and then evacuate the chamber by opening the valve to the pump (*see* **Note 8**).
4. Once the chamber has reached adequate vacuum, add an ice water slurry to the condenser sleeve. Lower the assembled apparatus onto the heated sand, ensuring that the bottom of the condenser is level and evenly submerged in the heated sand (*see* **Note 9**).
5. After the desired amount of matrix has been added, remove the assembled apparatus from the heated sand (*see* **Note 10**). Empty the slurry and allow the apparatus to equilibrate to room temperature before releasing the vacuum and removing the sample. Clean the apparatus using ethanol.

3.3 Lipid Imaging Mass Spectrometry

1. Prepare the matrix-coated tissue sample slide for imaging by first drawing fiducial markers using a silver metallic pen for the background and a fine pointed black Sharpie to create cross-hairs at four corners bracketing the sample. Place the slide in an MTP Slide Adapter and record an image of the marked plate using a flatbed scanner.
2. Introduce the MTP Slide Adapter containing the sample into the mass spectrometer. Optimize and save an ftmsControl acquisition method for positive ion mode PC analysis by tuning MALDI laser parameters, ion optic settings, and ICR cell analysis conditions (*see* **Note 11**). Ensure that the method is m/z calibrated. Matrix peaks, known endogenous compounds, and/or external standards are commonly used for m/z calibration.
3. Set up an image acquisition method using flexImaging. This process includes first "teaching" the fiducial marks from the instrument's optical camera in ftmsControl to the scanned sample image uploaded to flexImaging. Then select the appropriate ftmsControl acquisition method, select the desired step size (i.e., spatial resolution), and specify the areas to be imaged using the polygon region measurement tool.
4. Acquire the image and view the results using flexImaging. Ions of interest can then be selected for identification using procedures outlined below.

3.4 Ion/Ion Reactions Using Lipids Generated from Tissue

PC cations are tentatively identified using sum-composition nomenclature (expressed as total carbon:double bond [TC:DB]) via high resolution accurate mass (HRAM) measurements (*see* **Note 12**) [20–23]. However, identification of the fatty acyl tails present in the PC requires tandem mass spectrometry [24] and can be performed using the gas-phase ion/ion reactions discussed below (*see* **Note 13**).

1. PC cations of interest from the acquired ion image are chosen for identification via gas-phase charge inversion ion/ion reactions (Fig. 2a). MALDI-generated PC cations will be generated from regions of interest in the serial tissue section (Fig. 2b). The "position sample carrier" function in flexImaging can be used to accurately position the MALDI laser in ftmsControl.
2. Load the PDPA reagent sample solution into the ESI syringe and ensure reagent ion detection in negative ion mode (*see* **Note 14**).
3. The ion/ion reaction between the MALDI-generated PC cations and ESI-generated PDPA dianions is optimized by tuning various instrument ion optics (*see* **Note 15**). Ion/ion reactions can be integrated on solariX MS systems equipped

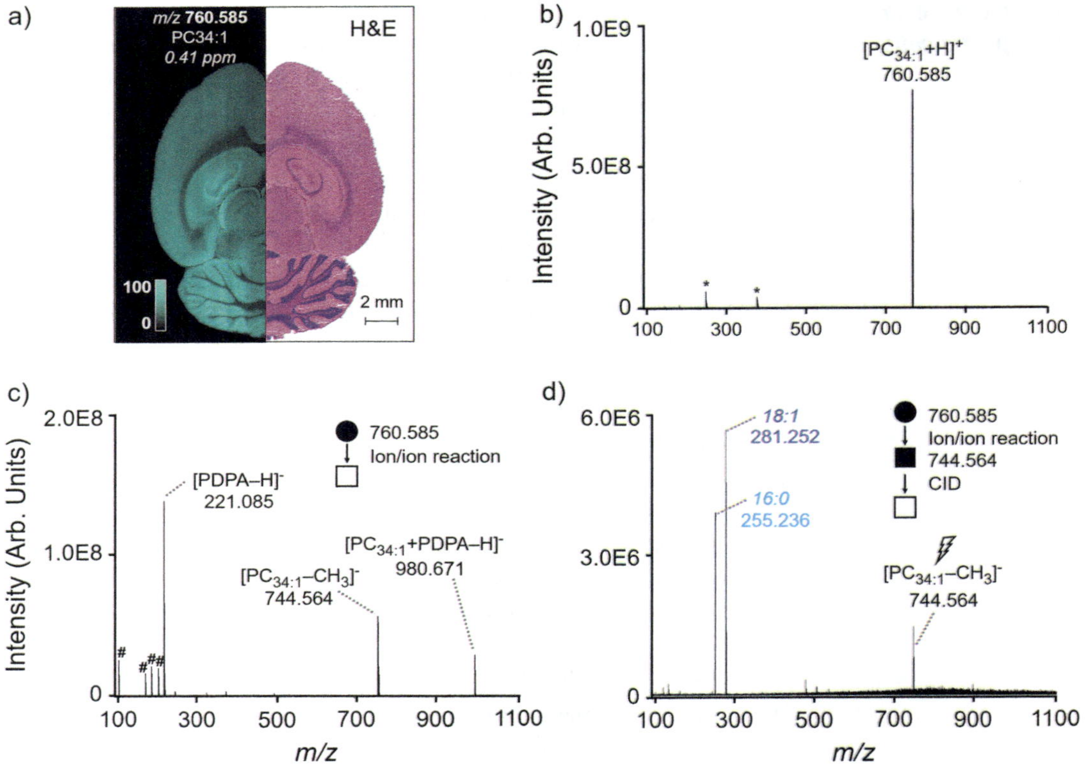

Fig. 2 (**a**) Ion image for *m/z* 760.585 in the rat brain, tentatively identified as $PC_{34:1}$ by an accurate mass measurement. A PDPA charge inversion ion/ion reaction of (**b**) $PC_{34:1}$ generated directly from the tissue surface (**c**) produces a PC/PDPA anionic complex and a demethylated PC anion. (**d**) Ion isolation of the demethylated PC anion followed by SORI CID produces fragment ions diagnostic for the 16:0 and 18:1 fatty acid tails in the lipid, allowing for the identification as $PC_{16:0\_18:1}$. Note that circles indicate positive ion mode analysis, squares indicate negative ion mode analysis, number signs denote fragment ions related to PDPA, asterisks denote harmonics/electronic noise, and the lightning bolt is used to denote the ion subjected to CID. All spectra are single scan measurements. (Reprinted with permission Specker JT, Orden SLV, Ridgeway ME, Prentice BM (2020) Identification of phosphatidylcholine isomers in imaging mass spectrometry using gas-phase charge inversion ion/ion reactions. Anal Chem 92:13192–13201 [8]. Copyright 2020 American Chemical Society)

with electron transfer dissociation (ETD) [25] by modifying the pulse program and ETD-binding rules, as described in our recent publication (*see* **Note 16**) [8]. Other gas-phase ion/ion reactions can be explored for different purposes and different reactants using this same instrumentation (*see* **Note 17**).

4. After optimization of the ion/ion reaction, a demethylated PC product anion should be detected (Fig. 2c) (*see* **Note 18**).
5. The desired demethylated PC product anion is then isolated in the ICR cell (*see* **Note 19**) and subjected to SORI CID (Fig. 2d) (*see* **Note 20**).

6. Following optimization of SORI CID conditions to observe the *sn*-1 and *sn*-2 fatty acid fragment anions, replicate measurements ($n = 10$ is common) should be made in desired regions of the tissue. These fragment ions allow for the identification of the fatty acyl groups in the original PC (*see* **Note 21**). In order to determine the *sn*-positions of the fatty acyl chains, comparison with synthetic lipid standards is necessary, as outlined below.
7. After measurements are completed, tissues can be stained using hematoxylin and eosin (H&E) and scanned using an optical microscope to validate morphological features in the tissue (Fig. 2a) [26].

3.5 Ion/Ion Reactions Using Lipid Standards

The *sn*-positions of fatty acid tails in a glycerophospholipid can typically be determined when performing CID on a deprotonated ion type by examining the relative intensities of the fatty acid fragment anions [24]. However, this is only true when the precursor glycerophospholipid is isomerically pure (*see* **Note 22**). PCs measured in the complex chemical environment of biological tissue samples are likely a mixture of *sn*-positional isomers [27–29]. Thus, a comparison with synthetic lipid standards is necessary to accurately determine the relative contribution of each isomer (*see* **Note 23**).

1. Prepare 10 mM solutions of lipid standards for each isomer in the identified lipid. Prepare samples for a calibration curve containing varying concentrations of each isomer using defined mixtures of the lipid standards (*see* **Note 24**). For example, manually spotted mixtures of $PC_{16:0/18:1}$ and $PC_{18:1/16:0}$ standards containing varying fractional amounts of each isomer are prepared for $PC_{34:1}$.
2. Mix the lipid standard solutions 1:1 (v/v) with aliquots of the DAN MALDI matrix solution. Use the dried-droplet method to deposit 1 μL aliquots of this mixture onto an MTP AnchorChip MALDI target.
3. The same ion/ion reaction and SORI CID method employed for on-tissue measurements should be used to interrogate the manually spotted calibration curve samples. Replicate measurements ($n = 10$) of each calibration point should be made. Data is then visualized in DataAnalysis.
4. The intensity of the 18:1 fatty acid fragment ion relative to the summed intensities of both the 16:0 and 18:1 fatty acid fragment ions will have a linear relationship with lipid isomer content (Fig. 3a). This calibration curve can be constructed in Excel and then used to determine the relative fraction of each $PC_{16:0\_18:1}$ isomer in different regions of the tissue (Fig. 3b, c) (*see* **Note 25**).

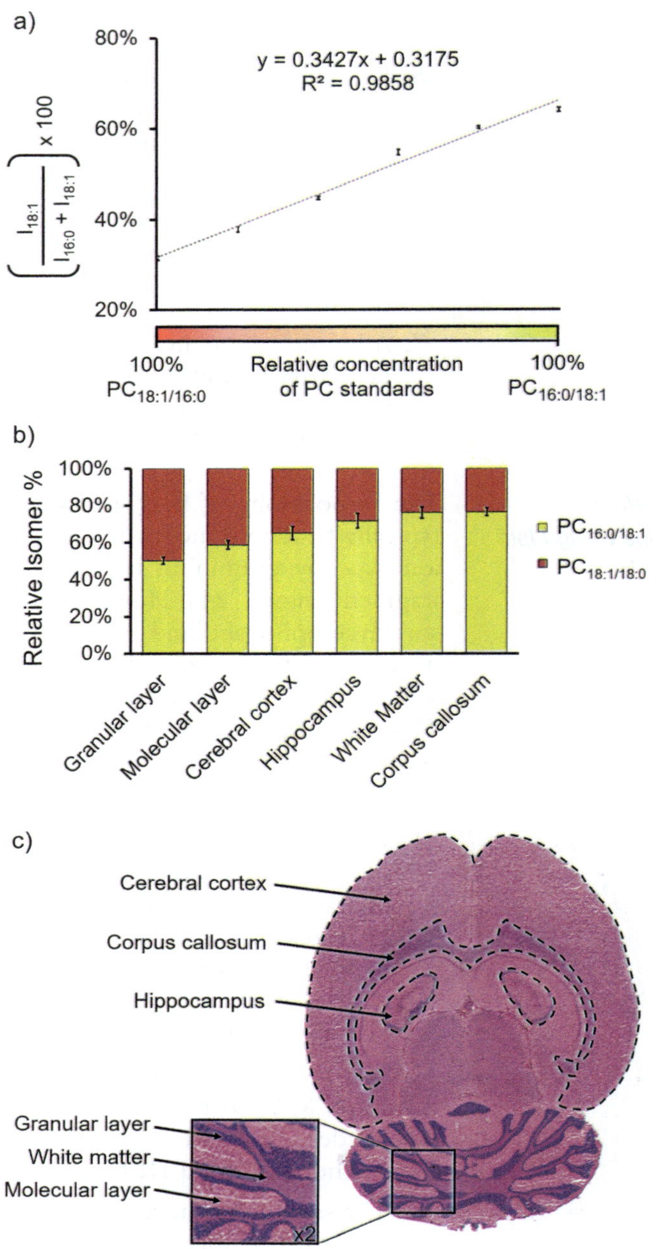

Fig. 3 (**a**) Mixtures of $PC_{16:0/18:1}$ and $PC_{18:1/16:0}$ lipid standards are used to calibrate the 18:1/(18:1 + 16:0) fragment ion intensity ratio obtained via CID of the $[PC_{34:1} - CH_3]^-$ anion produced via the PC/PDPA charge inversion reaction. (**b**) The ion/ion reaction method reveals that the relative isomeric content of $PC_{34:1}$ varies throughout rat brain. Each measurement in the standard curve and each region of the brain is the average of 10 spectra. (**c**) H&E-stained tissue showing regions of interest. (Reprinted with permission Specker JT, Orden SLV, Ridgeway ME, Prentice BM (2020) Identification of phosphatidylcholine isomers in imaging mass spectrometry using gas-phase charge inversion ion/ion reactions. Anal Chem 92:13192–13201 [8]. Copyright 2020 American Chemical Society)

4 Notes

1. Specimens stored in a −80 °C freezer need time to warm up to roughly −25 °C for optimal sectioning (this time will vary depending on the size and type of sample). While the sample tissue is warming, place other sectioning supplies (e.g., glass slides, paintbrushes, Teflon-coated slide) in the cryomicrotome chamber to cool. Tissues with higher fat content will require lower temperatures for adequate freezing and optimal sectioning.
2. Take care not to pour OCT on the bulk of the tissue that is to be sectioned, as it can cause ionization suppression in MALDI mass spectrometry. Avoid direct manual handling of the tissue so as not to thaw the organ. Instead, use forceps to manipulate the specimen.
3. Cryosectioning is a technique that requires patience and practice. If the tissue section adheres to the blade or antiroll plate, use the artist's paintbrushes to tease the tissue section free. If the tissue section strongly adheres to chamber surfaces, there may be excessive static electricity in the chamber or a mismatch between the object and chamber temperatures. An antistatic gun can be used to discharge a buildup of static electricity in the chamber. Ensure that the chamber temperature is colder than the object temperature to avoid a temperature mismatch. Small local adjustments in temperature (e.g., to the antiroll plate) can be made using a small piece of dry ice (to slightly cool a surface) or a finger (to slightly warm a surface).
4. Ensure that there are no folds in the tissue prior to mounting. If there is a slight curl to the tissue, face the curl downward onto the Teflon-coated slide so that it is flattened as the tissue section is picked up with the MALDI microscope slide.
5. If using a conductive slide, ensure that the tissue section is mounted on the conductive side (only one side of the slide is usually conductive). This can be verified by measuring the surface resistance of each side using a multimeter. The conductive side can be marked by etching a number in the corner of the slide using a diamond scribe. Serial slides can be consecutively numbered to keep track of the slide order. Numerical etches should be placed in the same corner of each slide and written so that the conductive surface can be easily identified. This is best done prior to placing the slides in the cryomicrotome.
6. Place the backside of the MALDI microscope slide containing the tissue section onto your palm to accelerate thaw-mounting. Ensure that all water content has been evaporated during the thaw-mounting process, as remaining water can refreeze in the

tissue and affect downstream analysis. Complete drying requires 15–60 s of thaw-mounting and will be evident when the tissue turns from a darker color to a more neutral matte color.

7. At 50 mTorr, DAN, 2,5-dihydroxyacetophenone (DHA), and 2,5-dihydroxybenzoic acid (DHB) sublimate well at 110 °C, sinapic acid sublimates well at 150 °C, α-cyano-4-hydroxycinnamic acid (CHCA) sublimates well at 170 °C, and 9-aminoacridine (9AA) sublimates well at 180 °C. Shake the sand bath to disperse sand evenly, allowing for uniform heating of the sand. After the temperature is set on the hot plate, it will typically take ~15–30 min to reach a stable temperature. The surface of the sand will typically be roughly 50 °C cooler than the surface setting of the hot plate and should be measured using a digital probe thermometer placed to the same depth in the sand bath as the condenser bottom discussed in later steps.
8. The vacuum gauge should quickly drop below 100 mTorr. If the pressure does not reach adequate pressure, ensure that the rubber O-ring is properly seated between the condenser sleeve and bottom. As the valve to the vacuum pump is opened, monitor the matrix coating on the bottom of the chamber. If the matrix particles are significantly disrupted as the valve is opened, there is likely a leak around the O-ring that is admitting air into the chamber. Leaks around the O-ring can also sometimes be audibly perceived as a high-pitched whistling.
9. Be consistent with the submersion depth to ensure reproducible matrix applications.
10. The application time will vary depending on the matrix (DHA will take ~3.5 min, DAN will take ~6 min, and CHCA, SA, and 9AA will take ~10–15 min). Weighing the MALDI microscope slide before and after matrix application allows for an accurate measure of added matrix. Typical optimal matrix coatings will weigh 1–3 mg. Sufficient matrix application can be monitored during the sublimation process by observation of a visible vapor layer roughly 1 cm above the bottom edge of the condenser tube.
11. ftmsControl has a built-in Help manual that details the function of many of the instrument parameters as well as suggested optimal ranges.
12. $PC_{34:1}$ is an example of TC:DB nomenclature, where 34 is the total number of carbon atoms in the fatty acyl chains and 1 is the total number of double bonds in the fatty acyl chains. HRAM identifications with better than 5 ppm mass accuracies are considered reliable and readily achieved with an accurately calibrated dataset. Tentative lipid identifications using accurate mass measurements can be made by searching any of several

lipid and metabolite databases [30]. However, care should be taken not to over-report the confidence of the identification. For example, *m/z* 760.585 can be tentatively identified as $PC_{34:1}$, but should not be reported as any of $PC_{16:0/18:1}$, $PC_{16:1/18:0}$, $PC_{(18:0/16:1)}$, or $PC_{18:1/16:0}$ based on intact mass alone.

13. While protonated PC ions are readily abundant in positive ion mode MALDI imaging mass spectrometry, CID of these ion types typically results exclusively in fragmentation of the choline head group and gives no information about the fatty acyl chains [24]. While negative ion mode CID results in fragment ions indicative of the fatty acid identities and positions present in the lipid, negative ion mode MALDI imaging mass spectrometry rarely produces PC anions, instead giving ion signals from more acidic glycerophospholipids (e.g., phosphatidylethanolamines) [21].
14. Stable ESI signal is key to ensuring facile ion/ion reaction tuning and is best obtained by optimizing the source gas parameters. A roughly tenfold excess of reagent ion signal compared to analyte ion signal is desirable to operate under conditions of pseudo-first order kinetics [31].
15. The analyte and reagent precursor ions should each be isolated using the quadrupole mass filter as they are sequentially injected into the hexapole ion trap (Fig. 2b). The hexapole ion trap RF potential and mirror RF potential are key to optimizing the reaction. These parameters define the pseudo-potential well depth in the reaction cell [32, 33]. If the RF potential is too high, the lower mass PDPA reactant dianions may be ejected due to a low mass cut-off (LMCO) effect. If the RF potential is too low, the higher mass product complex ions may be insufficiently confined in the ion trap [34]. The collision cell DC offset and lens potentials are also important ion/ion reaction parameters to optimize. These parameters facilitate trapping of the two ion populations. It is common to toggle back and forth between MALDI-only, ESI-only, and ion/ion reaction modes to confirm sufficient ion signal of both reactants as the reaction parameters are being optimized.
16. The modifications to the pulse programming and ETD-binding rules have enabled user-defined parameters to be entered directly in ftmsControl [8]. However, the polarity of the ESI source had to be manually defined using the SHEDS Command Interface. This was defined by programming the polarity command ("P_HV") of the ESI module ("ESI_") in NICE as negative ("1") using the "SetValue" feature. "GetSetValue" was used to confirm proper programming of the parameter.

17. Optimal ion/ion overlap for efficient reactions is best obtained when the reagent and analyte reactant ions are similar in *m/z* value [35]. At least one reactant should be multiply charged to avoid neutralization, and the reaction rate scales with the square of the ion charges [36]. Typical reagent ion injection and reaction times are 50–500 ms and 500–1500 ms, respectively.
18. The intact $[PC + PDPA - H]^-$ product ion complex spontaneously decomposes to the demethylated $[PC - CH_3]^-$ product ion via neutral loss of methylated PDPA. The energy to drive this spontaneous loss may originate from the coulombic recombination energy of the ion/ion reaction, the reaction conditions in the hexapole ion trap, and/or from the ion transfer conditions from the hexapole ion trap to the ICR cell [37]. It is also important to note that single proton transfer from the PC cation to the PDPA dianion is an undesired reaction pathway that competes with charge inversion and results in neutralization of the PC and a charge reduced $[PDPA - H]^-$ product anion.
19. The ion subjected to SORI CID should have a signal intensity of at least $1–5 \times 10^6$ in order to detect resulting fragment ion signals above the level of instrumental noise. Ion intensities of 10^7 will provide SORI CID fragment ion intensities of roughly 10^6, which are necessary to achieve satisfactory ion statistics for reliable downstream analysis.
20. Care should be taken to allow the pressure in the ICR cell to return below roughly 1×10^{-9} mbar between successive SORI CID experiments. This reduces the load on the vacuum system and avoids tripping an over-pressure interlock.
21. For example, *m/z* 760.585, tentatively identified as $PC_{34:1}$, was found to contain the fatty acyl chains equating to $PC_{16:0\_18:1}$ in rat brain tissue, where the underscore is used to denote ambiguity of the *sn*-position of the two fatty acyl groups.
22. The *sn*-positions of the fatty acyl groups in glycerophospholipid anions subjected to CID can typically be determined by examining the relative intensities of the fragment ions. Under low energy CID conditions, cleavage at the lipid *sn*-2 position is favored [24]. As a result, fatty acid fragment ions arising from the *sn*-2 position will generally be more abundant than those arising from the *sn*-1 position. However, PCs measured in the complex chemical environment of biological tissue samples are likely mixtures of *sn*-positional isomers, so these ratios will contain fragment ion populations from mixtures of precursor ions and will no longer be an accurate reflection of *sn*-position.

23. The procedure described herein allows for the differentiation of PC *sn*-positional isomers. However, additional structural isomers likely exist when accounting for the positions and orientations of the double bonds in the fatty acid tails (e.g., 6Z or 9Z). Alternative ion/ion reactions or other gas-phase approaches can be employed to identify these lipid structures [28, 29, 38, 39].
24. Determining the relative contributions of all isomers in a mixture using this method currently requires the generation of calibration curves containing each lipid isomer. This is dependent on the availability, affordability, and purity of synthetic lipid standards and may not be feasible for all experiments.
25. Relative standard deviations of on-tissue ratio measurements are typically <5%. The isomeric ratio can vary by at least 25% between different locations (i.e., substructures) in the tissue. Errors in this approach could be due to isomeric contamination in the synthetic lipid standards, which ranges from 95% to 99% purity.

Acknowledgments

This work was supported in part by the National Institutes of Health (NIH) under award R01 GM138660 (National Institute of General Medical Sciences [NIGMS]) and UF College of Liberal Arts and Sciences, the Office of Research, and the Department of Chemistry. The authors also thank Dr. Steven Van Order, Dr. Mark Ridgeway, Dr. Christian Berg, and Dr. Melvin Park of Bruker Daltonics for their assistance developing the ion/ion reaction software.

References

1. Liebisch G, Ejsing CS, Ekroos K (2015) Identification and annotation of lipid species in metabolomics studies need improvement. Clin Chem 61(12):1542–1544. https://doi.org/10.1373/clinchem.2015.244830
2. Fahy E, Subramaniam S, Brown HA, Glass CK, Merrill AH Jr, Murphy RC, Raetz CR, Russell DW, Seyama Y, Shaw W, Shimizu T, Spener F, van Meer G, VanNieuwenhze MS, White SH, Witztum JL, Dennis EA (2005) A comprehensive classification system for lipids. J Lipid Res 46(5):839–861. https://doi.org/10.1194/jlr.E400004-JLR200
3. Pauling JK, Hermansson M, Hartler J, Christiansen K, Gallego SF, Peng B, Ahrends R, Ejsing CS (2017) Proposal for a common nomenclature for fragment ions in mass spectra of lipids. PLoS One 12(11):e0188394. https://doi.org/10.1371/journal.pone.0188394
4. Kertesz V, Van Berkel GJ (2010) Liquid microjunction surface sampling coupled with high-pressure liquid chromatography–electrospray ionization-mass spectrometry for analysis of drugs and metabolites in whole-body thin tissue sections. Anal Chem 82(14):5917–5921. https://doi.org/10.1021/ac100954p
5. Van Berkel GJ, Kertesz V (2013) Continuous-flow liquid microjunction surface sampling probe connected on-line with high-performance liquid chromatography/mass spectrometry for spatially resolved analysis of small molecules and proteins. Rapid Commun

Mass Spectrom 27(12):1329–1334. https://doi.org/10.1002/rcm.6580
6. Prentice BM, Caprioli RM (2016) The need for speed in matrix-assisted laser desorption/ionization imaging mass spectrometry. J Postdoct Res 4(3):3–13
7. McLuckey SA, Mentinova M (2011) Ion/neutral, ion/electron, ion/photon, and ion/ion interactions in tandem mass spectrometry: do we need them all? Are they enough? J Am Soc Mass Spectrom 22(1):3–12. https://doi.org/10.1007/s13361-010-0004-9
8. Specker JT, Orden SLV, Ridgeway ME, Prentice BM (2020) Identification of phosphatidylcholine isomers in imaging mass spectrometry using gas-phase charge inversion ion/ion reactions. Anal Chem 92:13192–13201
9. Prentice BM, McLuckey SA (2013) Gas-phase ion/ion reactions of peptides and proteins: acid/base, redox, and covalent chemistries. Chem Commun 49(10):947–965. https://doi.org/10.1039/c2cc36577d
10. Foreman DJ, McLuckey SA (2020) Recent developments in gas-phase ion/ion reactions for analytical mass spectrometry. Anal Chem 92(1):252–266. https://doi.org/10.1021/acs.analchem.9b05014
11. Bonney JR, Prentice BM (2021) Perspective on emerging mass spectrometry technologies for comprehensive lipid structural elucidation. Anal Chem 93:6311. https://doi.org/10.1021/acs.analchem.1c00061
12. Stutzman JR, Blanksby SJ, McLuckey SA (2013) Gas-phase transformation of phosphatidylcholine cations to structurally informative anions via ion/ion chemistry. Anal Chem 85 (7):3752–3757. https://doi.org/10.1021/ac400190k
13. Rojas-Betancourt S, Stutzman JR, Londry FA, Blanksby SJ, McLuckey SA (2015) Gas-phase chemical separation of phosphatidylcholine and phosphatidylethanolamine cations via charge inversion ion/ion chemistry. Anal Chem 87(22):11255–11262. https://doi.org/10.1021/acs.analchem.5b02243
14. Schwartz SA, Reyzer ML, Caprioli RM (2003) Direct tissue analysis using matrix-assisted laser desorption/ionization mass spectrometry: practical aspects of sample preparation. J Mass Spectrom 38(7):699–708. https://doi.org/10.1002/jms.505
15. Kaletaş B, van der Wiel I, Stauber J, Lennard JD, Güzel C, Kros J, Luider T, Heeren R (2009) Sample preparation issues for tissue imaging by imaging MS. Proteomics 9 (10):2622–2633. https://doi.org/10.1002/pmic.200800364
16. Norris JL, Caprioli RM (2013) Analysis of tissue specimens by matrix-assisted laser desorption/ionization imaging mass spectrometry in biological and clinical research. Chem Rev 113 (4):2309–2342. https://doi.org/10.1021/cr3004295
17. Perry WJ, Patterson NH, Prentice BM, Neumann EK, Caprioli RM, Spraggins JM (2020) Uncovering matrix effects on lipid analyses in MALDI imaging mass spectrometry experiments. J Mass Spectrom 55(4):e4491. https://doi.org/10.1002/jms.4491
18. Hankin JA, Barkley RM, Murphy RC (2007) Sublimation as a method of matrix application for mass spectrometric imaging. J Am Soc Mass Spectrom 18(9):1646–1652. https://doi.org/10.1016/j.jasms.2007.06.010
19. Thomas A, Charbonneau J, Fournaise E, Chaurand P (2012) Sublimation of new matrix candidates for high spatial resolution imaging mass spectrometry of lipids: enhanced information in both positive and negative polarities after 1,5-diaminonapthalene deposition. Anal Chem 84(4):2048–2054. https://doi.org/10.1021/ac2033547
20. Cornett DS, Frappier SL, Caprioli RM (2008) MALDI-FTICR imaging mass spectrometry of drugs and metabolites in tissue. Anal Chem 80 (14):5648–5653. https://doi.org/10.1021/ac800617s
21. Zemski Berry KA, Hankin JA, Barkley RA, Spraggins JM, Caprioli RM, Murphy RC (2011) MALDI imaging of lipid biochemistry in tissues by mass spectrometry. Chem Rev 111 (10):6491–6512
22. Liebisch G, Vizcaíno JA, Köfeler H, Trötzmüller M, Griffiths WJ, Schmitz G, Spener F, Wakelam MJ (2013) Shorthand notation for lipid structures derived from mass spectrometry. J Lipid Res 54 (6):1523–1530. https://doi.org/10.1194/jlr.M033506
23. Bowman AP, Blakney GT, Hendrickson CL, Ellis SR, Heeren RMA, Smith DF (2020) Ultra-high mass resolving power, mass accuracy, and dynamic range MALDI mass spectrometry imaging by 21-T FT-ICR MS. Anal Chem 92(4):3133–3142. https://doi.org/10.1021/acs.analchem.9b04768
24. Murphy RC, Axelsen PH (2011) Mass spectrometric analysis of long-chain lipids. Mass Spectrom Rev 30(4):579–599. https://doi.org/10.1002/mas.20284
25. Kaplan DA, Hartmer R, Speir JP, Stoermer C, Gumerov D, Easterling ML, Brekenfeld A, Kim T, Laukien F, Park MA (2008) Electron transfer dissociation in the hexapole collision

cell of a hybrid quadrupole-hexapole Fourier transform ion cyclotron resonance mass spectrometer. Rapid Commun Mass Spectrom 22 (3):271–278. https://doi.org/10.1002/rcm.3356

26. Chaurand P, Schwartz SA, Billheimer D, Xu BJ, Crecelius A, Caprioli RM (2004) Integrating histology and imaging mass spectrometry. Anal Chem 76(4):1145–1155. https://doi.org/10.1021/ac0351264
27. Ma X, Chong L, Tian R, Shi R, Hu TY, Ouyang Z, Xia Y (2016) Identification and quantitation of lipid C=C location isomers: a shotgun lipidomics approach enabled by photochemical reaction. Proc Natl Acad Sci U S A 113(10):2573–2578. https://doi.org/10.1073/pnas.1523356113
28. Paine MRL, Poad BLJ, Eijkel GB, Marshall DL, Blanksby SJ, Heeren RMA, Ellis SR (2018) Mass spectrometry imaging with isomeric resolution enabled by ozone-induced dissociation. Angew Chem Int Ed Eng 57 (33):10530–10534. https://doi.org/10.1002/anie.201802937
29. Born M-EN, Prentice BM (2020) Structural elucidation of phosphatidylcholines from tissue using electron induced dissociation. Int J Mass Spectrom 452:116338. https://doi.org/10.1016/j.ijms.2020.116338
30. Sud M, Fahy E, Cotter D, Brown A, Dennis EA, Glass CK, Merrill AH, Murphy RC, Raetz CRH, Russell DW, Subramaniam SA (2007) LMSD: LIPID MAPS structure database. Nucleic Acids Res 35(1):D527–D532. https://doi.org/10.1093/nar/gkl838
31. McLuckey SA, Stephenson JL Jr, Asano KG (1998) Ion/ion proton-transfer kinetics: implications for analysis of ions derived from electrospray of protein mixtures. Anal Chem 70:1198–1202. https://doi.org/10.1021/ac9710137
32. Dehmelt HG (1967) Radiofrequency spectroscopy of stored ions I: storage. In: Bates DR, Estermann I (eds) Advances in atomic and molecular physics, vol 3. Academic Press, New York, NY, pp 53–72
33. March RE (1997) An introduction to quadrupole ion trap mass spectrometry. J Mass Spectrom 32:351–369
34. Prentice BM, McLuckey SA (2012) Analysis of high mass-to-charge ions in a quadrupole ion trap mass spectrometer via an end-cap quadrupolar direct current downscan. Anal Chem 84 (17):7562–7569. https://doi.org/10.1021/ac301741a
35. McLuckey SA (2010) The emerging role of ion/ion reactions in biological mass spectrometry: considerations for reagent ion selection. Eur J Mass Spectrom 16(3):429–436. https://doi.org/10.1255/ejms.1031
36. Wells JM, Chrisman PA, McLuckey SA (2001) Formation of protein–protein complexes in vacuo. J Am Chem Soc 123:12428–12429. https://doi.org/10.1021/ja0170403
37. Bu JX, Fisher CM, Gilbert JD, Prentice BM, McLuckey SA (2016) Selective covalent chemistry via gas-phase ion/ion reactions: an exploration of the energy surfaces associated with N-hydroxysuccinimide ester reagents and primary amines and guanidine groups. J Am Soc Mass Spectrom 27(6):1089–1098. https://doi.org/10.1007/s13361-016-1359-3
38. Randolph CE, Foreman DJ, Betancourt SK, Blanksby SJ, McLuckey SA (2018) Gas-phase ion/ion reactions involving tris-phenanthroline alkaline earth metal complexes as charge inversion reagents for the identification of fatty acids. Anal Chem 90 (21):12861–12869. https://doi.org/10.1021/acs.analchem.8b03441
39. Randolph CE, Blanksby SJ, McLuckey SA (2019) Towards complete structure elucidation of glycerophospholipids in the gas phase through charge inversion ion/ion chemistry. Anal Chem 92:1219. https://doi.org/10.1021/acs.analchem.9b04376

Chapter 2

MALDI-2 and t-MALDI-2 Mass Spectrometry Imaging

Klaus Dreisewerd, Tanja Bien, and Jens Soltwisch

Abstract

Matrix-assisted laser desorption ionization mass spectrometry imaging (MALDI-MSI) combined with laser-induced postionization for MALDI-2 enables the simultaneous registration of numerous classes of small molecules (e.g., secondary metabolites including sterols) as well as phospholipids, glycolipids, and glycans from tissue sections and from cell cultures with strongly boosted ion yields. Here, we describe methodological aspects that are key for optimizing the analytical sensitivity and spatial resolution of a MALDI-2 imaging experiment. We will include both top-illumination MALDI-2 as well as the recently introduced transmission (t-) mode MALDI-2 approach.

Key words MALDI imaging, MALDI-2, Transmission-mode MALDI-2, Sample preparation, Cell cultures

1 Introduction

Matrix-assisted laser desorption ionization mass spectrometry imaging (MALDI-MSI) is a label-free technique that is increasingly used to visualize the distribution of various classes of biomolecules such as phospho- and glycolipids, glycans, and tryptic peptides, and furthermore also exogenous substances like pharmaceuticals and their metabolic products, in thin tissue sections and in cell cultures [1–3]. The MALDI-MSI principle is depicted in Fig. 1. A matrix-coated sample (e.g., a thin tissue section) is "rastered" with the beam of a pulsed laser. At each irradiated pixel, a mixture of neutral and ionic analyte and matrix molecules is ejected. An electrical field extracts the ionic fraction of the MALDI plume into a mass analyzer, where individual ions are separated according to their mass-to-charge (*m/z*) ratio and are registered with an ion detector. Because the mass of a molecule (ion) is defined by its atomic composition, precisely measuring the mass provides information on the molecular identity of the compound. Most typically, protonated $[M + H]^+$ and alkali metal adduct species $[M + Na]^+/[M + K]^+$ are generated by MALDI in the positive ion mode,

Young-Jin Lee (ed.), *Mass Spectrometry Imaging of Small Molecules: Methods and Protocols*,
Methods in Molecular Biology, vol. 2437, https://doi.org/10.1007/978-1-0716-2030-4_2,
© The Author(s), under exclusive license to Springer Science+Business Media, LLC, part of Springer Nature 2022

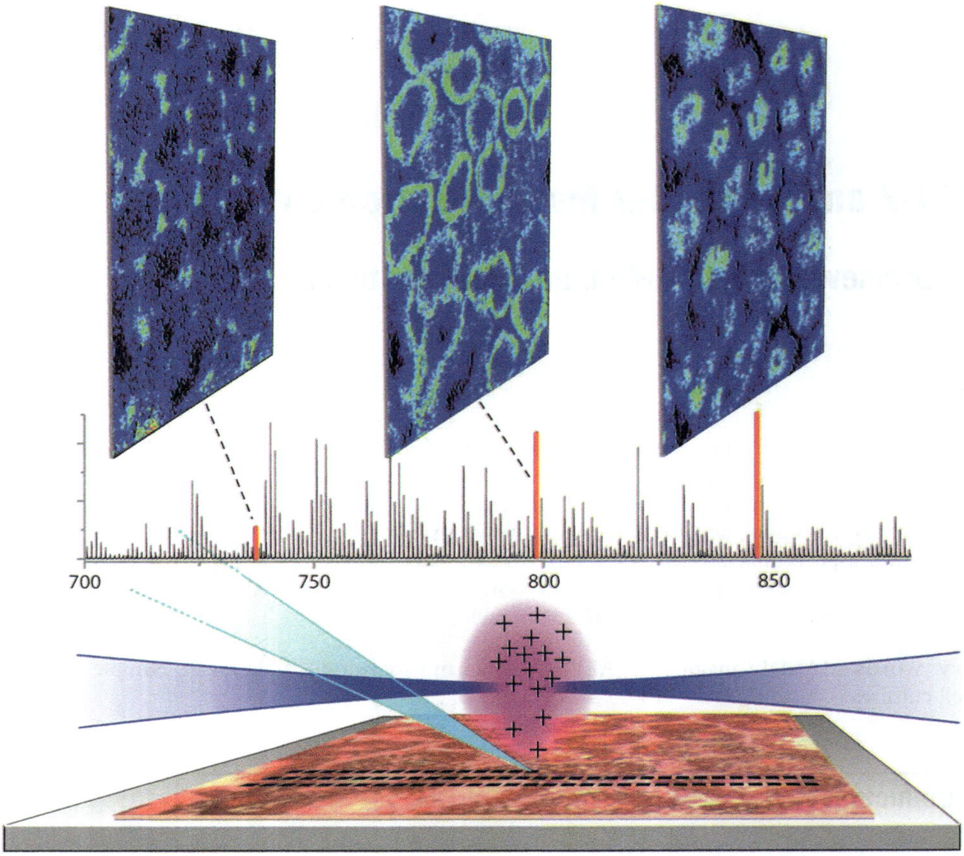

Fig. 1 The principles of MALDI and MALDI-2-MS imaging. A UV laser scans a sample (e.g., a tissue section), which is coated with a "MALDI matrix," and removes matrix and analyte molecules from each irradiated pixel. With MALDI-2, a pulsed "post-ionization laser" irradiates the material cloud produced by the first laser and, under fine vacuum conditions in the ion source, initiates a cascade of secondary MALDI-like ionization processes that add to the overall ion yields. From the mass spectra recorded for each individual pixel, a computer calculates molecular distribution images for the registered substances based on their individual mass (*m/z*) values. (Figure: Willi Kramer, University Hospital Münster)

whereas deprotonated species $[M - H]^-$ are commonly dominating the mass spectra acquired in the negative ion mode. Ion images reflecting the lateral distribution of the compounds in the sample are reconstructed from the generated set of mass spectra using dedicated software tools. A sufficient mass resolution and analytical sensitivity provided, several dozens to eventually hundreds of different substances can potentially be registered simultaneously in high-resolving MALDI imaging, for instance by use of Orbitrap, Fourier-Transform Ion Cyclotron Resonance (FTICR) or orthogonal-extracting time-of-flight (QTOF) instruments.

The lateral resolution of MALDI is determined by the quality of the sample preparation, which in particular includes the critical matrix coating step. Within this step, analyte compounds are extracted from the sample and cocrystallized with the matrix. Next to the mean crystal sizes, further key parameters defining the achievable spatial resolution comprise the focal laser spot size and the fidelity/precision of the sample stage movement. Most contemporary MALDI-MSI instruments now provide pixel sizes in the 5–20 μm range. New technology concepts show how this limit may be further advanced to the low, possibly even submicrometer level [4, 5]. Because of the minute amounts of material, which is available with such small pixel sizes, and because only charged molecules (ions) are registered by the mass analyzer, achieving an as high as possible ionization efficiency becomes exceedingly important in high-resolving MALDI-MSI applications. This is where standard MALDI, however, has two notable weak points: First, for many compound classes less than one out of a million of the desorbed molecules is concomitantly ionized [6, 7]. Second, particularly well-ionizing compounds (e.g., phosphatidylcholines (PC)) can scavenge charges from other compounds, a scenario that is referred to as ion suppression [8]. Several instrumental advancements have recently significantly extended the analytical abilities of MALDI-MS imaging [9, 10]. One particularly relevant one is the introduction of MALDI with laser-induced post-ionization, a technique that is generally referred to as MALDI-2 [11]. With MALDI-2 the ion yields for numerous classes of small molecules, phospho- and glycolipids, glycans (oligosaccharides), and pharmaceutical compounds can be boosted by up to 2–3 orders of magnitude [11–14].

Two prerequisites for exploiting the MALDI-2 effect are the employment of ion sources that are operated under elevated pressure conditions (a few mbar of nitrogen buffer gas) and the use of a pulsed ultraviolet (UV) postionization laser. Furthermore, the wavelength of this laser should fall below the two-photon threshold of the utilized MALDI matrix. For aromatic MALDI matrices, such as, for instance, 2,5-dihydroxybenzoic acid (DHB), α-cyano-4-hydroxy-cinnamic acid (CHCA), or 2,5-dihydroxyacetophenone (DHAP), this threshold is found around 310 nm. In particular, frequency-quadrupled Q-switched Nd:YAG lasers which emit at 266 nm constitute suitable laser sources [15, 16].

MALDI-2 was first developed on a hybrid QTOF mass spectrometer, a Synapt G2-S HDMS instrument from Waters/Micromass [11]. A schematic presentation of the postionization source developed on this instrument is shown in Fig. 2a. Successively, the technique was implemented by us and others on different Orbitrap

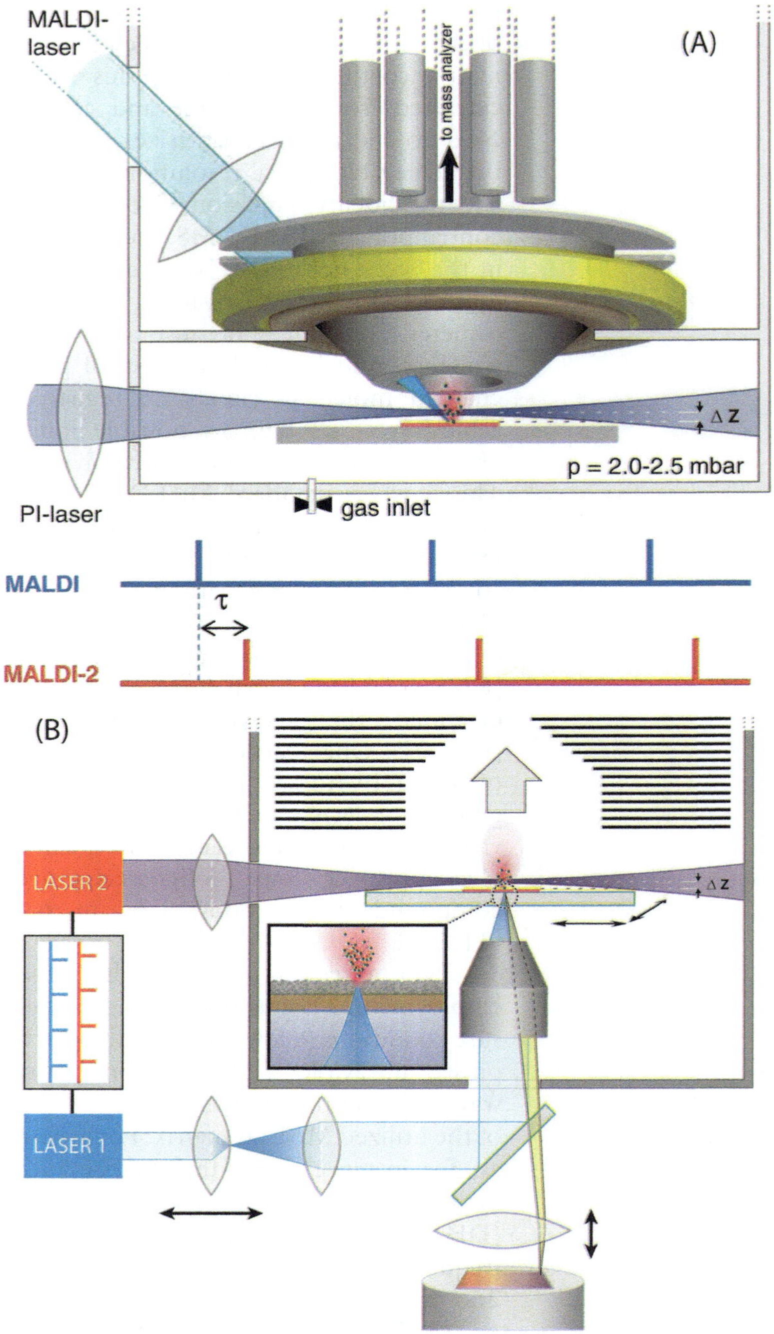

Fig. 2 Schematic drawings of two MALDI-2 ion sources, here sketched for (**a**) an ion source of a modified Synapt mass spectrometer, operated in top-illumination geometry, and (**b**) an ion funnel device operated in transmission-mode geometry, respectively. (Adapted from refs. 11 and 5, with permissions; © AAAS 2015, © Springer Nature 2019)

mass spectrometers (Thermo Fisher Scientific) [12, 14, 16]. In all cases, modified dual ion funnel MALDI/ESI injectors from Spectroglyph were used as the MALDI-2 ion source. Very recently, MALDI-2 has also been established in combination with atmospheric pressure (AP) MALDI [17]. The most recent addition to the MALDI-2 family is the timsTOF fleX MALDI-2, which was commercialized by Bruker Daltonics in 2020 [15]. The name TIMS stands for trapped ion mobility spectrometry. This online gas phase-based method provides another orthogonal separation dimension that is based on the collisional cross section (CCS) of the ions dragged through a nitrogen collision gas. As is demonstrated in Fig. 3 at the example of phospholipids and monohexosylceramides registered from rat testes, the TIMS–MSI combination can be exceedingly helpful for unravelling the high complexity of the lipidome and to separate isobaric or even isomeric molecules which may be not differentiable by their mass or in tandem MS experiments alone.

Transmission (t-) mode MALDI is a technique in which the laser beam is focused onto a matrix-coated sample in reversed irradiation or "backside illumination" geometry [5]. The principle of t-MALDI-2 is depicted in Fig. 2b and in Fig. 4, there at the example of analyzing cell cultures (*see* Subheading 3 below). A UV-transmitting microscopic objective with high numerical aperture is placed at a close distance (≤15–20 mm) to the matrix-coated sample. With t-MALDI-2, pixel sizes in the 1–2 μm range and below, if oversampling is used, are achieved [5]. As is demonstrated in Fig. 5 at the example of three cancer cell lines [18], single cells are readily differentiated within this resolution range, based on the recorded MALDI-MSI profiles, and even a subcellular resolution is indicated. The ability to visualize cell-to-cell heterogeneity on a molecular level is of high interest for many fields of cellular biology. Tracking cellular differentiation in carcinogenesis on a metabolic level would constitute just one important application example [19]. Different to top-illumination MALDI-2, t-MALDI-2 has at the time of preparing this book chapter not yet been commercialized.

A successful adaptation of MALDI-2 and t-MALDI-2 critically depends on the proper adjustment of a few key instrumental parameters, on the one hand, and on an optimized sample preparation, on the other [18, 20]. In the following, we will describe such optimized workflows as developed in our laboratory within the recent past for both top-illumination and t-mode MALDI-2-MSI of tissue sections and of cultured cells.

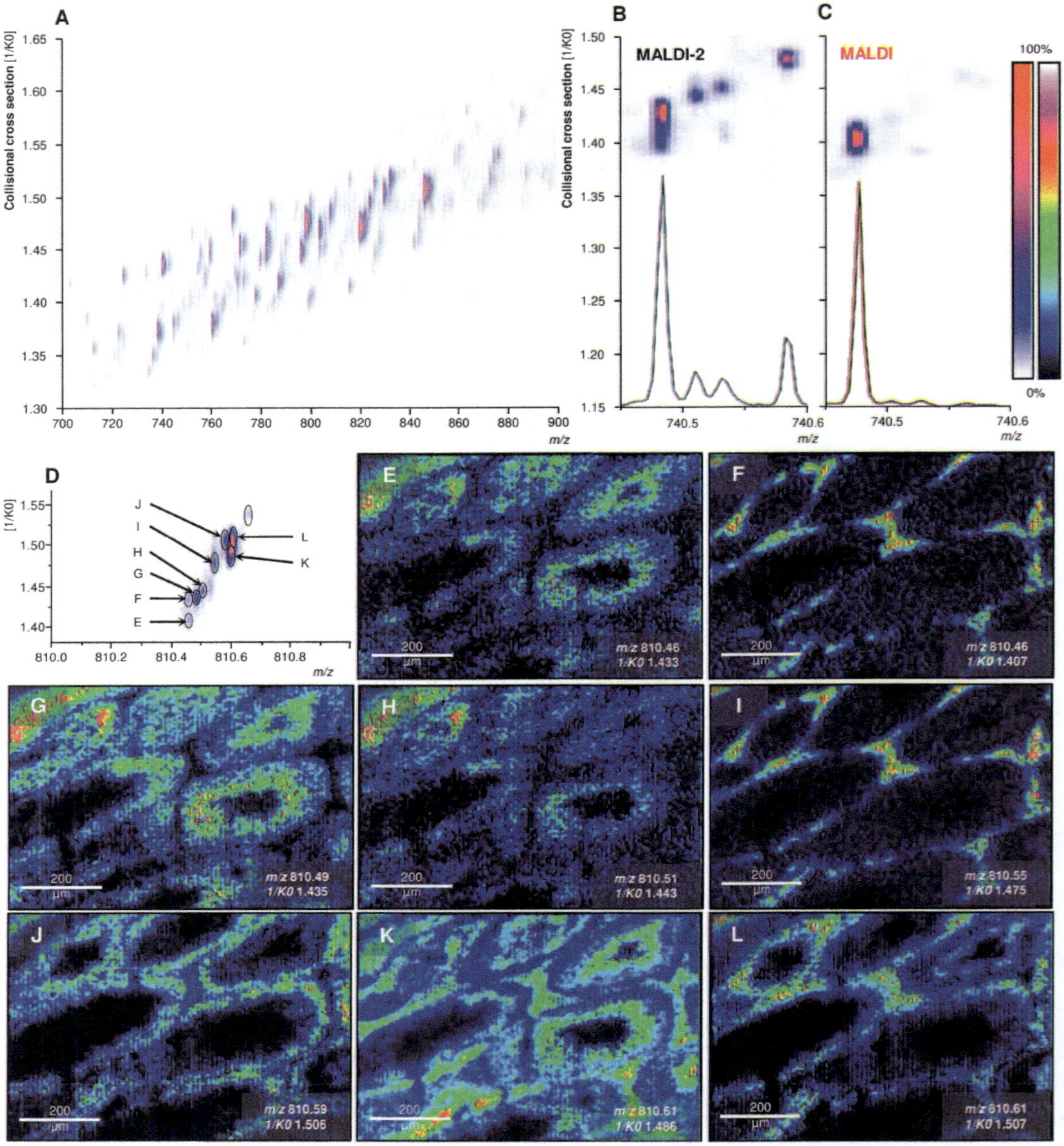

Fig. 3 MALDI-2-timsTOF fleX-MS imaging of lipids in rat testis. (**a**) Signal intensity/ *m/z* /ion mobility plot of the "phospholipid region." A comparison between (**b**) MALDI-2 and (**c**) conventional MALDI analyses shows that numerous features (mainly various phospholipids and different monohexosylceramides) are only detected with laser postionization. (**d**) The *m/z*-region between 810.0 and 811.0, displaying nine distinct features. (**e–l**) Ion images of eight of these features, some of which were only separated by use of TIMS. The data set consists of 20838 10 × 10 μm^2-wide pixels. (Reprinted from ref. 15 with permission; © ACS 2020)

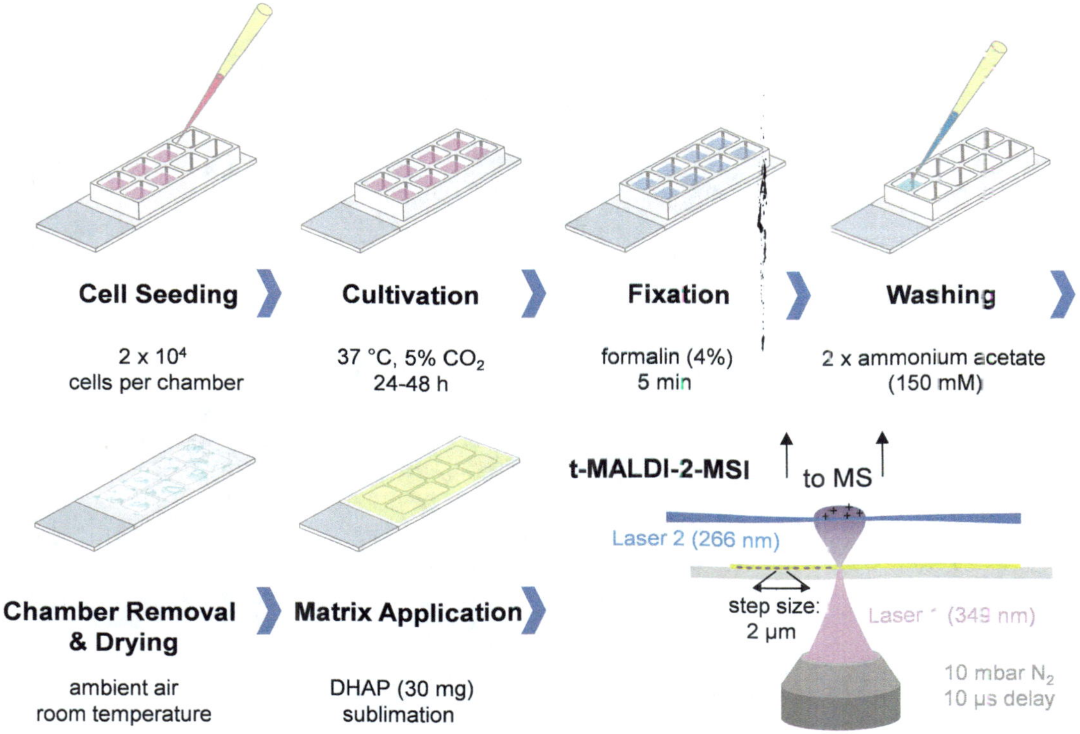

Fig. 4 Scheme of the optimized workflow developed for the t-MALDI-2-MSI analysis of single cells/cell cultures and simplified scheme of the t-MALDI-2 setup. (Reprinted and adapted from ref. 18 with permission; © ACS 2021)

2 Materials

2.1 Chemicals and Materials

2.1.1 Tissues

1. Heptane (*see* **Note 1**).
2. Indium tin oxide (ITO)-coated glass slides (75 mm × 25 mm; 70–100 Ω/sq (*see* **Note 2**)).
3. 2-Hydroxyethylcellulose (2–4%, average M_v ~1,500,000 g/mol).
4. Red phosphorus.
5. Ethanol (EtOH) or isopropanol for washing.
6. Hematoxylin and Eosin (H&E) solutions according to standard histological protocols.

2.1.2 Cell Cultures

1. 8-well chamber slides (Millicell EZ Slide, Merck).
2. Indium tin oxide (ITO)-coated glass slides (75 mm × 25 mm; 70–100 Ω/sq (*see* **Note 2**)).
3. Ethanol (EtOH, 70%).

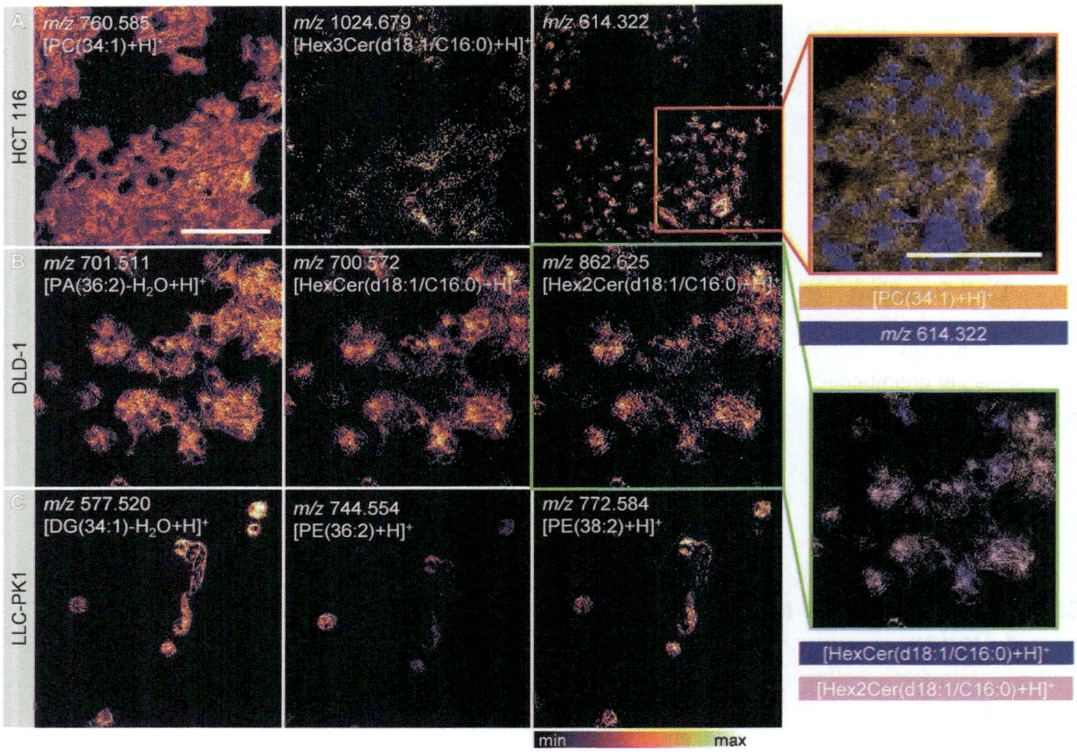

Fig. 5 T-MALDI-2 MS imaging of the three different cell culture lines (**a**) HCT 116, (**b**) DLD-1, and (**c**) LLC-PKI. Shown are the ion distribution images of selected glycosphingo- and phospholipids and of an unknown compound registered at *m*/*z* 614.322. The cells were grown to subconfluency and prepared with an optimized sample preparation protocol for high-resolution MALDI imaging. *DG* diacylglycerol, *HexCer* hexosylceramide, *PA* phosphatidic acid, *PC* phosphatidylcholine, *PE*: phosphatidylethanolamine; Scale bar: 100 μm applies to all images. (Reprinted from ref. 18 with permission; © ACS 2021)

4. Cell culturing flasks (75 mL).
5. Desired cell lines.
6. Cell culturing medium according to the used cell line (*see* **Note 3**).
7. Trypsin-EDTA (0.25%).
8. Phosphate-buffered saline (PBS).
9. Formaldehyde (4% in PBS).
10. Ammonium acetate (150 mM).

2.1.3 Matrix Spray Preparation

1. Deionized water.
2. Methanol (MeOH).
3. Acetonitrile (ACN).
4. 2,5-Dihydroxyacetophenone (DHAP).

Fig. 6 Snapshot of the home-made sublimation chamber with main elements depicted. (Credit: Tanja Bien)

2.1.4 Matrix Sublimation Preparation

1. Acetone.
2. 2,5-Dihydroxyacetophenone (DHAP).

2.2 Small Scale Instruments

1. Cryostat (Leica Jung Frigocut 2800N).
2. Ultrasonic sprayer (SimCoat with AccuMist nozzle; Sono-Tek).
3. Sublimation chamber (homemade; *see* Fig. 6 and **Note 4**).
4. Optical scanner.
5. Microscope.

2.3 Mass Spectrometers

1. TimsTOF fleX MALDI-2 (Bruker Daltonics) equipped with a Q-switched SmartBeam 3D laser emitting at 355 nm and a diode-pumped, actively Q-switched, frequency-quadrupled Nd:YAG laser (NL 204-1k-FH, EKSPLA, Vilnius, Lithuania; wavelength, 266 nm; pulse duration, 7 ns; maximum pulse duration, 1 kHz; maximum pulse energy, ~500 μJ) for laser postionization. The focal spot size of the scanning SmartBeam MALDI laser can be varied from 5 μm to larger squares. The postionization laser has a focal beam waist of about 100 μm.
2. Q Exactive plus Orbitrap mass analyzer (Thermo Fisher Scientific) equipped with a dual ion funnel dual-MALDI/ESI ion

injector from Spectroglyph. The source is equipped with a frequency-tripled Nd:YLF laser (Explorer One; Newport/ Spectra Physics) with 349 nm emission wavelength. By use of custom-made beam shaping, a focal spot size of ~9–10 μm by diameter (defined as the area of visible material ejection) was obtained in top-illumination geometry. For laser postionization, a Nd:YAG laser (PL2231–100, Ekspla) with 266 nm wavelength, 28 ps pulse duration, and 100 Hz pulse repetition frequency was used. The postionization laser has a focal beam waist of about 100 μm.

2.4 Software

1. Flex Imaging (Bruker).
2. TDF Viewer (Bruker).
3. SCiLS Lab MVS (SCiLS; *see* **Note 5**).
4. Image Insight software (Spectroglyph).
5. Custom control software for t-MALDI stage written in-house.

3 Methods

3.1 Tissue Preparation

Sample preparation is a crucial step in generating reproducible, artifact-free mass spectrometry imaging data. The quality of the prepared samples depends crucially on factors such as tissue fixation methods, the used embedding medium, the thickness to which sections are cut, the way samples are stored if not immediately analyzed, and not least the quality of the MALDI matrix coating.

1. Select a suitable embedding medium that does not interfere with MALDI-MS imaging by causing higher abundances of chemical background ion signals. Suitable materials are, for example, 2-hydroxy-ethylcellulose or carboxy-methylcellulose polymers with high average chain length, Shandon M-1 (Thermo Scientific), or gelatin (*see* **Note 6**).
2. Snap-freeze samples by dipping into liquid nitrogen (*see* **Note 7**).
3. If not immediately processed further, store the frozen samples at −20 °C. If samples need to be stored for longer times or constitute particular sensitive materials which tends to degrade or oxidize store them at −80 °C before further use. In addition, consider (vacuum) sealing the samples to avoid any contact with ambient air including moisture.
4. Avoid repetitive thawing and freezing, in particular if this is connected with condensation of water vapor. To avoid condensation, thaw samples in a dry atmosphere, for example under a stream of nitrogen gas.
5. Mount tissue blocks on the sample holder of a cryotome.

6. Section samples to a thickness between about 10 and 18 μm for front-side illumination and to 5–10 μm thickness for t-MALDI (*see* **Note 8**).
7. Thaw-mount samples on the conductive side of an ITO slide or microscopy slide suitable for cryo-sectioning (e.g., "Superfrost" slides from Thermo Scientific; *see* **Note 2**).
8. Apply fiducial markers around the tissue section, for example by scratching into the glass with a glass-marking pen or by using another suitable way of marking a position.
9. Record an optical image of the slide using a flatbed or dedicated slide scanner. Make sure to include the fiducial markers within the overview scan.
10. Coat samples with MALDI matrix by use of either the spraying or sublimation protocol.

3.2 Cell Cultures

A schematic illustrating of the cultivation and preparation of cell cultures for (t-) MALDI-2-MSI is presented in Fig. 4. Detailed information can be found in ref. 18.

1. Grow desired cell line in 75 mL culture flask in 20 mL of adequate cell culture media. Use a humidified atmosphere at 37 °C with 5% CO_2. Passage cells using 0.25% trypsin-EDTA as soon as they reach confluency.
2. Remove 8-well chambers from commercially available chamber slides. Sterilize chambers and ITO-coated glass slides with EtOH (70%) and allow the materials to dry at ambient conditions. Reassemble 8-well chambers with the ITO slides forming the base. Further information can be found in ref. 18.
3. Culture cells on the ITO slides equipped with 8-well growth chambers. Seed the desired amount of cells in 0.5 mL of growth media directly into each chamber of the slides and grow them to subconfluency (*see* **Note 9**).
4. Remove the cell culture media.
5. Optionally, add 4% formaldehyde in PBS for chemical fixation to retain cellular morphology. Incubate cells with 4% formaldehyde for 5 min and remove the solution immediately afterward (*see* **Note 10**).
6. Wash the cells in their chambers to remove media and 4% formaldehyde solution. Gently add the washing solution and remove it directly after a few seconds. Use 1 × 0.5 mL of PBS and 2 × 0.5 mL of ammonium acetate solution as washing agents.
7. Remove the chamber housing and let cells dry under ambient conditions over night.
8. Coat samples with MALDI matrix by use of the sublimation protocol.

3.3 Matrix Application

The two most widely used techniques for coating of samples with a uniform layer of a MALDI matrix are by spray-coating and by sublimation. For the first method, commercial pneumatic- or ultrasonic assisted devices are available from different vendors. Some researchers also use less expensive artistic airbrush systems, albeit these may tend to produce less reproducible coating results and are not recommended for high-resolution purposes. Matrix coating by sublimation is often followed by an additional recrystallization step in a humid atmosphere, in order to assist analyte extraction from the samples and their subsequent cocrystallization with the matrix. Depending on analyte type, for example a defined concentration of ethanol or another suitable solvent may be added. Depending on class of analyte, both spray-coating and sublimation may be used for tissue sections, whereas generally the sublimation technique will be more suitable for cell cultures (*see* **Note 11**).

3.3.1 Ultrasonic-Assisted Application of MALDI Matrix

1. Dissolve DHAP MALDI matrix to 15 mg/mL in ACN–MeOH–H_2O (8:1:1, v/v/v).
2. Fill the syringe of the syringe pump with one to a few milliliter of the solution.
3. Mount the sample plate on the stage of the SimCoat sprayer.
4. Adjust the distance of the spray nozzle outlet from the sample surface to 60 mm.
5. Adjust the flow rate of the syringe pump to 50 μL/min. Set the ultrasonic nebulizer working at 50% in "power mode" and using a nitrogen backing gas of ~70 hPa.
6. Perform robotic spraying by continuously moving the nozzle across the sample with a scan speed of 30 mm/s and a meandering pattern with a track spacing of 1.8 mm. Each cycle should include two passes across the whole sample area with a change of pattern tracks by 90° at half-time.
7. Use 20 spray cycles in total with a resting period of 10 s after each cycle to allow for solvent evaporation.

3.3.2 Sublimation Application of MALDI Matrix

1. Use a sublimation chamber (*see* Fig. 6 for a picture of a home-built set-up) for coating samples by sublimation. Sublimation temperature should be actively regulated by hard- and software (for example built in-house for home-built instrument). The overall setup consists of a sample holder construction, electronics, and a vacuum chamber.
2. Preheat the brass pan inside the sample holder, serving as the matrix reservoir, to 130 °C.
3. Dissolve DHAP MALDI matrix to 20 mg/mL in acetone.
4. Add 1.5 mL of DHAP matrix solution to the heated pan and wait until the solvent has fully evaporated.

5. Place the sample on a water-cooled surface of ~4 °C at a distance of 80 mm above the pan, with the sample pointing downward, facing the matrix reservoir.
6. Transfer the "sublimation stage," including matrix reservoir and sample, to a vacuum recipient and evacuate the chamber to ~5 × 10^{-3} mbar.
7. Allow for sublimation of the matrix and its redeposition on the cooled sample for 6 min.
8. Stop the process by switching off the heating for the matrix reservoir and by venting the sublimation chamber with nitrogen gas; successively remove the sample. With our apparatus, the described protocol resulted in a uniform DHAP matrix coverage of approx. 300 ± 20 μg/cm^2.
9. Optionally—carry out "recrystallization" of the matrix coating by placing the sample in a saturated atmosphere of H_2O–ethanol (1000:5, v/v) at 75 °C for 2.5 min (*see* **Note 12**).

3.4 Data Acquisition

3.4.1 MALDI-2 Imaging on the timsTOF fleX

1. Fix the histological slides with sample in the Bruker sample holder.
2. Deposit a small amount of red phosphorus to a position just next to the sample.
3. Mass-calibrate the timsTOF fleX mass spectrometer using the distribution of well-known red phosphorus signals and the corresponding mass list.
4. Generate MALDI-2-methods by optimizing laser parameters for the individual sample and analyte types at hand. In previous fundamental MALDI-2 studies [20, 21], the optimum value for the postionization laser energy was identified to ~300 μJ. An optimal inter-laser delay is generally found between 8 and 15 μs, depending somewhat on sample type and used ion polarity, and the inter-laser pulse delay time to about 10 μs (*see* **Notes 13** and **14**).
5. Define the desired measurement area in the flexImaging software.
6. Define step size and load the MALDI-method (e.g., positive versus negative ion mode polarity).
7. Start the MALDI-imaging run.

3.4.2 t-MALDI-2 Imaging on the Orbitrap Q Exactive Plus

1. Mount the ITO or uncoated histological glass slide with sample in the custom-made sample holder. The sample area of interest needs to be in the free aperture range of the transmission stage. Further details can be found in ref. 5.
2. To account for possible height differences between different types of slides and/or samples, adjust the optical system:

(a) stage height adjustable: Use the sample stage in *z*-direction to bring the sample into focus using the optical image. Because visual inspection and laser focus share the same focal depth, adjustment is done.

(b) stage height fixed: Adjust the position of the tube lens to bring the optical image into focus. By tuning the beam expander of the laser system, adjust the spot size of the desorption laser visible in the optical inspection to a minimum. Fine adjustment can be achieved using the TIC signal recorded from a homogenous area of the sample similar as for adjusting postionization laser parameters (*see* **Note 14**).

3. Optimize for MALDI-2 signal intensity by carefully adjusting the focusing lens of the postionization laser (*see* **Note 13**).
4. Adjust the "MALDI-1" desorption laser pulse energy to values slightly above the threshold at which typical MALDI-2 signals start to be registered in sizable numbers.
5. Adjust the postionization laser pulse energy to about 300 μJ (*see* **Note 14**).
6. Adjust the N_2 buffer gas pressure in the ion source to ca. 10 mbar.
7. Define the region to be measured and pixel size in the respective software.
8. Define the desired spectral resolution and appropriate (fixed) acquisition time for the FT-MS measurement and start the measurement with contact closure enabled (*see* **Note 15**).
9. Start the imaging run using the stage control software.
10. Use Image Insight software to import the Q Exactive mass spectral data and connect them with the Spectroglyph position file.

3.5 Image Processing

1. Export the data to the vendor independent imzML format.
2. Import the imzML data into SCiLS Lab software for data evaluation and image representation.
3. Optional: Perform hotspot removal and use desired ion signal normalization (e.g., TIC; *see* **Note 16**).
4. Use TDF Viewer (Bruker) to visualize mobilograms.
5. In addition to MALDI-2-MSI, in particular from tissue sections, standard hematoxylin and eosin (H&E) staining may be performed to visualize morphologies with a standard histological technique (*see* **Note 17**). This can be done by either washing the matrix from the section (e.g., by using ethanol or isopropanol) or by staining consecutive sections. An example of *N*-glycans analyzed from human cerebellum brain tissue is

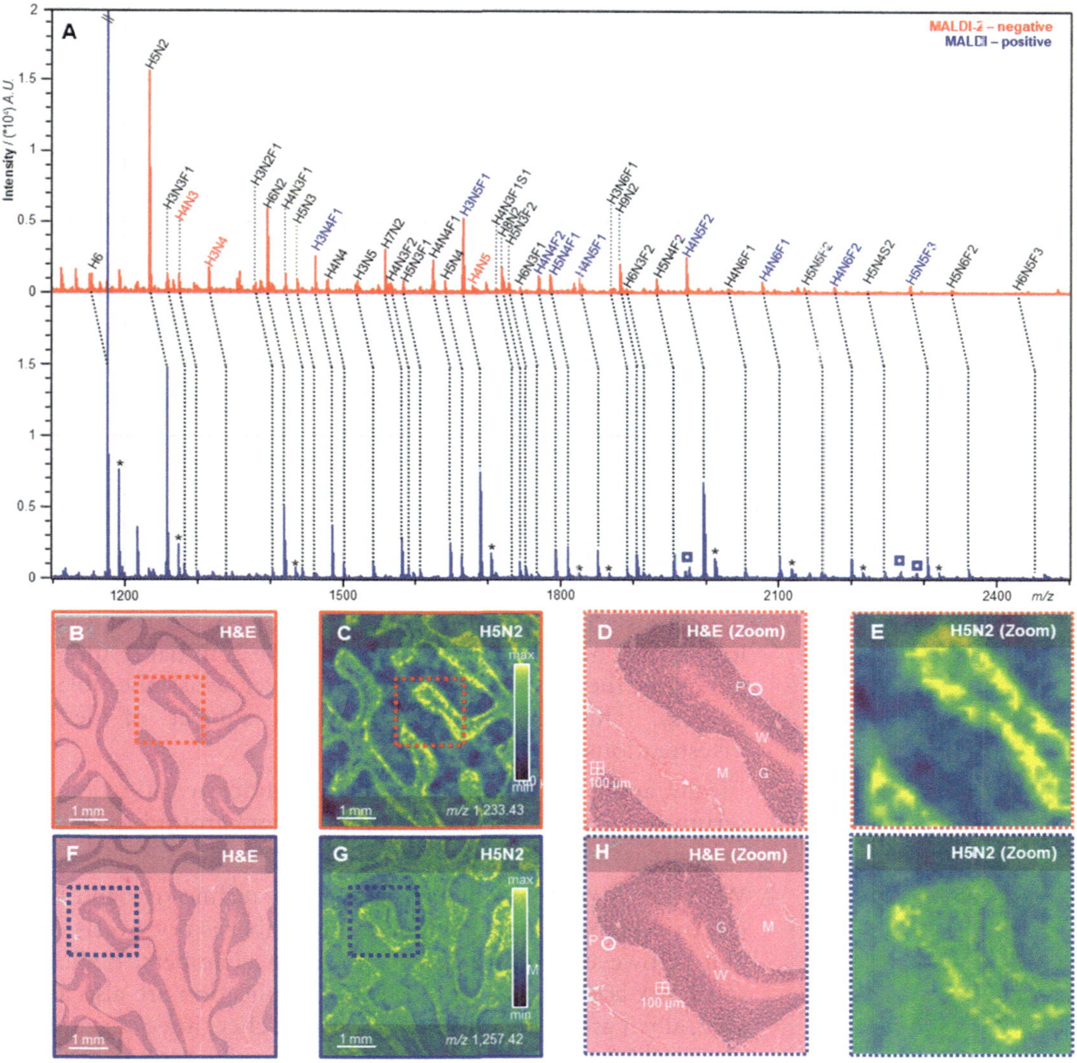

Fig. 7 Mass spectra and selected ion images of *N*-glycans in human cerebellum. Data were obtained with the timsTOF fleX MALDI-2 in the positive and negative ion modes, respectively. (**a**) Average spectra for negative ion-mode MALDI-2-MSI (red) and positive ion-mode MALDI-MSI (blue) with assigned *N*-glycan species. (**b**, **f**) H&E-stained consecutive section displaying cerebellar brain morphology. (**c**, **g**) Example image for the distribution of H5N2 in human cerebellum registered in the negative and positive ion modes, respectively. (**d**, **h**) Zoom in on histology with different morphological structures annotated (*P* purkinje cell, *g* granular layer, *M* molecular layer, *W* white matter). (**e**, **i**) Zoom in on example H5N2 images. (Reprinted from ref. 13 with permission; © ACS 2020, *see* **Notes 18** and **19**)

presented in Fig. 7. In this case H&E stains showing the brain morphology, including the visualization of single Purkinje cells in human cerebellum, were obtained from consecutive sections.

4 Notes

1. If tissues contain a high number of fat cells and are, therefore, rich in triacylglycerols, the mass spectral quality obtained from the sections can be severely reduced because of ion suppression [8] as well as "smearing" effects. To mitigate this problem, apply a washing step with heptane (3 × 1 mL) before matrix application [19]. The heptane wash efficiently removes triacylglycerols without affecting most membrane lipids (i.e., the fidelity on the position they are registered from is not impaired).
2. ITO slides are coated with a thin layer of indium tin oxide, which renders them electrically conducting. Because, with hybrid MALDI-2 mass spectrometers, the mass analysis step is decoupled from the ion generation, requirements on the extraction fields are reduced and the use of an ITO slide is not strictly necessary. Thus, with top-illumination MALDI-2, also more cost-effective standard histological slides (e.g., SuperFrost slides from Thermo Scientific) may be used without loss in MSI performance. With t-MALDI-2, initial results suggest that the ITO coating may however assist the ablation process. This feature is currently a subject of detailed investigations.
3. Ideally, and if possible for the cell lines under investigation, use serum-free culture media to minimize background ion signals in the mass spectra.
4. With some basic knowledge on vacuum technology and basic electronics, building a sublimation chamber is not overly difficult. However, commercial sublimation chambers, dedicated to matrix coating in a MALDI-MSI application are also available, albeit at substantial costs. Consider checking out standard chemical lab supplies for a more cost-effective "compromise".
5. There are two different versions of SCiLS Lab software. The first can be used exclusively with mass spectrometers from Bruker. Since a few years, the company also supports data formats as generated by the platforms of a few other vendors (Waters, Thermo Fisher Scientific) within SCiLS Lab MVS, where MVS stands for multi-vendor version(s). In particular, the vendor-independent data format imzML is supported. A possible alternative to SCiLS Lab is LipostarMSI from Molecular Horizon (Bettona, Perugia, Italy).
6. The role of the embedding medium is to produce a block that can be mounted in a cryotome. To generate artifact-free tissue sections, the embedding media should ideally enclose the tissue in a "snug-fit" way. Avoid embedding media such as "optimum cutting temperature (OCT)" polymer, which is widely used in

histology. OCT contains high concentrations of polymers that readily ionize under MALDI conditions—this can severely affect the quality of the mass spectral data. Helpful hints on preparing high-quality sections for MALDI-MSI can, for example, be found in ref. 22.

7. For many tissues, rapid dipping into lq N_2 is a convenient snap-freezing method. However, freezing artifacts like tissue cracks are then generally difficult to fully avoid. These artifacts may be only visible in high-resolution MSI experiments, though [5]. Precooling the material in the cold vapor phase just above the lq. N_2 reservoir prior to dipping them into liquid ultracold nitrogen may help to reduce such artifacts. Also check out the literature for further protocols if needed.

8. Most cryotomes allow to reliably cut tissue sections with thicknesses between about a few to several 10 μm. However, generating cryo-fixated tissue sections with a thicknesses below 10 μm is challenging and generally requires particular expertise. Sections thicker than 20 μm may also be used for top-illumination MALDI-2, but the lateral resolution of the MALDI-MSI measurement may then potentially be impaired, because analyte substances, which successively cocrystallizes with the MALDI matrix, may be extracted from more extended volume elements.

9. Cells grown on ITO-coated glass slides can show a different growth behavior than those grown in standard culture flasks or on uncoated glass slides. Growth conditions (e.g., cell density and culturing time) should therefore be optimized for each cell line under investigation, individually. For example, for Vero-B4 cells, 2×10^4 cells are the ideal amount of cells that, cultivated for 48 h in 8-well chambers, show subconfluency. For other cell types, these values may differ and optimal values may need to be determined individually.

10. Fixation with 4% formaldehyde helps to retain cellular morphology, albeit at some expense of chemical and membrane integrity, as amino group–containing proteins and lipids are cross-linked and the cell membrane gets permeabilized—which leads to a leakage of the cytosol, in particular in subsequent washing steps. Thus, small metabolites from the cytoplasm may potentially no longer be accessible to the MSI measurement. The chemical fixation step can be skipped if retaining the cellular morphology is less important.

11. Suitable protocols for other spraying and sublimation systems are available in the literature or from vendors supplying commercial devices.

12. If possible, avoid using "too wet" matrix application for cells, which can lead to delocalization of the outer leaflet plasma

membrane lipids. This can occur during spray-coating, recrystallization of sublimated matrix, or even in sublimation-coated preparations with samples that are exposed to ambient air for too long. Lipids from the inner leaflet of the plasma membrane are usually less affected.

13. A successful MALDI-2 experiments requires a careful unobstructed alignment of the optical system. Regular control measurements on a dedicated test sample are therefore advised. Realignment, adjustments, or replacement of the optical system, however, should only be done by fully trained personnel. Note that when eventually working with free laser beams it is mandatory to wear protective laser goggles.

14. At constant laser pulse energies, the highest TIC is generally obtained when the final focusing lens in the postionization laser path is positioned such that the maximum fluence (corresponding to the minimal focal beam width) is reached in the interaction zone of laser beam and MALDI particle cloud (plume).

15. For an area of 1 mm × 1 mm with a pixel size of 1 μm and a fixed C-trap injection time of 250 ms this results in a data acquisition time of approximately 3 days (at a data acquisition rate of ~3.7 pixel/s and 25 laser pulses/pixel). The high mechanic stability of the instrument and its superior positional reproducibility, however, allows for such long measurements without any notable degradation in MS image quality. To further improve the constancy of the mass accuracy of the MSI run over time, for example prominent matrix-derived ion signals that are generated by the postionization laser may be used as a lock mass. For higher mass-resolving power of the Q Exactive plus, increase the C-trap injection time. For example, for the highest possible mass-resolving power of 280,000 (defined for an *m/z* value of 200), use 1 s. This changes the data acquisition time to ~0.9 pixel/s.

16. With hotspot removal, the predefined quantiles of the pixel-intensities (usually 99%) are used to calculate the cutoff value for intensity visualization to enable a more intuitive use of the color scale. Normalization strategies can be used to account for systematic errors in data acquisition. It may however also lead to the formation of artifacts and should be handled with due care. Strategies for normalization and hot spot removal are for example described in ref. 23.

17. This can crucially support feature detection, for example when the data are discussed with a pathologist. A few dyes are also MALDI-2-MSI compatible, such as Hoechst 33342, which stains cell nuclei. Tissue sections or cell cultures may in these cases be stained before the matrix coating and imaged using a fluorescence microscope.

18. Although generally cryo-fixated tissues are highly preferable as samples for MALDI-MSI imaging, formalin-fixed, paraffin-embedded (FFPE) tissues may be usable for certain analyte classes such as the discussed glycans, but also enzymatically digested proteins. In these cases, dewaxing, and eventually antigen retrieval steps, are required.

19. Because of the higher sensitivity, traditionally MALDI-MS and -MSI of *N*-glycans is performed in the positive ion mode and by relying on the detection of $[M + Na]^+$ or $[M + K]^-$ adduct ions. With MALDI-2, the normally lower yields of $[M - H]^-$ can be improved by up to 3 orders of magnitudes. As demonstrated in ref. 13 this has, in particular, implications for an advanced structural analysis of the oligosaccharides, because collisional-induced dissociation (CID) experiments performed with deprotonated molecular precursors generally provide more structural details, including those obtained from cross-ring cleavages.

Acknowledgments

We acknowledge the invaluable input and support from numerous expert colleagues, many of whom are listed as coauthors on the cited papers. Financial support by the German Research Foundation (DFG; grants DR416/9-1, project number 208319078, to K. D.; DR416/12-1 and SO976/3-1, project number 290343045, to K.D. and J.S.; and SO976/5-1, project number 400912714, to J.S.) and the Interdisciplinary Center for Clinical Research (IZKF) of the Münster University Medical School (grant Drei2/018/17, to K.D. and J.S.) is gratefully acknowledged.

References

1. Buchberger AR, DeLaney K, Johnson J, Li L (2018) Mass spectrometry imaging: a review of emerging advancements and future insights. Anal Chem 90:240–265
2. He L (ed) (2015) Mass spectrometry imaging of small molecules. Methods in molecular biology. Humana Press/Springer, New York, NY
3. Cole M (ed) (2015) Imaging mass spectrometry: methods and protocols. Methods in molecular biology. Humana Press/Springer, New York, NY
4. Kompauer M, Heiles S, Spengler B (2017) Atmospheric pressure MALDI mass spectrometry imaging of tissues and cells at 1.4-μm lateral resolution. Nat Methods 14:90–96
5. Niehaus M, Soltwisch J, Belov ME, Dreisewerd K (2019) Transmission-mode MALDI-2 mass spectrometry imaging of cells and tissues at subcellular resolution. Nat Methods 16: 925–931
6. Dreisewerd K (2004) The descrption process in MALDI. Chem Rev 103:395–426
7. Lu IC, Chu KY, Lin CY, Wu SY, Dyakov YA, Chen JL, Gray-Weale A, Lee YT, Ni CK (2015) Ion-to-neutral ratios and thermal proton transfer in matrix-assisted laser desorption/ionization. J Am Soc Mass Spectrom 26:1242–1251
8. Boskamp MS, Soltwisch J (2020) Charge distribution between different classes of glycerophospolipids in MALDI-MS imaging. Anal Chem 92:5222–5230
9. Taylor MJ, Lukowski JK, Anderton CR (2021) Spatially resolved mass spectrometry at the single cell: recent innovations in proteomics

and metabolomics. J Am Soc Mass Spectrom 32:872–894
10. Dilmetz BA, Lee Y-R, Condina MR, Briggs M, Young C, Desire CT, Klingler-Hoffmann M, Hoffmann P (2021) Novel technical developments in mass spectrometry imaging in 2020: a mini review. Anal Sci Adv 2:225–237
11. Soltwisch J, Kettling H, Vens-Cappell S, Wiegelmann M, Müthing J, Dreisewerd K (2015) Mass spectrometry imaging with laser-induced postionization. Science 348:211–215
12. Ellis S, Soltwisch J, Paine MRL, Dreisewerd K, Heeren RMA (2017) Laser post-ionisation combined with a high resolving orbitrap mass spectrometer for enhanced MALDI-MS imaging of lipids. Chem Commun 53:7246–7249
13. Heijs B, Potthoff A, Soltwisch J, Dreisewerd K (2020) MALDI-2 for the enhanced analysis of *N*-linked glycans by mass spectrometry imaging. *Anal Chem* 92:13904–13911
14. Barré FPY, Paine MRL, Flinders B, Trevitt AJ, Kelly PD, Ait-Belkacem R, Garcia JP, Creemers LB, Stauber J, Vreeken RJ, Cillero-Pastor B, Ellis SR, Heeren RMA (2019) Enhanced sensitivity using MALDI imaging coupled with laser postionization (MALDI-2) for pharmaceutical research. Anal Chem 91:10840–10848
15. Soltwisch J, Heijs B, Koch A, Vens-Cappell S, Höhndorf J, Dreisewerd K (2020) MALDI-2 on a trapped ion mobility quadrupole time-of-flight instrument for rapid mass spectrometry imaging and ion mobility separation of complex lipid profiles. Anal Chem 92:8697–8703
16. Bednařík A, Bölsker S, Soltwisch J, Dreisewerd K (2018) An On-tissue Paternò-Büchi reaction for localization of carbon-carbon double bonds in phospho- and glycolipids by matrix-assisted laser desorption ionization mass spectrometry imaging. Angew Chem Int Ed 57:12092–12096
17. Niehaus M, Robinson KN, Murta T, Elia EA, Race AM, Yan B, Steven RT, Bunch J (2020) MALDI-2 at atmospheric pressure-parameter optimization and first imaging experiments. J Am Soc Mass Spectrom 31:2287–2295
18. Bien T, Bessler S, Dreisewerd K, Soltwisch J (2021) Transmission-mode MALDI mass spectrometry imaging of single cells: optimizing sample preparation protocols. Anal Chem 93:4513–4520
19. Bien T, Perl M, Machmüller AC, Nitsche U, Conrad A, Johannes L, Müthing J, Soltwisch J, Janssen KP, Dreisewerd K (2020) MALDI-2 mass spectrometry and immunohistochemistry imaging of Gb3Cer, Gb4Cer, and further glycosphingolipids in human colorectal cancer tissue. Anal Chem 92:7096–7105
20. Niehaus M, Soltwisch J (2018) New insights into mechanisms of material ejection in MALDI mass spectrometry for a wide range of spot sizes. Sci Rep 8:7755
21. Potthoff A, Dreisewerd K, Soltwisch J (2020) Detailed characterization of the post-ionization efficiencies in MALDI-2 as a function of relevant input parameters. J Am Soc Mass Spectrom 31:1844–1853
22. Dannhorn A, Kazanc E, Ling S, Nikula C, Karali E, Serra MP, Vorng J-L, Inglese P, Maglennon G, Hamm G, Swales J, Strittmatter N, Barry ST, Sansom OJ, Poulogiannis G, Bunch J, Goodwin RJA, Takats Z (2020) Universal sample preparation unlocking multimodal molecular tissue imaging. Anal Chem 92:11080–11088
23. Deininger SO, Cornett DS, Paape R, Becker M, Pineau C, Rauser S, Walch A, Wolski E (2011) Normalization in MALDI-TOF imaging datasets of proteins: practical considerations. Anal Bioanal Chem 401:167–181

Chapter 3

Ambient Mass Spectrometry Imaging of Small Molecules from Cells and Tissues

Jae Young Kim, Heejin Lim, and Dae Won Moon

Abstract

New methods to analyze cells and tissues in ambient condition without any harsh chemical fixation or physical freezing and drying are summarized in this report. The first approach, an atmospheric pressure mass spectrometry imaging method, is based on laser ablation in atmospheric pressure assisted by atmospheric plasma and nanomaterials such as nanoparticles and graphene to enhance laser ablation. The second one is based on secondary ion mass spectrometry (SIMS) imaging of live cells in solution capped with single-layer graphene to preserve intact and hydrated biological samples even under ultrahigh vacuum for SIMS bio-imaging in solution.

Key words Ambient mass spectrometric imaging, Atmospheric pressure mass spectrometry imaging, Secondary ion mass spectrometry, Single-layer graphene, Laser ablation, Nanomaterials

1 Introduction

Recent advances in mass spectrometric imaging (MSI) enable label-free molecular mapping in 3D samples with subcellular lateral resolution [1, 2]. Major limitations of applying MSI techniques such as matrix-assisted laser desorption ionization (MALDI) and secondary ion mass spectrometry (SIMS) to biomedical and material researches are harsh and laborious sample preparation procedures such as chemical fixation and dehydration [3–5] for the compatibility to ultrahigh vacuum environment.

Recently, we have investigated new methods to analyze cells and tissues in ambient condition without any harsh chemical fixation or physical freezing and drying. The first approach, an atmospheric pressure mass spectrometry imaging (AP-MSI) method, is based on laser ablation in atmospheric pressure assisted by

Jae Young Kim and Heejin Lim contributed equally with all other contributors.

Young-Jin Lee (ed.), *Mass Spectrometry Imaging of Small Molecules: Methods and Protocols*,
Methods in Molecular Biology, vol. 2437, https://doi.org/10.1007/978-1-0716-2030-4_3,
© The Author(s), under exclusive license to Springer Science+Business Media, LLC, part of Springer Nature 2022

atmospheric plasma and nanomaterials such as nanoparticles and graphene to enhance laser ablation [6–11]. This approach is unique, since it is applicable for alive cells and tissues. The second approach is based on SIMS imaging of live cells in solution capped with single-layer graphene [12]. We used single-layer graphene to preserve intact and hydrated biological samples even under ultra-high vacuum for SIMS bio-imaging in solution with subcellular spatial resolution.

We wish new techniques described in this chapter on ambient mass spectrometry imaging of small molecules based on laser beam and ion beam assisted by nanomaterials to be a useful way to innovative biomedical researches for various areas of cell biology, disease mechanism studies and drug development.

2 Materials

2.1 Materials for Atmospheric Pressure Mass Spectrometry Imaging (AP-MSI)

2.1.1 Mass Spectrometry System

1. Mass spectrometer (*see* **Note 1**).
2. Electrospray ionization source.
3. Optical table.
4. Long drive scope stage.
5. Inverted microscope.
6. Programmable scanning stage for mounting onto an inverted microscope.
7. Relay and wire.

2.1.2 Laser Desorption Source

1. Diode-pumped solid-state (DPSS) laser system: CW laser with 300 mW at 532 nm.
2. Graduated neutral density (ND) filter.
3. Dichroic beam splitter: selectively reflection of a wavelength of 532 nm.
4. Filter cube for dichroic beam splitter mounting (microscope parts).
5. Optical shutter.
6. 100% reflective mirrors.

2.1.3 AP-Ionization Source: Nonthermal AP Helium Plasma Jet

1. Quartz tube: 2 mm ID, 3 mm OD, 15 cm long.
2. Copper tape: 6 mm wide.
3. Stainless steel tube: 4.57 mm ID, 6.35 mm OD (1/4″ OD), 10 cm long.
4. Epoxy glue and tube fitting parts.
5. Power supply: sinusoidal waveform, amplitude: up to 20 kV, frequency: 25–30 kHz.

6. High voltage (HV) cable.
7. Helium gas cylinder (UHP grade: 99.999%), regulator, and mass flow controller (MFC).

2.1.4 AP-Ionization Source: Electrospray Device

1. Head part of a commercial desorption electrospray ionization (DESI) source (Prosolia Inc., USA).
2. Customized chamber.
3. DC Power supply: up to 5 kV.
4. Syringe and syringe pump.
5. Methanol/water solvent (MeOH/H_2O 1:1, v/v) with 1% acetic acid.
6. Nitrogen gas cylinder (UHP grade: 99.999%), regulator, and MFC.
7. Stainless steel tubing: 4.57 mm ID, 6.35 mm OD, 7 cm long.
8. The 60-degree curved glass capillary: 2 mm ID, 2.5 mm OD, 20 mm long.
9. Silicone tubing: 6 mm ID, 8 mm OD, 2 m long.
10. Tube bender and gas torch.
11. Tube fitting parts.
12. Silicone tube pieces.
13. Elastic and stretchable film (Parafilm M).

2.1.5 Airflow-Assisted Particle Transport System

1. Customized chamber.
2. Diaphragm pump.
3. Stainless steel tubing: 4.57 mm ID, 6.35 mm OD, 12 cm long.
4. The 60-degree curved glass capillary: 2 mm ID, 2.5 mm OD, 20 mm long.
5. Silicone tubing: 6 mm ID, 8 mm OD, 2 m long.
6. Helium gas cylinder (HP grade: 99.999%), regulator, and MFC.
7. Tube bender and gas torch.
8. Tube fitting parts.
9. Silicone tube pieces.
10. Elastic and stretchable film (Parafilm M).

2.1.6 Preparation of Hippocampal Brain Slices of Adult Mouse

1. Adult mouse: 6- to 8-week-old, model: ICR, C57BL/6.
2. CO_2 chamber for small animals, CO_2 gas cylinder, regulator, and gas flowmeter.
3. Tissue chopper.
4. Aquarium bubbler and tubing.

5. Scalpel, scissors, curved spatula, and forceps.
6. Glass disposable Pasteur pipette.
7. Large glass petri dish.
8. Microscope.
9. Incubator.
10. Sucrose artificial cerebral spinal fluid (sACSF): Ingredients—124 mM NaCl, 2.5 mM KCl, 1.25 mM KH_2PO_4, 26 mM $NaHCO_3$, 2 mM $MgSO_4$, 2.5 mM $CaCl_2$, 10 mM D-glucose, 4 mM D-sucrose.
11. Phosphate buffer saline (PBS) for washing.
12. Mixed gas: 95% O_2–5% CO_2, regulator, and gas flowmeter.
13. Citrate-gold nanoparticles (AuNPs), graphene oxide (GO) nanosheets, and reduced graphene oxide (rGO) nanosheets.
14. 0.1% Polyethylenimine (PEI, in 25 mM borate buffer).
15. Slide glass.

2.1.7 Preparation of Zebrafish Caudal Fin

1. Wild-type (WT) zebrafish: standard lengths (SL) between 25.0 and 30.0 mm.
2. Fish tank for zebrafish.
3. Tricaine: 0.6 mM.
4. Razor blade, brush, and forceps.
5. Deionized (DI) water.
6. Citrate-AuNPs: 25 nm, optical density 10.0 at 800 nm, 200 μL.
7. 0.1% PEI in 25 mM borate buffer.
8. Slide glass.

2.1.8 Preparation of Single Graphene Layer on Glass Substrate

1. Single-layer graphene covered copper (Cu) foil.
2. Poly(methyl methacrylate) (PMMA) solution: MW ~996,000, 4.0% (w/w) in anisole.
3. Spin-coater.
4. 0.1 M ammonium persulfate solution (APS).
5. DI water.
6. Slide glass.
7. Hot plate for annealing.
8. Acetone, methanol, and isopropanol.

2.1.9 Software for AP-MSI

1. Tune and Xcalibur: software running in Thermo Scientific mass spectrometer.
2. Software for driving the scanning stage.

3. FireFly (v. 3.0, Prosolia, Indianapolis, USA): software for MSI data construction (imzML format).
4. BioMAP or MSiReader: software for MSI viewer and data analysis.

2.2 Materials for SIMS Imaging of Untreated Wet Cells in Solution Capped by Single-Layer Graphene

2.2.1 Cell Culture Wet Substrate

1. 6-in. silicon wafer.
2. Fabrication equipment (*see* Subheading 3.2.1 fabrication of a cell culture wet substrate).
3. Neutralized collagen solution (80% v/v collagen type I, 10% v/v 10× phosphate-buffered saline (PBS), 10% v/v 0.1 M sodium hydroxide).
4. Culture medium (RPMI-1640, L-glutamine, 10% v/v FBS, 50 μg/mL penicillin–streptomycin, 100 μg/mL neomycin).
5. High-density polyethylene film (HDPE).
6. Optical adhesive.
7. Ultraviolet (UV) lamp at a wavelength of 356 nm.
8. Glass substrates of size 6 × 13 mm^2.
9. A549 human lung carcinoma cells.

2.2.2 Graphene Capping

1. Single-layer graphene (SLG) grown on a Cu foil through chemical vapor deposition.
2. Aluminum foil loop with a 3 mm diameter hole in the center.
3. 0.1 M ammonium persulfate solution (APS).
4. 75 and 150 mM ammonium acetate solution (AAS).

3 Methods

3.1 Methods for Atmospheric Pressure Mass Spectrometry Imaging (AP-MSI)

This method section largely consists of four components: (1) Construct the AP-MSI system (Subheadings 3.1.1 through 3.1.6), (2) prepare the specimen (Subheadings 3.1.7 through 3.1.9), (3) data acquisition (Subheadings 3.1.8 through 3.1.10) and (4) convert the raw data to imzML format (Subheading 3.1.11) and construct chemical images (Subheading 3.1.12). The overall AP-MSI system is shown in Fig. 1 (schematics) and Fig. 2 (photograph). The AP-MSI system consists of a mass spectrometer (Subheading 3.1.1), a 532-nm CW laser (Subheading 3.1.2), an atmospheric pressure (AP) helium plasma jet device (Subheading 3.1.3), airflow-assisted particle transport equipment (Subheading 3.1.4), and a sampling and scanning stage (Subheading 3.1.6). Instead of AP-plasma jet equipment and airflow-assisted particle transport equipment, the electrospray ionization equipment (Subheading 3.1.5) can be used for postionization. For the tissue sample preparation, three methods are introduced here based on a short-

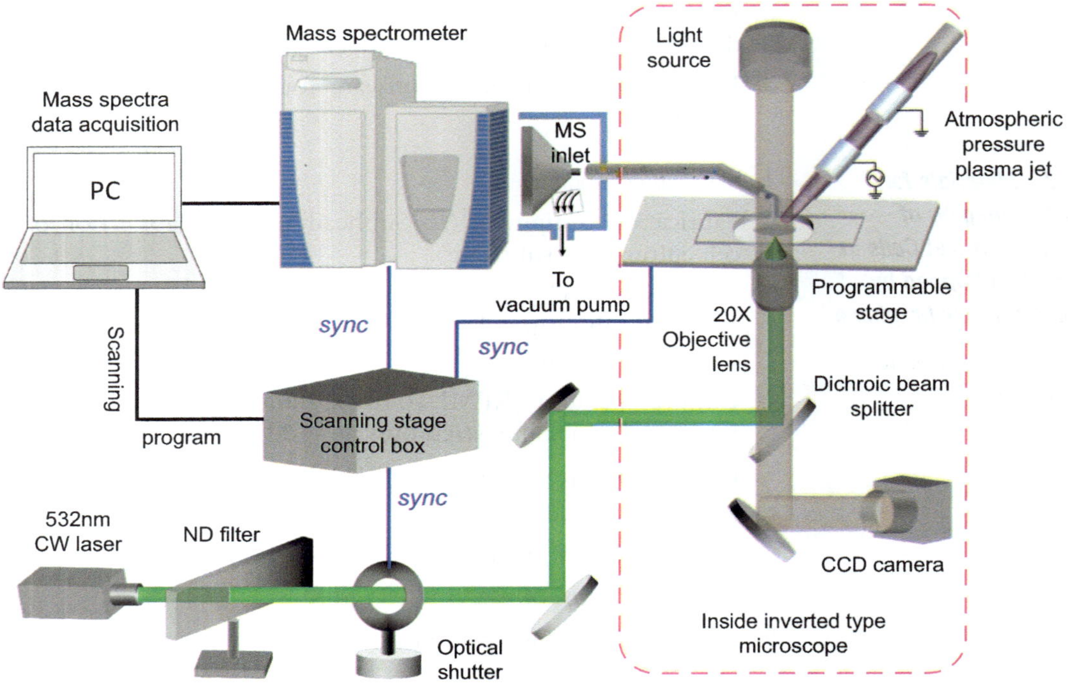

Fig. 1 Schematic description for whole AP-MSI system consists of a mass spectrometer, a scanning stage, a 532-nm CW laser, an AP plasma device, and airflow-assisted particle transport equipment. (Modified and reprinted with permission from ref. 9. Copyright 2019 American Chemical Society)

term living tissue slices well-established as pseudo two-dimensional models for the study of neurons, pathology, and electrophysiology [13, 14]: mouse hippocampus tissues treated with nanomaterials (Subheading 3.1.7) or on a single-layer graphene substrate (Subheading 3.1.8), and zebrafish fin (Subheading 3.1.9).

3.1.1 Basic Arrangement of Mass Spectrometer System

1. Place an optical table next to the table with the mass spectrometer (Fig. 2).
2. Install the long drive scope stage on the optical table and place an inverted type microscope on it (*see* **Note 2**).
3. Install the microscope a programmable stage on the inverted microscope.
4. Adjust the height of the mass spectrometer's table or optical table so that the height of the mass inlet of the mass spectrometer is approximately 2–3 cm higher than that of the programmable stage mounted on the inverted microscope.

3.1.2 Installation of Laser Desorption Source

To desorb the specimen with focused laser light, the objective lens of an inverted microscope is positioned under the specimen, and the AP-plasma jet device is positioned above the specimen. Thus,

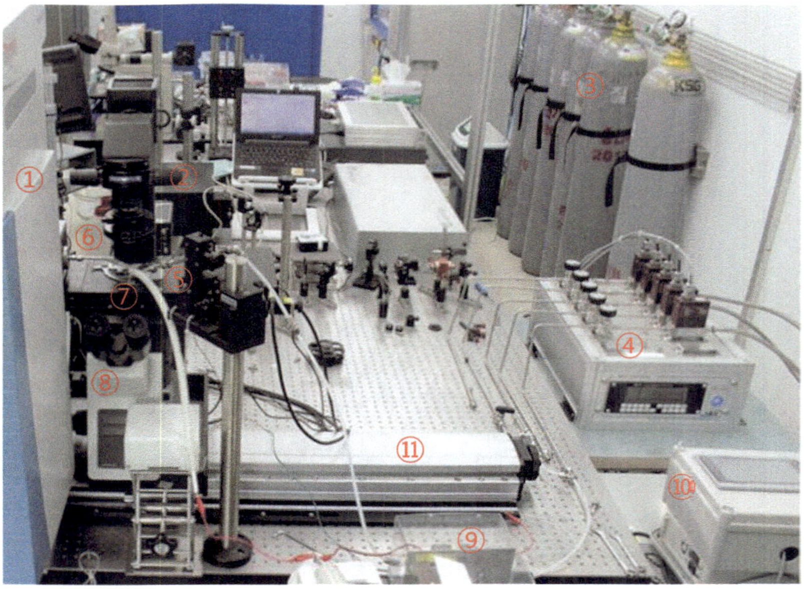

Fig. 2 A photograph of the entire system for AP-MSI (1) QE Orbitrap mass spectrometer, (2) CW laser system, (3) gas cylinders, (4) 5 ch. gas flow controller, (5) nonthermal atmospheric pressure plasma jet, (6) airflow-assisted particle transfer system, (7) auto scanning stage (sample stage), (8) inverted microscope, (9) HV driving circuit and HV probe, (10) joystick of auto scanning stage, and (11) long drive scope stage for microscope shift. (Modified and reprinted with permission from ref. 6. Copyright 2017 Springer Nature)

the laser light and plasma medium are completely separated during the sampling step; so even if they work at the same time, they do not affect each other's processes (Fig. 3).

1. Place the DPSS laser system on the optical table.
2. Using multiple sets of 100% reflective mirrors, create an optical path so that the laser beam can be aligned vertically and horizontally.
3. Introduce the CW laser beam to the rear port of the inverted microscope and align the laser beam to the center of the rear port.
4. Mount a dichroic beam splitter onto the microscope using a filter cube (*see* **Note 3**).
5. Insert a graduated neutral density (ND) filter on the optical path to control the input laser power.
6. Insert an optical shutter device on the optical path.

3.1.3 Installation of AP-Ionization Source: Nonthermal AP Helium Plasma Jet

A nonthermal AP helium plasma jet device is a typical dielectric barrier discharge device with a small discharge current [15–17].

1. Prepare the body of the plasma jet apparatus by attaching the 1/4″ stainless steel tube to the quartz tube using epoxy glue.

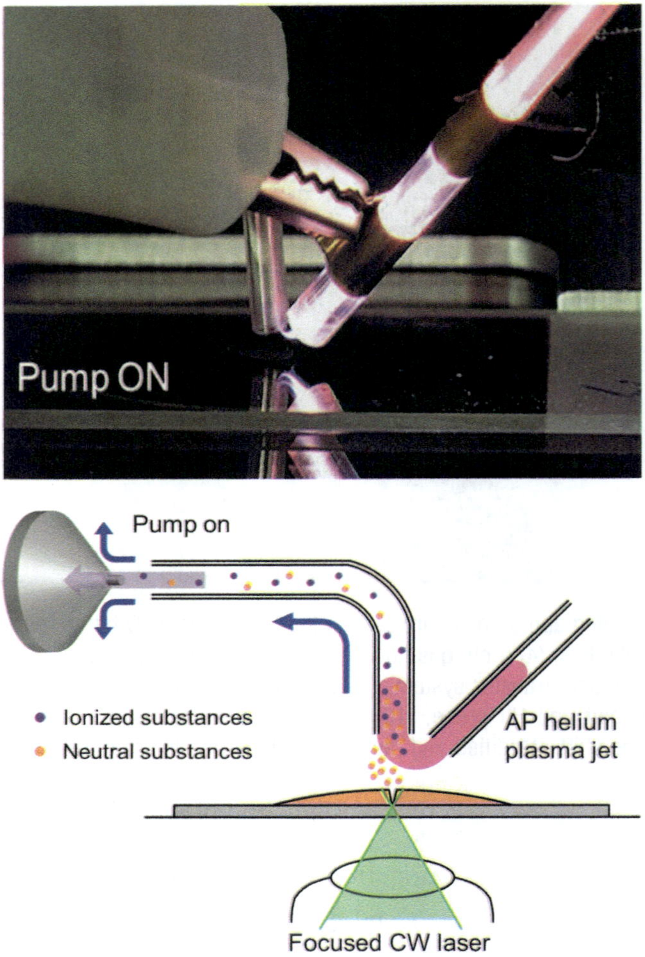

Fig. 3 A photograph and schematics during MS analysis on analysis stage. These show the configuration of AP-desorption and ionization sources, sampling stage, air-flow particle-transfer equipment, and mass spectrometer in order for transmission-mode laser desorption and AP helium plasma postionization. (Modified and reprinted with permission from ref. 6. Copyright 2017 Springer Nature)

2. Wrap one copper tape at a distance of 5 mm from the end of the quartz tube to form a powered electrode, and wrap another copper tape at a distance of 7 mm from the powered electrode to form the ground (GND) electrode.
3. Connect the plasma device to the *x*–*y*–*z* axis controllable support with an angle of incidence of 30°.
4. Connect the stainless steel tube of the plasma jet apparatus to the helium gas line including MFC using tube fitting parts.
5. Connect the HV cable and GND cable of the high voltage power supply into the powered electrode and GND electrode of the plasma jet apparatus, respectively (Electric feedthrough).

3.1.4 Setup of Airflow-Assisted Particle Transport System

1. Design and fabricate the customized chamber as a default ionization source to install the particle-transfer tube and a diaphragm pump as shown in Fig. 4 (*see* **Note 4**).
2. Using a tube bender, bend the stainless steel tube with 30° to make the particle-transfer tube.
3. Bend a glass capillary (i.d. 2 mm, o.d. 2.5 mm, length 20 mm) with 60-degrees using a gas torch and connect it to the transfer tube with pieces of silicone tubes and a highly elastic and stretchable film (Parafilm M).

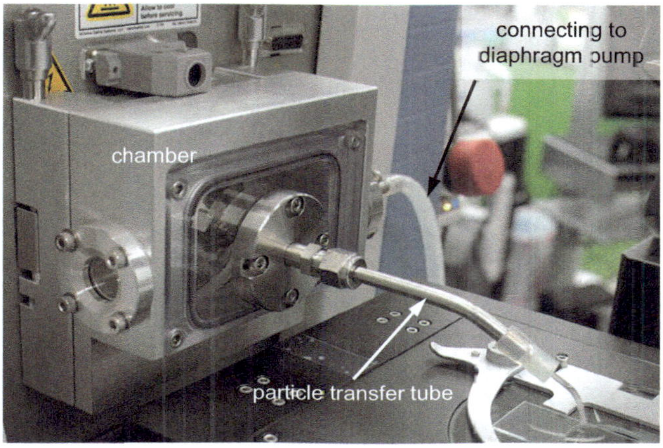

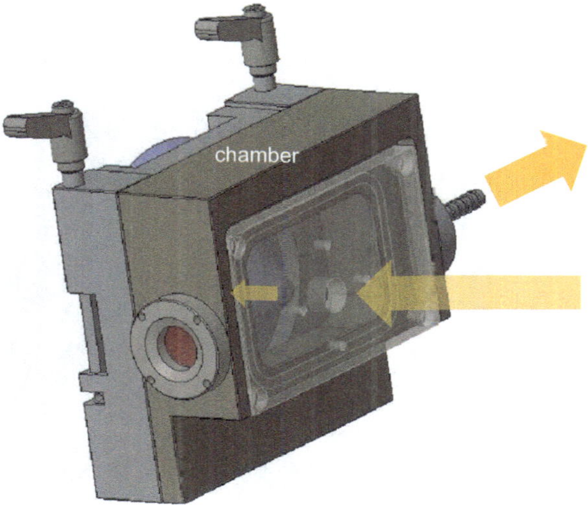

Fig. 4 A photograph of airflow-assisted particle-transfer equipment consists of particle-transfer tube, chamber, and diaphragm pump and a drawing of the customized chamber. The pump is connected to the side hole of the chamber and generated the airflow toward the MS inlet inside the particle transfer tube. (Modified and reprinted with permission from ref. 6. Copyright 2017 Springer Nature)

4. Connect the particle-transfer tube to the center hole of the chamber with a tube fitting.
5. Connect the pump to the side hole of the chamber via silicone tubing with o.d. 6.35 mm (1/4 in.). A flow rate of 30 slm is provided in the transfer tube.
6. Install the airflow-assisted particle transport equipment into the mass inlet (Fig. 4).

3.1.5 Installation of AP-Ionization Source: Electrospray Ionization Equipment

1. Design and build a simple chamber to be connected to the mass spectrometer inlet that can be equipped with a particle-transfer tube, electrospray ionization source, and diaphragm pump (Fig. 5).

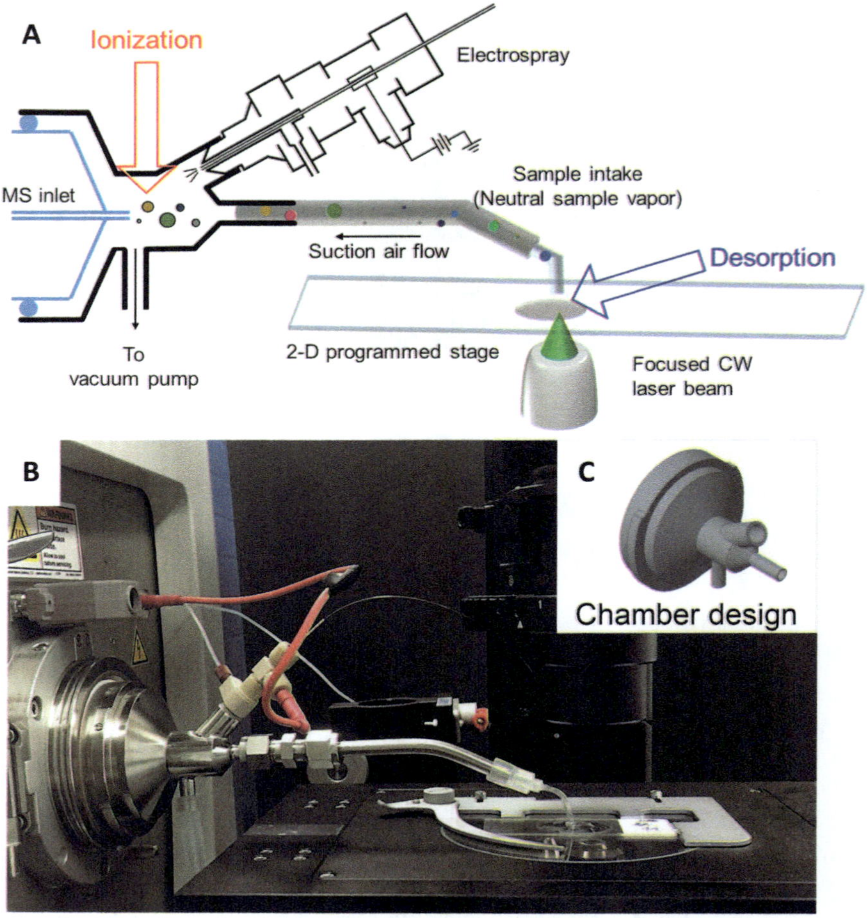

Fig. 5 A modified AP-MSI system with a combination of 532-nm CW laser, electrospray source, and airflow-assisted particle-transfer equipment. Upper image: a photograph of analysis stage in the modified AP-MS system and design of a customized chamber. Lower image: schematics of AP desorption and ionization procedure on analysis stage and particle-transport. (Modified and reprinted with permission from ref. 10. Copyright 2019 AIP Publishing)

2. Install the electrospray device, a head part of a commercial DESI source, in the simple chamber.
3. Connect the 3.5 kV DC voltage to the electrospray.
4. Connect the nitrogen gas feedline to the electrospray and introduce N_2 gas at 580 sccm.
5. Using syringe and syringe pump, apply methanol–water solvent ($MeOH–H_2O$ 1:1, v/v) with 1% acetic acid at 10 μL/min into the electrospray to produce charged droplets.
6. Using a tube bender, bend the stainless steel tube with 30° to make the particle-transfer tube.
7. Bend a glass capillary (i.d. 2 mm, o.d. 2.5 mm, length 20 mm) with 60-degrees using a gas torch and connect it to the transfer tube.
8. Connect the particle-transfer tube to the center hole of the chamber with a tube fitting.
9. Connect the pump to the bottom hole of the chamber via silicone tubing with o.d. 6.35 mm (1/4 in.). A flow rate of 30 slm is provided in the transfer tube.
10. Install the electrospray ionization equipment into the mass inlet (Fig. 5).

3.1.6 Setup of Programmable Scanning Stage

1. Mount a programmable motorized *X–Y* scanning stage on the inverted optical microscope (*see* **Note 5**).
2. To electrically connect the scanning stage of **step 1** with the mass spectrometer and optical shutter to transmit and receive logic signals, do the following (Fig. 1, *see* **Note 6**).
3. Connect the logic output terminal of the scanning stage's control box to a relay with wire.
4. Connect one contact of the relay to the analog input terminal of the mass spectrometer and connect the other contact of the relay to the input of the optical shutter for laser.

3.1.7 Preparation of Hippocampal Tissue Slices of Adult Mouse with Nanomaterial Treatments

1. Sacrifice an adult mouse using CO_2 chamber (*see* **Note 7**).
2. Using surgical equipment (scalpel and scissors), cut the head of the mouse and open the skull to extract the brain.
3. Isolate hippocampus from the brain using curved spatula and scalpel and wash it with sACSF.
4. Slice the isolated hippocampus transversely at a thickness of 200 μm using a tissue chopper (*see* **Note 8**).
5. Place the sliced hippocampal tissues in a large glass Petri dish filled with sACSF solution and separate from each other manually with a glass disposable Pasteur pipette.

6. Aerate the hippocampal tissue slices with sACSF bubbled by 95% O_2–5% CO_2 using an aquarium bubbler at 32 °C for 1 h inside the incubator.
7. Immerse the hippocampal tissue slices in 5 mL of sACSF solution with nanomaterials, such as citrate-gold nanoparticles (AuNPs), graphene oxide (GO) nanosheets, and reduced graphene oxide (rGO) nanosheets, at 32 °C for 1 h inside the incubator (*see* **Note 9**).
8. After incubation with the nanomaterials, wash the tissue slices 10 times with sACSF.
9. Drop 0.1–0.2 mL of 0.1% PEI in 25 mM borate buffer onto slide glass and place the tissue slice on the slide before the PEI is completely dry.

3.1.8 Preparation of Hippocampal Tissue Slices on Graphene Coated Glass Substrate

1. Spin-coat PMMA solution onto the surface of the graphene covered Cu foil at 4000 rpm for 45 s and dry in air.
2. Prepare the PMMA-supported graphene by removing the Cu foil using 0.1 M APS.
3. Rinse the PMMA-supported graphene with deionized water several times and transfer it onto a glass slide.
4. Dry the glass slide in air for 30 min and anneal it at 180 °C for 3 h.
5. Dissolve the PMMA polymer by immersion in acetone for 30 min, rinse by methanol for 10 min and by isopropanol for 10 min, respectively, and dry it.
6. Follow the same steps in Subheading 3.1.7 except **steps** 7 and **8**.

3.1.9 Preparation of Zebrafish Caudal Fin

1. Maintain zebrafish at 25 °C in separate tanks (one fish per tank).
2. Anesthetize zebrafish in 0.6 mM Tricaine and amputate the caudal fin using a razor blade (*see* **Note 10**).
3. Submerge the amputated fins in 5 mL of deionized (DI) water with citrate-AuNPs (optical density 10.0 at 800 nm, 200 μL) for 1 h.
4. After 1 h of treatment with the AuNPs, rinse the fin 10 times with DI water and place it on 0.1% PEI-coated slide glass (*see* **Note 11**).

3.1.10 Data Acquisition

This subsection will describe how to perform a mass spectrometry experiment on a lab-built AP-MSI system including the Q-Exactive hybrid quadrupole-Orbitrap mass spectrometer (QE Orbitrap; Thermo Fisher Scientific; Q Exactive, San Jose, CA, USA). Similar data analysis can be performed for other types of imaging mass spectrometers following the vendor-specific user manual. The following data acquisition should be proceeded after mass calibration using the ESI source and Tune software.

1. Place a slide with the tissue specimen on the 2-D programmable stage mounted on the inverted optical microscope.
2. Observe the entire area of the specimen through a microscope and specify the area to be analyzed in the specimen using the stage's drive software or an external joystick.
3. In the stage software, define the area to create the MS image by entering coordinate values for the *X*-axis raster speed and the *Y*-axis spacing.
4. Turn on the laser source and adjust the output power of the CW beam to the specimen desorption condition by controlling the graduated ND filter in front of the laser (*see* **Note 12**).
5. Turn on the optical shutter located between the ND filter and the rear port of the microscope in the laser optical path, activate it in remote mode, and check whether it can be operated by the stage software.
6. Apply helium gas with a gas flow rate of 0.5 slm to the AP-plasma ionization source and generate the plasma jet by applying AC voltage (5 kV_p and a frequency of 27 kHz).
7. Turn on the diaphragm pump connected to the chamber of the particle transfer equipment.
8. Use the Tune software to save the adjustment file containing the scan parameters, ionization source parameters, and acquisition information.
9. Set up a new device method using Xcalibur software. This method should include the desired mass range, resolution, polarity, and adjustment file saved in **step 8**.
10. Set up a new sequence using Xcalibur software. Includes the name of the data file, the destination folder for the file, the saved device method, and the saved location file. Repeat this for all tissue sections you want to run (*see* **Note 13**).
11. Select all samples to be analyzed and select "Run order."

3.1.11 MS Raw Data Conversion to imzML Format

1. Open the FireFly software, select the converted raw data files and load them into the software.
2. Identify the *x*-axis and *y*-axis of expecting MS imaging and define all user parameters appropriately.
3. Click "Convert" to generate an MSI file (imzML format).
4. View MS images for specific m/z values using BioMAP [18] or MSiReader [19] to ensure the ions of interest are derived from the tissue specimen.

3.1.12 Chemical Mapping Information of MSI and Tentative Mass Assignment

1. Find as many ion peaks as possible that show spatial distributions similar to the outline of the specimen by comparing ion images with optical images.
2. Search for biomolecules that may exist in the tissue specimen using the Human Metabolome Database (www.hmdb.ca).
3. Perform the assignment of ion peaks by matching the theoretical mass value of the molecules with the measured mass value.
4. Confirm the ion assignment with comparison of the mass fragment patterns of the tandem MS data with the corresponding standard materials (*see* **Note 14**).

3.2 Methods for SIMS Imaging of Untreated Wet Cells in Solution Capped by Single-Layer Graphene

3.2.1 Fabrication of a Cell Culture Wet Substrate (Fig. 6)

1. Deposit low-stress silicon nitride (Si_3N_4), 100 nm in thickness, on both sides of a silicon wafer using plasma-enhanced chemical vapor deposition (Fig. 6a).
2. Pattern the Si_3N_4 thin film on the front of the silicon wafer into a 2 × 2 or 3 × 3 array of 40 × 40 μm^2 microholes, with each hole diameter of 1 μm and spaced at intervals of 2.2–3 μm, by using photolithography (Fig. 6b, m–o) (*see* **Note 15**).
3. Pattern the Si_3N_4 thin film on the back of the silicon wafer into a large window of 2.5 × 2.5 mm^2 in the center (Fig. 6c).
4. Remove photoresist film from both sides of the silicon wafer by plasma ashing (Fig. 6d).
5. Etch the back of the silicon wafer in 30% potassium hydroxide (KOH) solution at 80 °C to make a reservoir for cell culture media, and stop the etching when the microholes in the Si_3N_4 thin membrane is exposed (Fig. 6e, j, k) (*see* **Note 16**).

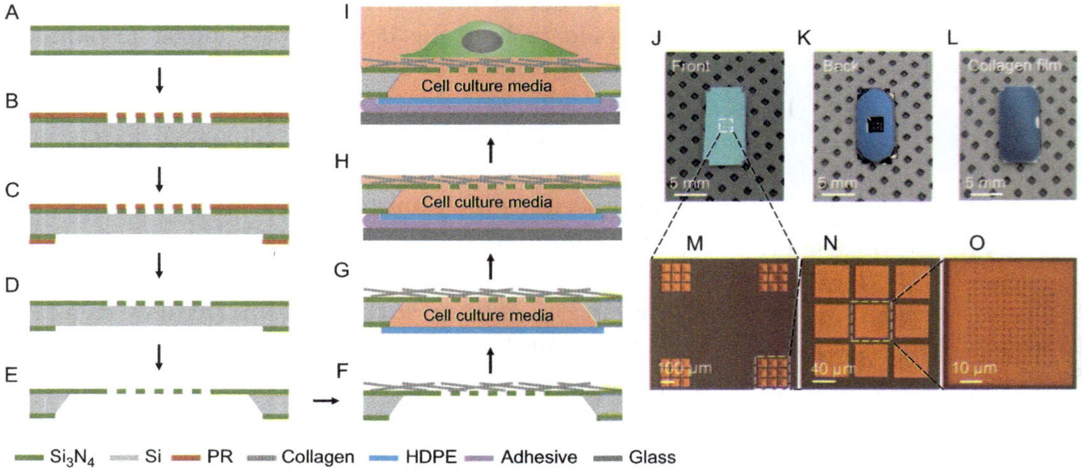

Fig. 6 Fabrication of a cell culture wet substrate. (**a–i**), Schematic workflow of the fabrication. (**j**, **k**) The front (**j**) and back (**k**) sides of the patterned substrate with a microhole array Si_3N_4 membrane. (**l**) Collagen film coated on the top of the substrate. (**m–o**) a microhole array Si_3N_4 membrane. (Modified and reprinted with permission from ref. 12. Copyright 2021 Springer Nature)

6. Rinse the silicon wafer in distilled water for 20 min.
7. Cut the silicon wafer into each fabricated substrates of size $6 \times 12\ mm^2$.
8. Drop 100 μL neutralized collagen solution onto the entire surface on the front of a fabricated substrate and dry overnight to form an ~1-μm-thick collagen film over the microholes (Fig. 6f, l) (*see* **Note 17**).
9. Rinse the collagen film substrate in distilled water 3 times 1 h each time and dry in air.
10. Inject one drop of cell culture medium of 5 μL into the reservoir on the back of the fabricated substrate.
11. Put an HDPE film of size $4 \times 4\ mm^2$ over the drop (Fig. 6g) (*see* **Note 18**).
12. Apply an optical adhesive between the back of the fabricated substrate and a glass substrate, and cure the adhesive using a UV lamp at a wavelength of 356 nm for 1 min to seal the culture media inside and complete assembly (Fig. 6h) (*see* **Note 19**).
13. Store the completed wet substrate in cell culture medium overnight.
14. Rinse the wet substrate in PBS twice.
15. Culture cells on the wet substrate in cell culture medium overnight (Fig. 6i).

3.2.2 Graphene Capping (Fig. 7)

1. Remove graphene on the back of a Cu foil using O_2 plasma reactive ion etching for 13 s (*see* **Note 20**).
2. Cut the Cu foil into $5 \times 5\ mm^2$ squares.
3. Place a flat aluminum foil loop with a 3 mm diameter hole in the center on the front of the Cu foil (*see* **Note 21**).

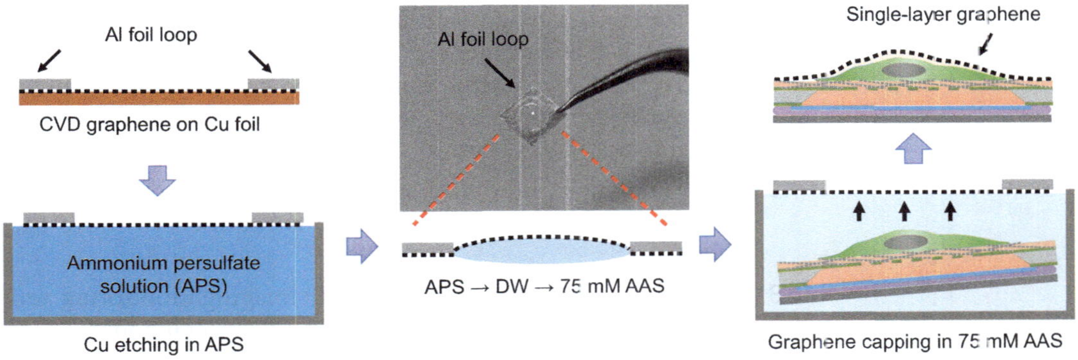

Fig. 7 Schematic workflow of graphene capping

4. Etch the Cu foil in a 0.1 M APS for 4 h to leave single-layer graphene floating on the surface (*see* **Note 22**).
5. Rinse the graphene in distilled water for 1 h (*see* **Note 23**).
6. Move graphene onto the surface of a 75 mM AAS.
7. Rinse the wet substrate culturing the cells in a 150 mM and then a 75 mM AAS once each.
8. Use the wet substrate to scoop up the graphene floating on the 75 mM AAS surface to transfer the graphene onto the cells.
9. Dry the graphene-covered untreated cells at 20 °C for 10 min before inserting the cells into the high-vacuum chamber of ToF-SIMS (*see* **Note 24**).

3.2.3 ToF-SIMS Imaging

1. Conduct ToF-SIMS analysis. Here are recommended instrument parameters for ToF-SIMS 5-100 instrument (ION-TOF, Münster, Germany): pulsed 30 keV Bi_3^+ primary ion beam in the delayed extraction mode, primary ion dose of 6.02×10^{11} ions/cm^2, analysis time of 10 min, both positive and negative ion images over a 500×500 μm^2 area with 256×256 pixels.
2. Perform internal mass calibration for all of the ToF-SIMS spectra using the peaks of CH_3^+, Na^+, $C_2H_3^+$, $C_3H_5^+$, and $C_4H_7^+$ for the positive ion mode, and C^-, C_2^-, C_3^-, and C_4^- for the negative ion mode before further analysis.
3. Inspect the graphene quality from all sets of microhole arrays and select the best set without noticeable damages in graphene for SIMS imaging.
4. Conduct ToF-SIMS imaging for graphene-covered untreated cells (*see* **Note 25**).

4 Notes

1. Obtaining MSI does not require a very high performance mass spectrometer. However, since the specimen is analyzed in the solid state, mass assignment based on the retention time is not available, it is recommended to use a mass spectrometer with tandem MS (MS/MS) function. The mass spectrometer employed is a Q-Exactive hybrid quadrupole-Orbitrap mass spectrometer (QE Orbitrap; Thermo Fisher Scientific).
2. When attaching and detaching the HESI source for mass calibration, the long drive scope stage is required to move the microscope.
3. A dichroic beam splitter allows optical imaging monitoring and laser desorption of a biological specimen simultaneously.

4. The way the Tune software recognizes the type of ion source is by connecting two pins of the ion source's serial port into one. For AP-MSI analysis, the serial port is set-up to recognize the customized chamber as a nanoelectrospray ionization (NSI) source. In addition, note that the ion source is only activated when the safety interlock engages.
5. The rastering should be performed by only moving this programmable stage while keeping at the fixed positions for the particle-transfer tube, the atmospheric pressure plasma jet device, and the laser focus.
6. Logic signals programmed and generated by the stage software were simultaneously fed to the scan stage, mass spectrometer, and optical shutter to trigger synchronized operation. Logic signals generated at the beginning and end of individual scan-lines are simultaneously transmitted to the mass spectrometer and optical shutter, enabling MS analysis of a specific area defined by synchronized multiple line scans.
7. All experiments must be approved and performed in accordance with the researcher's affiliated institution for animal use and care guidelines.
8. It is difficult to obtain uniform hippocampus tissue slice less than 200 μm without freezing or paraffin-fixation. In addition, the thickness more than 200 μm is also required to maintain short-term living tissue state (viable for 1–2 weeks).
9. Nanomaterials such as AuNPs and GO/rGO nanosheets that induce photothermal conversion in response to the selected wavelength of the laser light are commercially available and can be appropriately selected. The synthesis methods of these nanoparticles have been described in detailed in our previous works [6, 7]. Besides these nanoparticles, there are many nanomaterials that lead to photothermal conversion in response to laser light of a specific wavelength that can be applied in this kind of research.
10. When the zebrafish is quickly moved to the tank after the caudal fin amputation, the zebrafish comes alive (the anesthesia wears off) and the caudal tail begins to regenerate.
11. Since these tissue samples did not need to be completely dried, the analysis could be then started within 20 min.
12. The specimen desorption condition depends on the thickness of the specimen and the nanoparticle-treatment condition. As a result of repeated experiments, the output power of a CW laser is sufficient for 250 mW for a 200 μm-thick specimen.
13. In the sequence function of Xcalibur software, a single data file is mass spectrometry information for a single *x*-line of the resultant MSI, and the number of data files determines the

vertical size of the MSI. The data files in the sequence must be repeated so that it is performed up to the desired *y*-axis position. Depending on the size of the analysis area, it can be hundreds of the data files.

14. In the proposed AP-desorption/ionization method, in-source fragmentation may occur [20] leading to potentially incorrect annotation. The mass assignment in this study is limited to "tentative."
15. Fabricate multiple sets of microhole arrays on a cell culture wet substrate to increase the likelihood of successful SIMS imaging without damages in graphene.
16. Find the precise time for etching and Make sure not to etch overly until the whole Si_3N_4 membrane is exposed. Etching takes 6.5–7 h for a 650 μm thick silicon wafer. When only areas of microholes are exposed, they are seen as tiny pinholes by a naked eye (Fig. 6k).
17. Neutralized collagen solution should be used immediately after PBS and NaOH are added for collagen gel coating.
18. Make sure no air bubbles form.
19. After exposing the adhesive to UV for 2 s, apply it to the glass substrate first and put it together with the fabricated substrate.
20. Find the precise setting for reactive ion etching to remove graphene on the back completely and not to damage graphene on the front.
21. Make the aluminum foil loop flat by pressing it between two slide glasses.
22. After Cu etching, check the quality of graphene without damages and shrinkage.
23. Graphene can be safely moved between solutions by using the aluminum foil loop as a grip. Graphene floats stably on the solution surface suspended in the hole of the aluminum foil loop in the air (Fig. 7).
24. Take extra care in handling single-layer graphene because damages in graphene such as shrinkage, wrinkles, and cracks can adversely affect SIMS imaging.
25. ToF-SIMS imaging can start 20 min after graphene capping because it takes 10 min to dry graphene-covered cells and another 10 min to insert them into the high-vacuum chamber and get all the settings ready. We found that the SIMS images in the first 10 min of SIMS imaging show the intrinsic distribution in intact cell membranes [12].

Acknowledgments

This work was supported by the National Research Foundation of Korea (NRF) grant funded by the Korea government (MSIT) (NRF-2021R1C1C2010928 and 2021R1A2C1008704) and the BK21 FOUR project funded by the Ministry of Education, Korea (4199990113966).

References

1. Passarelli MK, Pirkl A, Moellers R, Grinfeld D, Kollmer F, Havelund R, Newman CF, Marshall PS, Arlinghaus H, Alexander MR (2017) The 3D OrbiSIMS—label-free metabolic imaging with subcellular lateral resolution and high mass-resolving power. Nat Methods 14 (12):1175–1183
2. Kompauer M, Heiles S, Spengler B (2017) Autofocusing MALDI mass spectrometry imaging of tissue sections and 3D chemical topography of nonflat surfaces. Nat Methods 14(12):1156–1158
3. Kourkoutis LF, Plitzko JM, Baumeister W (2012) Electron microscopy of biological materials at the nanometer scale. Annu Rev Mater Res 42:33–58
4. Svitkina T (2007) Electron microscopic analysis of the leading edge in migrating cells. Methods Cell Biol 79:295–319
5. Jungnickel H, Laux P, Luch A (2016) Time-of-flight secondary ion mass spectrometry (ToF-SIMS): a new tool for the analysis of toxicological effects on single cell level. Toxics 4(1):5
6. Kim JY, Seo ES, Kim H, Park J-W, Lim D-K, Moon DW (2017) Atmospheric pressure mass spectrometric imaging of live hippocampal tissue slices with subcellular spatial resolution. Nat Commun 8(1):1–13
7. Kim JY, Seo ES, Lim HJ, Kim H, Park J-W, Shin HH, Lim D-K, Moon DW (2018) Nanomaterials and continuous wave laser-based efficient desorption for atmospheric pressure mass spectrometric imaging of live hippocampal tissue slices. RSC Adv 8(15):8021–8025
8. Kim JY, Lee SY, Kim H, Park J-W, Lim D-K, Moon DW (2018) Biomolecular imaging of regeneration of zebrafish caudal fins using high spatial resolution ambient mass spectrometry. Anal Chem 90(21):12723–12730
9. Kim JY, Lim H, Lee SY, Song C, Park J-W, Shin HH, Lim D-K, Moon DW (2019) Graphene-coated glass substrate for continuous wave laser desorption and atmospheric pressure mass spectrometric imaging of a live hippocampal tissue. ACS Appl Mater Interfaces 11(30):27153–27161
10. Kim JY, Lee SY, Moon DW, Park J-W, Lim D-K, Song C (2019) Atmospheric pressure mass spectrometric imaging of bio-tissue specimen using electrospray-assisted CW laser desorption and ionization source. Biointerphases 14(4):041001
11. Lee SY, Lim H, Moon DW, Kim JY (2019) Improved ion imaging of slowly dried neurons and skin cells by graphene cover in time-of-flight secondary ion mass spectrometry. Biointerphases 14(5):051001
12. Lim H, Lee SY, Park Y, Jin H, Seo D, Jang YH, Moon DW (2021) Mass spectrometry imaging of untreated wet cell membranes in solution using single-layer graphene. Nat Methods 18:316. https://doi.org/10.1038/s41592-020-01055-6
13. Buskila Y, Breen PP, Tapson J, Van Schaik A, Barton M, Morley JW (2014) Extending the viability of acute brain slices. Sci Rep 4(1):1–7
14. Opitz-Araya X, Barria A (2011) Organotypic hippocampal slice cultures. J Vis Exp (48):2462
15. Lu X, Jiang Z, Xiong Q, Tang Z, Hu X, Pan Y (2008) An 11 cm long atmospheric pressure cold plasma plume for applications of plasma medicine. Appl Phys Lett 92(8):081502
16. Li Q, Li J-T, Zhu W-C, Zhu X-M, Pu Y-K (2009) Effects of gas flow rate on the length of atmospheric pressure nonequilibrium plasma jets. Appl Phys Lett 95(14):141502
17. Sands BL, Huang SK, Ganguly BN (2009) Current scaling in an atmospheric pressure capillary dielectric barrier discharge. Appl Phys Lett 95(5):051502
18. BioMap. http://www.maldi-msi.org
19. MSiReader. https://msireader.wordpress.ncsu.edu/
20. Xu Y-F, Lu W, Rabinowitz JD (2015) Avoiding misannotation of in-source fragmentation products as cellular metabolites in liquid chromatography–mass spectrometry-based metabolomics. Anal Chem 87(4):2273–2281

Chapter 4

Single-Cell Metabolomics with Rapid Determination of Chemical Formulas from Isotopic Fine Structures

Laith Z. Samarah, Akos Vertes, and Christopher R. Anderton

Abstract

Metabolomic measurements can provide functional readouts of cellular states and phenotypes. Here, we present a protocol for single-cell metabolomics that permits direct untargeted detection of a broad number of metabolites under ambient conditions, without the need for sample processing, and with high confidence in the discovery and identification of the molecular formulas for detected metabolites. This protocol describes combining fiber-based laser ablation electrospray ionization (f-LAESI) with a 21 Tesla Fourier transform ion cyclotron resonance mass spectrometer (21T-FTICR-MS) to obtain high confidence molecular formula information about detected metabolites. The f-LAESI source utilizes mid-infrared laser ablation through a sharp optical fiber tip, affording direct ambient analysis of cells without the need for sample processing. Using the 21T-FTICR-MS as a mass analyzer enabled measurement of the isotopic fine structure (IFS) for numerous metabolites simultaneously from single cells, and the IFSs were in turn computationally processed to rapidly determine the corresponding elemental compositions. This metabolomics technique complements other single cell omics measurement methods, helping to resolve complex molecular interactions that take place within cells unattainable from single cell transcriptomic and proteomics methods.

Key words Mass spectrometry, FTICR, FTMS, Ultrahigh mass resolution, LAESI, Single-cell analysis, Spatially resolved mass spectrometry

1 Introduction

Metabolites can provide a functional readout of cellular states and phenotypes, serving as real-time indicators of biochemical activity [1]. Over the last few decades, metabolomics has exploded in areas like nuclear magnetic resonance (NMR) and mass spectrometry (MS), where these techniques have the capability of studying an extensive collection of small molecules present within cells or tissues of interest [2, 3]. These studies aim to complete our understanding of the complex molecular interactions that take place within biological systems, filling in knowledge gaps unattainable

Young-Jin Lee (ed.), *Mass Spectrometry Imaging of Small Molecules: Methods and Protocols*,
Methods in Molecular Biology, vol. 2437, https://doi.org/10.1007/978-1-0716-2030-4_4,
© The Author(s), under exclusive license to Springer Science+Business Media, LLC, part of Springer Nature 2022

from genomic and proteomic studies [1, 4]. MS-based metabolomic methods are ubiquitously employed for bulk characterization of metabolites extracted from homogenized tissue and cell lysates [5]. These common tissue-scale or cell-population-level analyses, however, generate data blind to the metabolic state of individual cells, and they yield abundance data that averages out aspects of cellular heterogeneity [6, 7]. Moreover, chemical species present only in rare cells are not even detected due to the dilution caused by the presence of all the other cells.

The bioanalytical chemistry research community has increasingly set its sights on adapting MS-based metabolomics to a single-cell scale. Single-cell omics differs from bulk omics in that analysis and/or extraction is constrained and targeted to individual cells within a tissue or a culture. However, the ability to obtain single-cell metabolomics data is a nontrivial endeavor, complicated, in part, by the low sample volume (e.g., ~1 pL for a mammalian cell) [8], high turnover rates [9], and the determination of metabolite identities usually based on only accurate mass data [10]. Specifically, per the latter issue, many small molecules with different elemental compositions have close to identical accurate masses. Fourier transform ion cyclotron resonance (FTICR)-MS has the mass resolving power to minimize molecular interferences, increase the MS-peak capacity, and measure the isotopic fine structure (IFS) of metabolites [3]. Capturing the IFSs provides increased confidence in the identification of detected metabolites.

Our approach of combining fiber-based laser ablation electrospray ionization (f-LAESI) with a 21 Tesla FTICR (21T-FTICR) mass spectrometer allows us to overcome many obstacles associated with single-cell metabolomics. The f-LAESI approach utilizes a mid-infrared (IR) laser ablation by a sharp optical fiber tip [3], and this permits us to perform direct in situ analysis of cells without the need for sample processing. Since all key metrics of FTICR-MS performance improve with increased magnetic field strength (e.g., mass resolution, acquisition speed, dynamic range, etc.) [11], we exploit the unique 21T-FTICR mass spectrometer at the Environmental Molecular Science Laboratory (EMSL) [12] to capture IFS for numerous metabolites simultaneously from single cells. The IFSs are in turn computationally processed to rapidly determine the corresponding elemental compositions.

2 Materials

2.1 Reagents and Chemicals

1. Electrospray solution for positive-ion mode ionization: 0.1 % (v/v) acetic acid in 1:1 (v/v) methanol–HPLC-grade water.
2. Electrospray solution for negative-ion mode ionization: 2:1 (v/v) methanol–chloroform.

3. 1-Methyl-2-pyrrolidinone.
4. 4% (v/v) nitric acid solution in HPLC-grade water.
5. 2.5 % carboxymethylcellulose (CMC).
6. Wide-base (100 mm) Pyrex beaker.

2.2 Biological Samples and Associated Tools

1. Soybean [*Glycine max*, (L.) Merr from Williams 82 seeds] plants with root nodules infected by *Bradyrhizobium japonicum* (USDA 110) for 21 days [13].
2. Fresh *Allium cepa* bulb.
3. Cryostat microtome.
4. Fine-point tweezers.
5. Microscope glass slides.
6. Embedding base molds.
7. Specimen stage.

2.3 f-LAESI Components

2.3.1 Laser Pulse Delivery

1. Optical breadboard mounted on a vibration-isolated support frame.
2. Optical parametric oscillator driven by Nd:YAG laser (IR Opolette HE 2940; Opotek, Carlsbad, CA, USA) to produce laser pulses with 2940 nm wavelength.
3. External laser pulse energy attenuator that can be controlled with a continuous rotation mount.
4. Five-axis fiber coupler.
5. Planoconvex calcium fluoride lens (50 mm focal length).
6. Focusing lens tube.
7. Bare fiber chuck (BFC300, Siskiyou Corporation).
8. GeO_2-based glass optical fiber, 250 and 450 μm core diameter with outer thermoplastic and polyamide coatings (HP Fiber, Infrared Fiber Systems, Inc., Silver Spring, MD, USA) (*see* **Note 1**).
9. Ceramic probe holder (MXP-150, Siskiyou Corporation, Grants Pass, OR).
10. Five-axis micromanipulator (MX110, Siskiyou Corporation).
11. Magnetic base for mounting five-axis micromanipulator (MGB-3.0, Siskiyou Corporation).
12. Sapphire scribe (Ideal 45-358, Precision Fiber Products, Inc., CA, USA).
13. Photodiode laser-pulse-energy sensor.
14. Thermal paper or mid-infrared laser detector card.
15. Protective spectacle from mid-IR (2940 nm) radiation.

2.3.2 Electrospray

1. Tapered-tip stainless steel emitter, 50-μm inner diameter (MT320-50-5-5, New Objective Inc., Woburn, MA).
2. Metal union.
3. Conductive perfluoroelastomer ferrule.
4. Fittings.
5. Needle port.
6. Fused silica tubing.
7. Gastight syringe.

2.3.3 Visualization System

1. Components to assemble two long-distance video microscopes with 7 × precision zoom optic: 2.0 × TV Tube (56-626, Edmund Optics, Barrington, NJ) (quantity: 2), Manual upper zoom module (56-627, Edmund Optics) (quantity: 2), Coaxial focusable lower module (13-415, Edmund Optics) (quantity: 2).
2. White Mi-LED fiber optic LED illuminator (89-740, Edmund Optics).
3. 0.125″ Core quartz fiber optic light guide (38-955, Edmund Optics).
4. 10 × infinity corrected long-working-distance objective lens (M Plan Apo 10 ×, Mitutoyo Co., Kanagawa, Japan).
5. 5 × infinity corrected long-working-distance objective lens (M Plan Apo 5 ×, Mitutoyo).
6. Two CCD or CMOS cameras.
7. Dynamically damped post.

2.3.4 Sample Stage Components

1. Motorized XYZ translation stages.
2. Peltier cooling stage components (*see* **Note 2**): Thermoelectric cooler (TEC) controller, Thermistor, Computer liquid cooling system, AC Power Supply Adapter, Liquid coolant, Liquid heat exchanger block, Cold Plate, Washer (quantity: 2), High Performance TEC.

2.4 Ultrahigh Mass Resolution Mass Spectrometer

1. A 21 T magnet (designed and made by Agilent Technologies, Yarnton, UK). The magnet is a 22-ton, 2.53 m long × 2.44 m wide × 3.36 m high system, with a room temperature bore of 112 mm. The cryostat of the magnet is directly pumped from a single reservoir system holding ~4050 L of liquid helium. The magnet itself features a multicoil, actively shielded design using both NbTi and Nb_3Sn superconducting wire [12].
2. A homebuilt hybrid mass spectrometer. System designed for the 21 T magnet used a Velos Pro dual linear quadrupole ion trap mass spectrometer (ThermoFisher Scientific, San Jose, CA, USA) for the front end of the hybrid platform [12]. The

hybrid system used a long linear quadrupole ion trap (75 mm long) to transfer ions from the Velos Pro to the ICR cell (it also served as a collision cell for tandem MS), and was operated at 1–3 mTorr of N_2. Two short rf-only quadrupole ion guides were used to transmit ions from the Velos Pro to the external collision cell, and then from the collision cell to the conductance limit of the first ultra-high vacuum stage. Another two quadrupole ion guides were used for ions to traverse the three stages of ultrahigh vacuum, each separated by conductance limits. The first three quadrupole ion guides are driven at ~3 MHz (300–800 $V_{p\text{-}p}$), and the last one, which transmits ions through the magnetic field gradient into the analyzer cell, was driven at ~4.4 MHz (up to 500 $V_{p\text{-}p}$).

3. Ions were measured using a custom analyzer cell developed at Pacific Northwest National Laboratory. The "window cell" is a harmonized ICR cell that utilizes external shimming to closely approximate an ideal quadrupolar trapping electric field [14].
4. The Velos Pro tune software (ThermoFisher Scientific, San Jose, CA, USA) was modified to enable integration of the additional power supplies of the hybrid systems and a FT acquisition computer. The ICR cell control was integrated into the Velos Pro tune software, and it enabled automated tuning of ion transfer to the ICR cell. Stored waveform inverse Fourier transform excite waveforms were generated by the FT acquisition computer and amplified by a linear power amplifier (T&C Power Conversion, Inc., Rochester, NY, USA).
5. Time-domain data processing was performed on the FT acquisition computer, and spectra were stored as .raw files.

2.5 Data Analysis

1. Thermo Xcalibur 2.2 (ThermoFisher Scientific, San Jose, CA, USA).
2. Ecipex for theoretical IFS generation (https://cran.r-project.org/web/packages/ecipex/index.html).
3. Formularity software for computational IFS matching [15] (https://omics.pnl.gov/software/formularity).
4. ChemCalc for molecular formula calculation (http://www.chemcalc.org/mf_finder).

3 Methods

Herein, we describe how to prepare samples for single cell analysis, fabricate the optical fiber, set up the probing source, utilize the mass spectrometer to obtain ultrahigh mass accuracy and mass resolution data, and perform data analysis to obtain chemical formulas from IFS information. While obtaining IFS of metabolites is possible on other FT-based instruments (e.g., Orbitrap systems), these approaches are limited to measuring small *m/z* molecules and complex mixtures should be avoided. As such, the use of FTICR-MS is recommended for IFS measurements of single cells (i.e., a complex mixture of biomolecules), because it provides the highest mass resolving power and mass accuracy of all mass analyzers [11] and it has a higher dynamic range than Orbitrap based systems.

3.1 Sample Preparation for Single Cell Analysis

1. Remove the nodules from the roots with fine tweezers, place them in base molds prefilled with 2.5% CMC. Cool the molds in a cryostat microtome at −10 to −15 °C until the embedding medium is frozen.
2. Mount the frozen block on a specimen stage and prepare 30 μm thick cryosections, thaw-mount them on glass slides, and keep them frozen in a dry-ice-filled container. If possible, fresh cryosections should be prepared and analyzed immediately instead of storing them prior to analysis.
3. For analysis of single epidermal cells from an *A. cepa* bulb, remove the outer layer (leaf base) and cut out a small piece of the inner layer using a sterile scalpel. Peel off a monolayer of epidermal cells using sterile fine point tweezers, and place it flat on the microscope slide.

3.2 Fabrication of Optical Fiber

1. Strip the outer thermoplastic and polyamide coatings covering the last ~2 cm ends of the optical fiber by placing them in a small glass beaker containing 10 mL of 1-methyl-2-pyrrolidinone and heating it to 200 °C. When the outer coating is dissolved and the glass cladding is exposed, rinse the optical fiber ends with isopropanol and dab dry with a lint-free napkin.
2. Cleave the two ends with a sharp sapphire scribe to form straight and roughly uniform surfaces.
3. Pour the 4% (v/v) nitric acid solution into the wide-base glass beaker to a liquid height of 5–10 mm. For optimal etching results, place the beaker on a vibration isolated tabletop for minimal vibrations (e.g., optical table).
4. Insert the optical fiber into a fiber chuck and allow ~5 cm of the fiber end to extend out of the opposite side of the chuck. Position the extended (~5 cm) portion vertically by securing the fiber chuck with the aid of a clamp that is in turn attached to a translation stage oriented in the vertical axis.
5. Slowly lower the extended fiber edge into the nitric acid solution using the translation stage until a liquid bridge forms between the surface of the acid solution and the fiber end. After the tip spontaneously detaches from the liquid surface, rinse it with deionized water to remove the acid residue.

6. Inspect the etched tip with a microscope. If it is etched to a diameter commensurate with the size of the studied cell, proceed to the next step. If the etching results in a tip diameter larger than the desired value, the same tip can be re-etched as described above, or the tip can be cleaved with the sapphire scribe and **steps 4** and **5** can be repeated.

3.3 f-LAESI Source Setup and Laser-Pulse Delivery

1. All f-LAESI components are mounted on a vibration-isolated optical table (Fig. 1).
2. Position the electrospray emitter on the same axis with, and ~8–10 mm from, the mass spectrometer inlet (*see* **Note 3**).
3. After etching, release the fiber from the chuck and secure its etched distal end with a ceramic probe holder. Position the holder at ~45° angle relative to the electrospray-inlet axis using a five-axis micromanipulator.
4. Hold the blunt (proximal) end with a fiber chuck and allow ~3–5 mm of the fiber to extend out of the chuck. Secure the latter into the five-axis fiber coupler by facing the proximal end to the focusing lens at a distance of ~50 mm (focal length).
5. Position the optical fiber tip 8–10 mm below the electrospray-inlet axis and 2–5 mm ahead of the emitter tip position (*see* **Note 3**).
6. Mount the first long-distance microscope (microscope 1) with the 10 × infinity-corrected objective lens and connected to a CCD camera on the dampened post and perpendicular to the electrospray-inlet axis. Mount the second long-distance microscope (microscope 2) with the 5 × infinity-corrected objective lens and connected to a CCD camera at ~15–30° angle from the electrospray-inlet axis (*see* **Note 4**). Position the two microscopes to focus on the fiber tip and to place it in the center of both fields of view. Adjust the magnification as required (*see* **Note 5**).
7. For visual aid during the laser-beam coupling process, switch on the red alignment diode that is built into the laser (*see* **Notes 6** and **7**).
8. Measure the laser pulse energy with the laser-pulse-energy sensor and set it at 0.5–1.3 mJ post-attenuation and in the path between the focusing lens and the proximal fiber end (*see* **Note 8**).
9. Begin a coarse alignment of the laser beam with the proximal end of the fiber by holding a piece of thermal paper onto the surface of the latter. Fire the laser and observe the location of the burn mark relative to the fiber end surface. If no burn mark can be seen on the thermal paper, adjust the distance between the focusing lens and the surface of the proximal end to be equal to the lens focal length (~50 mm). Observe the location

Fig. 1 f-LAESI setup on 21T-FTICR-MS. The top image shows the entire setup of the fiber LAESI system on a vibration-isolated optical table. Here, the laser (unit with a red casing) is coupled into the proximal end of the fiber optic with the bare fiber chuck and the five-axis fiber coupler. The distal end of the fiber is held in place with a ceramic probe holder and can be moved and finely adjusted using the five-axis micromanipulator. The bottom image shows a zoomed in region sampling area. Here, the sample is placed on the Peltier-cooled stage, the laser pulse is delivered through the fiber tip, and the ablated material is ionized via electrospray ionization before being measured by the 21T-FTICR-MS

of the burn mark relative to the fiber edge and fine-adjust the position of the latter using the translation knobs on the coupler until the two are aligned.

10. Set the Peltier stage temperature to 3–5 °C and raise its position with the aid of the motorized translation system until it is a few millimeters below the fiber tip (*see* **Note 9**).

11. Carefully place the sample, for example, onion epidermal-cell monolayer, on the cooling stage.
12. Raise the stage slowly until the cells become roughly in focus but before the fiber tip touches the sample surface. Both the sample surface and the fiber tip should be in the field of view and in roughly the same focus when visualized by the two microscopes.
13. Move the sample stage in the XY plane until the center of the targeted cell and the fiber tip are in the same plane orthogonal to the *Z* axis.
14. Pull the fiber tip out of the field of view in the *Z* direction and further raise the sample stage until the cells become in optimal focus. Obtain an optical image of the cell before laser ablation using the camera connected to the vertical microscope (*see* Fig. 2).

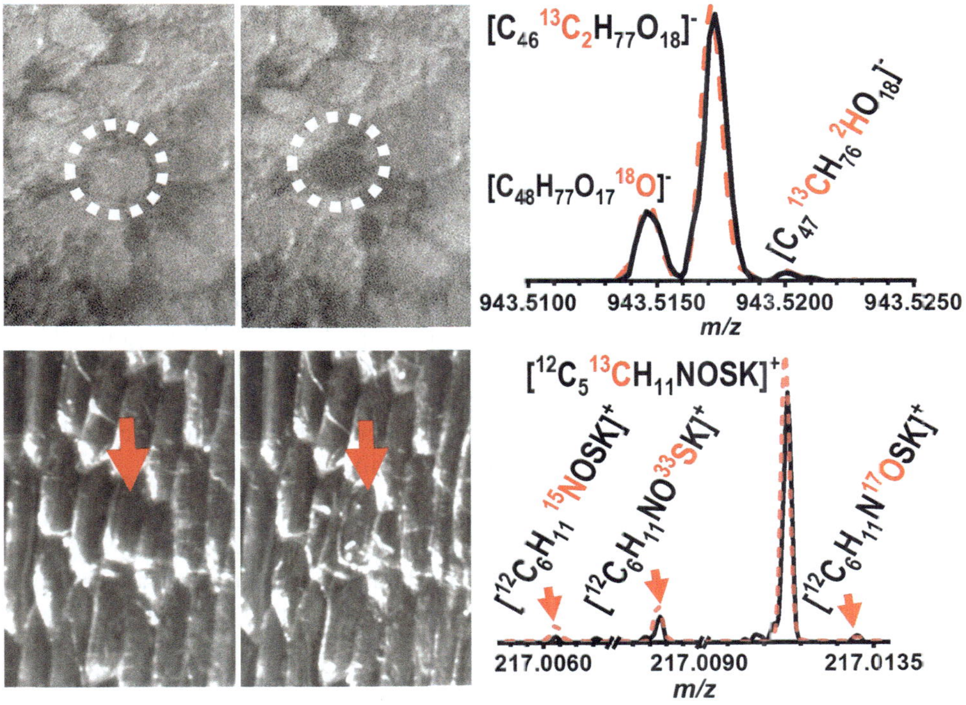

Fig. 2 Single-cell IFS. Top left panel shows microscope images of an infected cell from a soybean root nodule before (left) and after (right) laser ablation with the fiber tip. Top right panel shows an example of single-cell IFS at higher *m/z* range for dehydrosoyasaponin I. (Adapted with permission from [3]. Copyright 2020, American Chemical Society). Bottom left panel shows microscope images of an epidermal cell from *A. cepa* before (left) and after (right) laser ablation. Bottom right panel shows an example of a single-cell IFS at lower *m/z* range for alliin. Experimental and simulated IFS are displayed in solid black and dashed red lines, respectively

15. Carefully lower the fiber tip in the *Z* direction until it makes light physical contact with the cell surface (*see* **Note 10**).
16. Fire the laser at a pulse repetition rate of 1–20 Hz and keep it on while observing the targeted cell. If no ablation occurs, fine-adjust the proximal fiber edge position with the translation knobs on the coupler until the target is ablated (*see* **Note 11**).
17. If the laser-pulse delivery results in the ablation of more than one cell, lower the rate of energy deposition into the target by decreasing the pulse repetition rate. It is also possible that the laser pulse energy may need to be further attenuated.
18. After successful laser ablation of a single cell, pull the fiber out of the field of view in the *Z* direction and obtain an image of the ablation area (*see* Fig. 2). Inspect the "before" and "after" ablation images. Image processing, such as image subtraction, can be used to aid in estimating the amount of removed material.

3.4 FTICR-MS Data Acquisition

1. Ensure the mass spectrometer is operating correctly and is calibrated (*see* **Note 12**).
2. Pump the electrospray solution at 500 nL/min and apply a voltage of +3.3 and −2.5 kV for positive and negative ion mode, respectively (*see* **Note 13**).
3. Place the sample on the Peltier stage and follow **steps 12–15** in Subheading 3.3.
4. The ablation was synchronized with the opening of the ion trap on the hybrid mass spectrometer platform using a data acquisition board (NI USB-6341; National Instruments, Austin, TX, USA) and a custom-developed LabVIEW program. The TTL pulse sent from the Velos Pro would trigger the laser flash lamp of the mid-IR laser, and then another trigger would be sent after a 675 ns delay that opened the Q-switch (delivering a laser pulse through the optical fiber tip).
5. A 50 ms ion accumulation time was set on the Velos Pro to fill the trap with ions resulting from a single laser shot.
6. For "broadband" MS data acquisition, spectra were collected from *m/z* 175 to 1200 with different mass analyzer transient times ranging from 0.096 to 6.144 s, which translated to mass resolutions of ~40,000 to ~1,800,000 at *m/z* 400, respectively (*see* **Note 14**).
7. For "narrowband" MS data acquisition, intervals of 50 *m/z* windows were collected separately between *m/z* 180 and 1500 and stitched together to produce a comprehensive mass spectrum (*see* **Note 15**). The transient time was gradually increased as higher-*m/z* ions were recorded. This increases the sensitivity

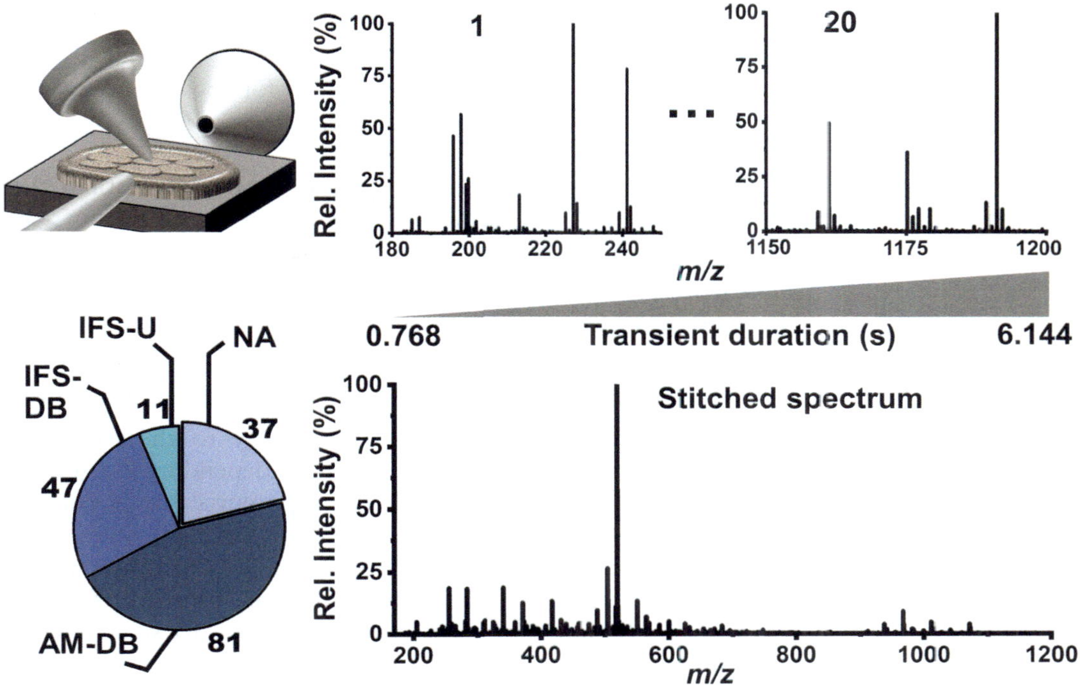

Fig. 3 Maximizing the number of metabolite assignments by accurate mass and IFS from single cells. Top left panel shows a schematic of f-LAESI-MS for single cells, each analyzed at a different 50-unit long *m/z* range. The top right panel represents 2 out of 20 mass spectra acquired at consecutive 50-unit *m/z* ranges from 20 single cells over $180 \leq m/z \leq 1200$. The transient time was increased as "narrowband" spectra were recorded for higher-*m/z* ions. Bottom panel shows a "stitched" spectrum from the individual "narrowband" spectra. Bottom left panel shows a graphical representation in the form of a pie chart to illustrate an example of the number of assigned and unassigned compounds in the stitched spectrum from a single cell by accurate mass and IFS. Note, background-related *m/z* peaks, for example, from ESI, are not included in the pie chart. *AM-DB* peaks assigned by database matches based on accurate mass only, *IFS-DB* peaks assigned by database matches based on accurate mass and IFS, *IFS-U* peaks assigned for unknown (newly discovered) compounds based on accurate mass and IFS without database matches, *NA* no assignments

and number of measured IFSs, because it reduces the number of ions being analyzed at a time and the mass resolution gradually increases at higher *m/z* values (*see* Fig. 3).

3.5 Metabolite Assignments by Ultrahigh Mass Accuracy and Isotope Fine Structure

3.5.1 Accurate Mass and IFS Library Generation for Known Compounds

1. Generate a library of molecular formulas and their accurate masses for sample-related compounds by downloading the corresponding Tab-delimited text files from online databases. For example, PlantCyc (https://www.plantcyc.org/data/search) offers lists of molecular formulas and accurate masses for compounds, for example, metabolites and lipids, in several plant species, whereas the yeast metabolome database (http://www.ymdb.ca/) provides the same for many microorganisms. Expectedly, the greater the library volume, the more hits can return from the search.

2. Add molecular formulas and accurate masses for common electrospray contaminants.
3. Compute and generate theoretical IFS for the compounds in the library using the software Ecipex.

3.5.2 Mass Spectra Processing and Rapid Assignments of Known Compounds by IFS

1. For a greater number of compound annotations, stitch the "narrowband" spectra into a comprehensive spectrum (*see* Fig. 3).
2. Create a peak list of m/z values with their corresponding intensities from the stitched spectrum.
3. Perform internal calibration of the peak list to further lower the mass error to less than 1 ppm using the software Formularity.
4. Set the search parameters in Formularity to the following values, but they may also be changed depending on several factors, including prior knowledge of potential compounds in the sample: elemental composition (Nitrogen ≤ 10, Sulfur ≤ 5, and Phosphorus ≤ 5), major peaks $S/N \geq 3$, minor peaks minimum $S/N \geq 1.5$, minimum peak presence/absence score $= 1$, and minimum peak probability to score equals 0.001.
5. Cross-match between the experimental and theoretical IFS by searching against the generated library.
6. IFS matching can be also done manually by plotting and comparing the simulated spectra to the experimental data using Thermo Xcalibur 2.2 (*see* Fig. 2).

3.5.3 Elemental Composition Assignments for Unknown Compounds by IFS

1. Enter the m/z values for unknown compounds that returned no hits in the Formularity database search into the web-based molecular formula calculator ChemCalc (http://www.chemcalc.org/mf_finder).
2. Set the mass error to <1 ppm, and, depending on the m/z value and ion-mode polarity, specify the highest possible number for C, H, N, O, S, P, Cl, Na, and K. Other elements may be included.
3. Search for possible molecular formulas that are consistent with the computational constraints. Generate simulated IFS for the calculated molecular formulas using Thermo Xcalibur 2.2.
4. Overlay the experimental and simulated IFS for the unknown compounds. Manually inspect the degree of matching by taking into account the mass accuracy and the relative isotopic abundance.
5. The total number of assigned peaks based on ultrahigh mass accuracy and IFS for known and unknown compounds can be compared to the unassigned ones in order to evaluate the coverage of this approach (*see* Fig. 3).

4 Notes

1. Optical fibers with 250 μm core diameter can be etched to tips with smaller diameters than 450 μm fibers. *G. max* root nodule cells are significantly smaller than *A. cepa* epidermal cells. Thus, the former and the latter were ablated with 250 and 450 μm core diameter fibers, respectively.
2. There are a number of reports describing fabrication of Peltier-cooled sample stages, including the one followed in this work [16].
3. The distance values reported here may need to be adjusted for optimal signal.
4. If there are spatial constraints for mounting the oblique long-distance microscope (microscope 2), a handheld digital microscope (Dino-Lite, Torrance, CA, USA) can be mounted instead.
5. For wider field of view, a 5 × long-working-distance objective can be mounted on the vertical microscope (microscope 1) instead of the 10× objective.
6. Wear appropriate protective eyewear for 2940 nm radiation.
7. The red light may be observed at the distal end of the fiber, but this does not necessarily indicate successful coupling of the laser pulses.
8. The sensor can be damaged if placed in the focal spot of the lens.
9. There is a tradeoff between the low stage-temperature and water vapor condensation on the sample surface. The optimal temperature will vary depending partly on the relative humidity, which, in our case, varied between 10% and 20%.
10. Light contact between the cell and the fiber tip may affect the cell prior to ablation. An alternative approach that was previously utilized to prevent this potential cell damage was to set the distance between the cell-surface and the fiber tip to ~2 × tip radius of curvature. At this distance, the laser pulse energy can be confined into the cell by evanescent waves [17].
11. The absence of ablation may not be solely due to suboptimal laser-pulse coupling into the fiber. This can also be a sign of insufficient laser-pulse repetition rate and/or energy. The values provided here for these parameters have worked for our samples, but can vary depending on several factors, for example, cell type, size, tensile strength, and fiber-tip diameter.
12. Mass calibration should be performed at least weekly using Pierce LTQ ESI positive and negative ion calibration solutions.

13. The electrospray solution flow rate and applied voltage on the ESI emitter need to be adjusted to obtain a stable electrospray and optimal signal from single cells.
14. Early work demonstrated a transient time of at least 0.768 s (mass resolution of ~330,000) were required for ISF measurements of disaccharides, for example [9].
15. One laser pulse per spectrum per cell is collected. To generate a comprehensive view of all the possible metabolites in a cell population, analyses of multiple cells is required using the "narrowband" MS data acquisition approach.

Acknowledgments

This material was supported by the US Department of Energy (DOE), Office of Biological and Environmental Research (OBER) under award number DESC0013978. Much of this research was performed at the Environmental Molecular Sciences Laboratory, a DOE Office of Science User Facility sponsored by the OBER and located at PNNL. PNNL is operated for DOE by Battelle Memorial Institute under contract DE-AC05-76RL01830.

References

1. Patti GJ, Yanes O, Siuzdak G (2012) Metabolomics: the apogee of the omics trilogy. Nat Rev Mol Cell Biol 13(4):263–269. https://doi.org/10.1038/nrm3314
2. Gomez-Casati DF, Zanor MI, Busi MV (2013) Metabolomics in plants and humans: applications in the prevention and diagnosis of diseases. BioMed Res Int 2013:792527. https://doi.org/10.1155/2013/792527
3. Samarah LZ, Khattar R, Tran TH, Stopka SA, Brantner CA, Parlanti P, Veličković D, Shaw JB, Agtuca BJ, Stacey G, Paša-Tolić L, Tolić N, Anderton CR, Vertes A (2020) Single-cell metabolic profiling: metabolite formulas from isotopic fine structures in heterogeneous plant cell populations. Anal Chem 92 (10):7289–7298. https://doi.org/10.1021/acs.analchem.0c00936
4. Fiehn O (2002) Metabolomics—the link between genotypes and phenotypes. Plant Mol Biol 48(1):155–171. https://doi.org/10.1023/A:1013713905833
5. Gemperline E, Keller C, Li L (2016) Mass spectrometry in plant-omics. Anal Chem 88 (7):3422–3434. https://doi.org/10.1021/acs.analchem.5b02938
6. Altschuler SJ, Wu LF (2010) Cellular heterogeneity: do differences make a difference? Cell 141(4):559–563. https://doi.org/10.1016/j.cell.2010.04.033
7. Zenobi R (2013) Single-cell metabolomics: analytical and biological perspectives. Science 342(6163):1243259. https://doi.org/10.1126/science.1243259
8. Zhang L, Foreman DP, Grant PA, Shrestha B, Moody SA, Villiers F, Kwak JM, Vertes A (2014) In Situ metabolic analysis of single plant cells by capillary microsampling and electrospray ionization mass spectrometry with ion mobility separation. Analyst 139 (20):5079–5085. https://doi.org/10.1039/C4AN01018C
9. Stopka SA, Samarah LZ, Shaw JB, Liyu AV, Veličković D, Agtuca BJ, Kukolj C, Koppenaal DW, Stacey G, Paša-Tolić L, Anderton CR, Vertes A (2019) Ambient metabolic profiling and imaging of biological samples with ultrahigh molecular resolution using laser ablation electrospray ionization 21 Tesla FTICR mass spectrometry. Anal Chem 91(8):5028–5035. https://doi.org/10.1021/acs.analchem.8b05084

10. Veličković D, Chu RK, Carrell AA, Thomas M, Paša-Tolić L, Weston DJ, Anderton CR (2018) Multimodal MSI in conjunction with broad coverage spatially resolved MS2 increases confidence in both molecular identification and localization. Anal Chem 90(1):702–707. https://doi.org/10.1021/acs.analchem.7b04319
11. Marshall AG, Guan S (1996) Advantages of high magnetic field for Fourier transform ion cyclotron resonance mass spectrometry. Rapid Commun Mass Spectrom 10(14):1819–1823. https://doi.org/10.1002/(SICI)1097-0231(199611)10:14<1819::AID-RCM686>3.0.CO;2-Z
12. Shaw JB, Lin T-Y, Leach FE, Tolmachev AV, Tolić N, Robinson EW, Koppenaal DW, Paša-Tolić L (2016) 21 Tesla Fourier transform ion cyclotron resonance mass spectrometer greatly expands mass spectrometry toolbox. J Am Soc Mass Spectrom 27(12):1929–1936. https://doi.org/10.1021/jasms.8b05198
13. Samarah LZ, Tran TH, Stacey G, Vertes A (2020) In vivo chemical analysis of plant sap from the xylem and single parenchymal cells by capillary microsampling electrospray ionization mass spectrometry. Anal Chem 92(10):7299–7306. https://doi.org/10.1021/acs.analchem.0c00939
14. Tolmachev AV, Robinson EW, Smith RD, Leach FE, Futrell JH, Paša-Tolić L (2012) A conceptual approach for FT-ICR cell harmonization utilizing external shim electrodes. Int J Mass Spectrom 325–327:45–50. https://doi.org/10.1016/j.ijms.2012.06.027
15. Tolić N, Liu Y, Liyu A, Shen Y, Tfaily MM, Kujawinski EB, Longnecker K, Kuo L-J, Robinson EW, Paša-Tolić L, Hess NJ (2017) Formularity: software for automated formula assignment of natural and other organic matter from ultrahigh-resolution mass spectra. Anal Chem 89(23):12659–12665. https://doi.org/10.1021/acs.analchem.7b03318
16. Von Seggern CE, Gardner BD, Cotter RJ (2004) Infrared atmospheric pressure MALDI ion trap mass spectrometry of frozen samples using a peltier-cooled sample stage. Anal Chem 76(19):5887–5893. https://doi.org/10.1021/ac049575v
17. Li H, Smith BK, Shrestha B, Márk L, Vertes A (2015) Automated cell-by-cell tissue imaging and single-cell analysis for targeted morphologies by laser ablation electrospray ionization mass spectrometry. In: He L (ed) Mass spectrometry imaging of small molecules. Springer New York, New York, pp 117–127. https://doi.org/10.1007/978-1-4939-1357-2_12

10. Ma ... Ko ... H, Chi ... L, ...
Re ... ., Weston ...
Mc ... ald ... MSI ... m ...
... and spatially ... ed ...
nde ... ca ... d ... welco ...
landscapes. An ... Ch ...
https://doi.org/10.102 ...
ac ...
11. Marsh ... S, Chan ...
... mine ... Michael ...
... WO ... ba ...
C ... M ... S ... ectr ...
... 2/j/ ... org/10.1002 ...
... 0 ... 1 ... 1 ...
12. Me ... ... 1 ...
B ... M ... 1 ... n ...
H ... H ... Hog ... Lisi ...
... mass ...
... ...
... NMR ... 21 ...
... ... 7 ... 1 ...
13. Sa ... ... T ...
... (2013) ...
... ...
... ...
... ...
... 3 ... 7 ...
... ...

Chapter 5

TOF-SIMS Imaging of Biological Tissue Sections and Structural Determination Using Tandem MS

Sebastiaan Van Nuffel and Alain Brunelle

Abstract

Over the past couple of years, imaging mass spectrometry (IMS) has arisen as a powerful tool to answer research questions in the biomedical field. Imaging mass spectrometry allows for label-free chemical imaging by providing full molecular information. The IMS technique best positioned for cell and tissue analysis is time-of-flight secondary ion mass spectrometry (ToF-SIMS) because it has the best spatial resolution of all the molecular IMS techniques and can detect many biochemical species and especially lipids with high sensitivity. Because one must rely on the mass and isotopic pattern of an ion in combination with positive correlations with lower mass fragments to help identify its structure, one major problem during ToF-SIMS experiments is the ambiguity when assigning a molecule to a certain mass peak. The solution are instruments with tandem MS capabilities as was already the case for many MALDI-ToF instruments more than a decade ago. It has been a few years since instruments with this capability were introduced and a number of interesting publications have been produced highlighting the advantages in biological SIMS work. Here, we present a protocol describing how tandem MS can be used to elucidate the structure of unknown or ambiguous mass peaks in biological tissue samples observed during ToF-SIMS imaging based on our experiences.

Key words SIMS, Time-of-flight, Orbitrap, Tandem MS, Imaging mass spectrometry, Lipids, Fatty acids

1 Introduction

There are a variety of articles and books, which explain the basics of ToF-SIMS imaging, and we refer the reader to those for a more profound understanding [1–3]. However, we will provide a brief summary for the sake of clarity. Simply put, secondary ion mass spectrometry (SIMS) refers to the use of primary ions, for example Bi_3^+ ions produced by a Bi liquid metal ion gun (LMIG), to generate secondary ions from solid samples. Firing primary ions at the sample causes a collision cascade that causes molecules from the top 1–3 monolayers of material at the sample surface to desorb and

Young-Jin Lee (ed.), *Mass Spectrometry Imaging of Small Molecules: Methods and Protocols*,
Methods in Molecular Biology, vol. 2437, https://doi.org/10.1007/978-1-0716-2030-4_5,
© The Author(s), under exclusive license to Springer Science+Business Media, LLC, part of Springer Nature 2022

ionize. Typically, these ions are then sent to a mass analyzer using a static electric field and their mass-to-charge (*m/z*) ratio is determined via a time-of-flight (ToF) measurement. Because the primary ions can be focused onto the sample as a narrow beam, it is possible to raster the beam across the sample to generate a hyperspectral 2D image where every pixel is a full mass spectrum. One of the advantages of an imaging mass spectrometry technique such as ToF-SIMS is the fact that the imaging is label-free, because individual molecules can be identified by characteristic mass peaks that show up in the ToF-SIMS mass spectrum. Using ToF-SIMS imaging, it is possible to visualize the spatial distribution of small molecules (<2000 Da) in 2D with a spatial resolution comparable to that of a light microscope. If equipped with a sputter gun such as an Ar gas cluster ion beam (GCIB), it is also possible to acquire 3D image stacks with <10 nm depth resolution by removing material through sputtering in between 2D ToF-SIMS imaging experiments. Because of these molecular imaging abilities, ToF-SIMS is extensively used for characterization in materials science and increasingly the life sciences.

SIMS can also provide structural information, because SIMS generally produces both molecular ions and fragment ions. This fragmentation pattern allows one to determine the structure of unknown compounds. However, a sample can contain many unknown chemistries and the secondary ions of the different compounds present in the sample are all formed together after the impact of the primary ion, which is why a "panoramic" detector with high transmission such as a ToF mass analyzer is typically used for SIMS. This property allows for label-free detection, but is a double-edged sword because it also means that a typical ToF-SIMS mass spectrum can be considered a summation of the spectra of the individual compounds present. In the case of ToF-SIMS imaging, it is possible to determine which mass peaks are positively correlated using principal component analysis or some other multivariate analysis technique. However, spatial colocation does not necessarily mean that these mass peaks all originate from one compound, particularly in the case of complex systems such as biological ones. This is further complicated by the fact that different compounds can produce similar fragmentation patterns with many of the same fragment ions. This is especially a problem when investigating biological materials, because biomolecules are often composed of the same building blocks. In addition, the mass resolution attainable with a ToF mass analyzer is typically limited to 10,000, which means not all peaks are necessarily resolved. One mass peak can thus harbor multiple ions. All this obfuscates which mass peaks belong together, that is, originate from the same compound. Finally, the secondary ion intensity and fragments produced using desorption-ionization techniques such as ToF-SIMS are highly dependent on the chemical environment of the compounds. This so-called matrix effect has made it very difficult to fingerprint and library approaches have proven largely ineffectual for ToF-SIMS. Even in simple systems, it is often necessary to work with reference materials in order to be certain of the identity of a mass peak of interest.

For these reasons, the manufacturers of ToF-SIMS instruments have come up with different strategies to implement MS/MS capabilities, as was already the case for many MALDI-ToF instruments more than a decade ago. MS/MS or Tandem MS is a technique where the ions of a particular *m/z*-ratio are selected and are then made to undergo fragmentation (usually through collision-induced dissociation or CID) before being mass analyzed again. While the first mass spectrum (MS1) is a summation of the molecular and fragment ions generated by all chemistries, the second mass spectrum (MS2) is a "pure" mass spectrum of single precursor ion if there are no isobaric interferences. In other words, MS2 allows one to unambiguously identify a mass peak in MS1. Because the fragmentation for MS2 is achieved in the gas phase, there are also no matrix effects causing reproducibility issues and this means it is now possible to use direct spectrum-to-spectrum matching against a reference library of previously observed MS/MS spectra. Even if there is more than one precursor ion species in the *m/z*-ratio window selected for MS/MS, it will still be relatively easy to deconvolute MS2. In the first ToF-SIMS instrument to demonstrate MS/MS capabilities, collisions of the secondary ions are achieved with argon gas in a field free region before the reflectron but after the exit of the buncher that samples the secondary ion stream, effectively performing high-energy CID [4, 5]. Each precursor and its corresponding product ions continue to travel at the same velocity and are selected using a timed ion gate so the MS/MS configuration can be considered a ToF/ToF. A second manufacturer built an instrument in which a precursor ion can be deflected into a CID cell [6, 7]. The resulting fragment ions are mass separated in a separate linear TOF with a second pulse counting detector, producing the MS/MS spectrum. A great advantage here is that MS/MS and traditional TOF-SIMS data are collected simultaneously from the same analytical area in a single experiment, allowing for MS/MS imaging at a high speed. A third instrument is a so-called Hybrid SIMS, combining a conventional ToF-SIMS instrument with an Orbitrap™, which allows for MS/MS through higher-energy collisional dissociation (HCD) with a mass resolution of up to 240,000 at *m/z* 200 and a mass accuracy below 1 ppm [8, 9]. This kind of mass resolution and accuracy allows one to determine the chemical formula of each fragment ion in the MS/MS spectrum without any ambiguity.

The following protocol describes how tandem MS can be used to elucidate the structure of unknown or ambiguous mass peaks in biological tissue samples observed during ToF-SIMS imaging. It is largely based on a previously published article, where we describe the combination of MALDI-ToF, ToF-SIMS, and Hybrid SIMS to identify markers for pulmonary arterial hypertension in human lung tissue [10]. However, the protocols are modified in order to be applicable to as wide a range of biological tissue samples possible.

2 Materials

2.1 Chemicals and Materials

1. Dry ice.
2. Optimal cutting temperature (OCT) compound.
3. Glass microscope slides and coverslips.
4. Indium tin oxide (ITO) coated glass microscope slides.
5. 4% paraformaldehyde solution.
6. Hematoxylin solution.
7. 0.5% Erythrosine B solution.
8. Ethanol.

2.2 Instrumentation

1. −80 °C freezer.
2. Cryomicrotome.
3. Optical microscope with a digital camera.
4. Freeze-dryer.
5. Vacuum desiccator.
6. Microscope slide scanner.
7. A ToF-SIMS instrument equipped with a Bi LMIG and Ar GCIB that has MS/MS capabilities (*see* **Note 1**).

3 Methods

3.1 Tissue Sectioning

This is a protocol for the preparation of tissue sections for both ToF-SIMS measurements and complementary histology. The protocol assumes to already have a tissue block of fresh-frozen tissue stored in a −80 °C freezer or a liquid nitrogen tank (*see* **Note 2**). The authors have applied this protocol to various tissue types of different organisms with little or no variation.

1. Remove frozen tissue block from storage and put the tissue block on dry ice to transport it to the cryotome.
2. Apply OCT to the back of the tissue block to hold it in place during sectioning (*see* **Note 3**).
3. Section tissue at −20 °C using a cryostat system. The section thickness tends to be limited by the cryostat system and is usually 5–15 μm (*see* **Note 4**).

4. For each tissue sample, obtain three serial sections. Thaw-mount sections 1 and 3 on ITO-coated glass microscope slides (*see* **Note 5**). Thaw-mount section 2 (middle of the series) on a glass microscope slide.
5. After thaw-mounting, immediately put the slide in a microscope slide box cooled on dry ice.
6. Store the microscope slide box in a −80 °C freezer.

3.2 Tissue Staining

This is a protocol for the histological staining of the adjacent tissue section on the glass microscope slide. This allows a pathologist to easily identify areas of interest for the ToF-SIMS measurements. This specific protocol was optimized for staining human lung tissue sections.

1. Defrost tissue sections that were thaw-mounted on glass microscope slides as described in Subheading 3.1.
2. Fix the sections chemically in 4% paraformaldehyde solution at room temperature (10 min).
3. Rinse in tap water (30 s).
4. Stain in hematoxylin solution (1 min).
5. Rinse in tap water (30 s).
6. Counterstain in 0.5% erythrosine solution (3 s).
7. Rinse in tap water (30 s).
8. Rinse in 70% ethanol (1 min).
9. Rinse in pure ethanol (1 min).
10. Air-dry the tissue.
11. Mount a cover slip and seal with nail polish.
12. Take pictures of the sample using an optical microscope with a digital camera.

3.3 Sample Preparation Prior to ToF-SIMS

ToF-SIMS requires ultra-high vacuum (UHV) so the high water content of cells and tissues complicates the sample preparation of biological materials for ToF-SIMS imaging. This part describes the preparation of freeze-dried tissue sections from frozen tissue sections for ToF-SIMS imaging (*see* **Note 6**). The authors have applied this protocol to various tissue types of different organisms with little or no variation.

1. Remove frozen tissue sections that were thaw-mounted on the ITO-coated glass microscope slides as described in Subheading 3.1 from storage and put them on dry ice.
2. Freeze-dry samples using a tray freeze-dryer (*see* **Note 7**). If no freeze-dryer is available, dehydrate the tissue sections in a vacuum desiccator for at least 30 min.

3. Scan the ITO coated glass microscope slides with a microscope slide scanner. Alternatively, use an optical microscope with a digital camera.
4. Store the dehydrated tissue sections in a vacuum desiccator prior to SIMS imaging.

3.4 ToF-SIMS Imaging

After the stained images and the images of the corresponding freeze-dried tissue sections are compared, regions of interest can be identified for ToF-SIMS imaging. ToF-SIMS imaging is performed using the Bi liquid metal ion gun (LMIG) as a primary ion beam in order to achieve a high spatial resolution ($\leq$1 μm pixel size).

1. Mount section 1 and introduce the sample holder in the vacuum chamber of the mass spectrometer. The image recorded by the slide scanner can be registered in the ToF-SIMS software's navigator in order to easily find small histological structures.
2. Optimize the 25 keV bismuth LMIG for molecular imaging using **steps 3–9**.
3. Select Bi_3^+ clusters and operate the gun the high-current bunched mode, which allows for higher mass resolution and has a beam spot of 1–2 μm.
4. Measure the pulsed primary ion current for Bi_3^+ (about 0.4 pA) using the Faraday cup provided on each sample holder and a cycle time of 100 μs (*see* **Note 8**).
5. Using a so-called A-Grid, a metallic tuning grid where the A-letter is drawn, the primary ion beam is aligned with the ToF-SIMS camera image.
6. Move the stage to the sample and optimize the distance (or *Z*-height) between the sample and the extraction cone.
7. Charging of the sample is compensated with the low-energetic (about 20 eV) electrons of the flood gun. Optimize the charge compensation on the sample surface.
8. Iteratively adjust beam alignment, *Z*-height and charge compensation on the sample surface for optimal imaging performance.
9. Save all parameters before starting an imaging experiment.
10. Internally calibrate the mass spectra using known mass peaks (*see* **Note 9**).
11. Image the area of interest. Areas larger than 500 × 500 μm^2 can be imaged by moving the sample stage and image stitching 500 × 500 μm^2 patches or smaller together.

3.5 Hybrid SIMS MS/MS

After analyzing the ToF-SIMS ion image data acquired in Subheading 3.3 using the commercial software, unknown or ambiguous mass peaks of interest can be selected for tandem MS. For tandem MS, the Hybrid SIMS's Thermo Scientific Q Exactive HF Orbitrap is used in this method chapter in combination with the Ar gas cluster ion beam (GCIB) as a quasi-continuous primary ion beam. A typical Hybrid SIMS MS/MS spectrum of a mass peak of interest detected in human lung tissue is provided in Fig. 1.

1. Mount tissue section 3 and introduce the sample holder in the vacuum chamber of the mass spectrometer. The image recorded by the slide scanner can be registered in the software's navigator in order to easily locate the same areas analyzed on tissue section 1 in Subheading 3.3.
2. Set up the Ar GCIB for tandem MS using 5–20 keV Ar_n clusters with $n = 1500$–3500 and a spot size of roughly 3 μm.
3. Measure the primary ion current (about 12 pA at 15% duty cycle) using the Faraday cup on the sample holder.
4. The Orbitrap analyzer should be calibrated using silver cluster ion patterns from a silver foil standard.

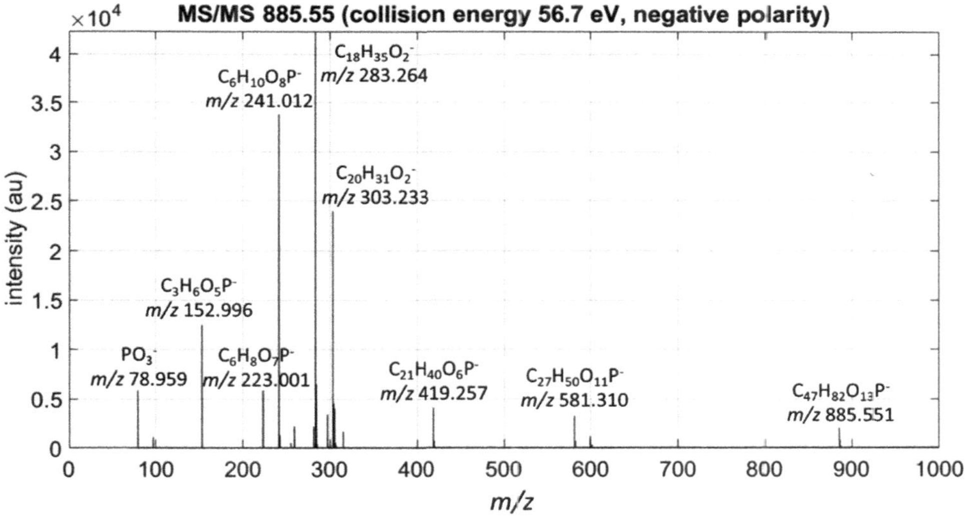

Fig. 1 The negative polarity MS/MS fragment ion spectrum (collision energy of 56.7 eV) of the *m/z* 885.551 precursor ion from human arterial tissue recorded using the IONTOF Hybrid SIMS instrument. The MS/MS fragment ion spectrum unambiguously identifies this precursor ion as a PI C18:0/C20:4 species, given the $C_6H_8O_7P^-$ ion at *m/z* 223.001 and the $C_6H_{10}O_8P^-$ ion at *m/z* 241.012, which are characteristic fragments for the inositol phosphate headgroup, and the $C_{18}H_{35}O_2^-$ ion at *m/z* 283.264 and the $C_{20}H_{31}O_2^-$ ion at *m/z* 303.233 corresponding to stearic and arachidonic acid, respectively. This mass spectrum also agrees with the commonly occurring product ions for a PI C18:0/C20:4 species published online in LIPIDMAPS, showing spectral matching is now possible. (Reproduced from Van Nuffel S. et al. Anal. Chem. 2020, 92:12079–12087 https:/pubs.acs.org/doi/10.1021/acs.analchem.0c02815 (ref. 10). Copyright 2020 American Chemical Society)

5. Move the stage to the sample areas of interest and acquire spectra with the same field of view used during ToF-SIMS imaging (*see* **Note 10**).
6. Record the MS/MS spectrum of the ion of interest using a mass window of 0.4 *m/z* and a collision energy of 40–60 eV.

4 Notes

1. In this protocol, the ToF-SIMS imaging is performed using a TOF-SIMS IV instrument (IONTOF GmbH, Münster, Germany). For the MS/MS analyses, the Hybrid SIMS is used (IONTOF GmbH, Münster, Germany), which is a combination of a typical ToF-SIMS instrument with an orbital trapping mass spectrometer (Q Exactive HF, ThermoFisher Scientific GmbH, Bremen, Germany). ToF-SIMS ion images are recorded and analyzed using the SurfaceLab 7 (IONTOF) software.
2. Since the goal is to detect and localize chemical compounds contained in the tissue, care should be taken not to alter its chemical composition during the sample preparation. Consequently, the use of fresh-frozen tissue is recommended as well as avoiding the introduction of any new chemistry through for example chemical fixation, sucrose infiltration, etc. Fresh-frozen means the tissue block should be excised and snap-frozen within 30–60 min. A small piece of tissue can be flash frozen in a cryogenic liquid such as ethane or isopentane. Otherwise, liquid nitrogen is used.
3. When the tissue block is small, it is possible to embed the sample using a cryomold and an appropriate embedding material to help preserve its shape and make it easier to manipulate the sections. Carboxymethyl cellulose (CMC) and gelatin compounds are more suitable than OCT, because OCT produces a strong signal in positive polarity that obscures lipid peaks and also contains a large amount of salts which could lead to ion suppression. When OCT is used to hold the block in place, care must be taken to avoid cross contamination.
4. The thickness is not really of importance to the ToF-SIMS measurement itself, because ToF-SIMS is a surface analysis technique with an information depth of c. 1 nm and only c. 20 nm of subsurface damage. However, one wants to slice as thin as possible in order to ensure the adjacent sections used for replicates or histology are similar.
5. The tissue sections are thaw-mounted on conductive ITO coated glass microscope slides, because dielectric substrates can negatively affect the ion extraction in a typical ToF-SIMS instrument.

6. Most biological sample preparation techniques for ToF-SIMS involve dehydration of the sample, because a dry sample can be kept at room temperature, allowing convenient sample handling and the use of complementary techniques such as optical microscopy or atomic force microscopy prior to ToF-SIMS analysis. Alternatively, the sample can be imaged in a frozen-hydrated state by performing the ToF-SIMS measurement with a cryogenically cooled stage at a temperature <-120 °C below which recrystallization does not occur [11]. This is referred to as cryo-ToF-SIMS and is conceptually analogous to cryo-EM.
7. When freeze-drying, it is imperative that the samples do not thaw before the pressure drops below that of the triple point. If the freeze-dryer has no temperature control, put in a steel plate that has been cooled in the -80 °C freezer and place the microscope slides on top of it.
8. This value allows one to calculate the primary ion dose density or "fluence" before each image acquisition and keep it below the static limit of SIMS ($\leq 10^{13}$ ions·cm^{-2}).
9. We refer the reader to ref. 12 for a recommended procedure of ToF-SIMS mass calibration.
10. A 500×500 μm^2 area can for example be rastered using 25×25 pixels with a 20 μm spot size. The primary ion dose density is typically in the order of 1×10^{15} primary ions per cm^2.

References

1. Vickerman JC, Briggs D (2013) ToF-SIMS: materials analysis by mass spectrometry. SurfaceSpectra, Manchester
2. Fearn S (2015) An introduction to time-of-flight secondary ion mass spectrometry (ToF-SIMS) and its application to materials science. Morgan & Claypool Publishers, San Rafael
3. Tuck M, Blanc L, Touti R, Patterson NH, Van Nuffel S, Villette S, Taveau J-C, Römpp A, Brunelle A, Lecomte S, Desbenoit N (2021) Multimodal imaging based on vibrational spectroscopies and mass spectrometry imaging applied to biological tissue: a multiscale and multiomics review. Anal Chem 93:445–477. https://doi.org/10.1021/acs.analchem.0c04595
4. Fletcher JS, Kotze HL, Armitage EG, Lockyer NP, Vickerman JC (2013) Evaluating the challenges associated with time-of-fight secondary ion mass spectrometry for metabolomics using pure and mixed metabolites. Metabolomics 9:535–544. https://doi.org/10.1007/s11306-012-0487-4
5. Wehrli PM, Lindberg E, Angerer TB, Wold AE, Gottfries J, Fletcher JS (2014) Maximising the potential for bacterial phenotyping using time-of-flight secondary ion mass spectrometry with multivariate analysis and tandem mass spectrometry. Surf Interface Anal 46:173–176. https://doi.org/10.1002/sia.5505
6. Fisher GL, Hammond JS, Larson PE, Bryan SR, Heeren RMA (2016) Parallel imaging MS/MS TOF-SIMS instrument. J Vac Sci Technol B 34:03H126. https://doi.org/10.1116/1.4943568
7. Fisher GL, Bruinen AL, Ogrinc Potočnik N, Hammond JS, Bryan SR, Larson PE, Heeren RMA (2016) A new method and mass spectrometer design for TOF-SIMS parallel imaging MS/MS. Anal Chem 88:6433–6440. https://doi.org/10.1021/acs.analchem.6b01022

8. Pirkl A, Moellers R, Arlinghaus H, Kollmer F, Niehuis E, Makarov A, Horning S, Passarelli M, Havelund R, Rakowska P, Race A, Shard AG, West A, Marshall P, Newman CF, Alexander M, Dollery C, Gilmore IS (2016) A novel hybrid dual analyzer SIMS instrument for improved surface and 3D-analysis. Microsc Microanal 22:340–341. https://doi.org/10.1017/S1431927616002555
9. Passarelli MK, Pirkl A, Moellers R, Grinfeld D, Kollmer F, Havelund R, Newman CF, Marshall PS, Arlinghaus H, Alexander MR, West A, Horning S, Niehuis E, Makarov A, Dollery CT, Gilmore IS (2017) The 3D OrbiSIMS—label-free metabolic imaging with subcellular lateral resolution and high mass-resolving power. Nat Methods 14:1175–1183. https://doi.org/10.1038/nmeth.4504
10. Van Nuffel S, Quatredeniers M, Pirkl A, Zakel J, Le Caer J-P, Elie N, Vanbellingen QP, Dumas SJ, Nakhleh MK, Ghigna M-R, Fadel E, Humbert M, Chaurand P, Touboul D, Cohen-Kaminsky S, Brunelle A (2020) Multimodal imaging mass spectrometry to identify markers of pulmonary arterial hypertension in human lung tissue using MALDI-ToF, ToF-SIMS, and hybrid SIMS. Anal Chem 53:19. https://doi.org/10.1021/acs.analchem.0c02815
11. Linner JG, Livesey SA, Harrison DS, Steiner AL (1986) A new technique for removal of amorphous phase tissue water without ice crystal damage: a preparative method for ultrastructural analysis and immunoelectron microscopy. J Histochem Cytochem 34:1123–1135. https://doi.org/10.1177/34.9.2426340
12. Green FM, Gilmore IS, Seah MP (2006) TOF-SIMS: accurate mass scale calibration. J Am Soc Mass Spectrom 17:514–523. https://doi.org/10.1021/jasms.8b02645

Part II

New Types of Laser Absorption Methods in Mass Spectrometry Imaging

Chapter 6

Mass Spectrometry Imaging of Biological Tissues by Laser Desorption Ionization from Silicon Nanopost Arrays

Laith Z. Samarah and Akos Vertes

Abstract

Mass spectrometry imaging (MSI) plays an expanding role in the label-free spatial mapping of hundreds of molecules simultaneously. Currently, matrix-assisted laser desorption ionization (MALDI) is among the most widely adopted MSI techniques. However, matrix application can impact the fidelity of spatial distributions, and matrix selection and related spectral interferences in the low mass range can lead to biased molecular coverage. Nanophotonic ionization from silicon nanopost arrays (NAPA) is an emerging matrix-free MSI platform with enhanced sensitivity for several molecular classes, for example, neutral lipids and biooligomers. Here, we describe a protocol with minimal sample preparation for NAPA-MSI of metabolites, lipids, and biooligomers from biological tissues.

Key words Mass spectrometry imaging, Nanopost arrays, Nanophotonic ionization, Matrix-free laser desorption ionization

1 Introduction

Acquiring information on the spatial distributions of molecules in biological tissues is a valuable approach for elucidating biomedical phenomena, including histological–biochemical correlations, colocalization of chemical species, regulation of biochemical pathways, intercellular molecular transport, and localization of drugs and their metabolites. Unlike other molecular imaging modalities, such as magnetic resonance imaging (MRI) and immunohistochemistry, mass spectrometry imaging (MSI) can reveal the spatial distributions for various types of molecules simultaneously and in an untargeted and label-free fashion [1].

Since its inception, matrix-assisted laser desorption ionization (MALDI) has ascended to be among the top MSI methods due to several attractive qualities, including high sensitivity, soft ionization, availability of a wide range of organic matrices, adjustable molecular coverage, and the robustness of matrix application

Young-Jin Lee (ed.), *Mass Spectrometry Imaging of Small Molecules: Methods and Protocols*, Methods in Molecular Biology, vol. 2437, https://doi.org/10.1007/978-1-0716-2030-4_6,
© The Author(s), under exclusive license to Springer Science+Business Media, LLC, part of Springer Nature 2022

techniques [2, 3]. However, the presence of matrix creates several limitations in MALDI-MSI, such as spectral interferences for low-*m/z* ions [4], and, in case of suboptimal matrix application, spot-to-spot heterogeneity can affect the fidelity of the acquired spatial distributions [5–7].

To address these limitations, matrix-free ionization methods have been developed, including laser desorption ionization (LDI) based on inorganic substrates [8–12], nanophotonic [13–15] and plasmonic platforms [16, 17], and some atmospheric-pressure ionization techniques [18–20]. Recently, nanophotonic ionization from silicon nanopost arrays (NAPA) has emerged as a matrix-free MSI platform capable of mapping the spatial distributions for various molecular classes in biological tissues and cell populations [14, 21–23]. Resonant nanophotonic ion production in NAPA is contingent in part on the physical dimensions of the nanoposts, such as their diameter and aspect ratio, which can be tailored using nanofabrication processes, for example, electron beam [15] or deep ultraviolet projection lithography [3]. In some applications, the molecular coverage for some inorganic matrices and substrates has been shown to be complementary to MALDI with conventional organic matrices [24, 25]. For example, the superior coverage of certain neutral lipids by LDI from NAPA relative to MALDI has been demonstrated in multimodal imaging of biological tissues using the two imaging modalities [26]. Here, we describe a protocol for NAPA-MSI of metabolites, lipids, and biooligomers in biological tissues.

2 Materials

2.1 Biological Samples

1. Soybean plants [*Glycine max*, (L.) Merr from Williams 82 seeds] with root nodules infected by *Bradyrhizobium japonicum* (USDA 110) [27].

2.2 NAPA Imaging Chips

1. NAPA imaging chips (5×5 mm^2) with nanoposts 1000 nm in height, 150 nm in diameter, and a periodicity of 337 nm [3, 15].

2.3 Sample-Preparation

1. 2.5% Carboxymethylcellulose (CMC).
2. Cryomicrotome.
3. Desiccator with diaphragm vacuum pump.
4. Embedding base molds.
5. Specimen stage.
6. Fine-tip brush for handling thin tissue cryosections.

2.4 Mass Spectrometer and Associated Supplies

1. MALDI-mass spectrometer.
2. Conductive double sided adhesive carbon tape (BZB5012947, SPW Industrial, CA, USA).

2.5 Software for Data Analysis

1. Thermo Xcalibur 2.0 (ThermoFisher Scientific, CA, USA).
2. ImageQuest (1.1.0) (ThermoFisher Scientific, CA, USA).

3 Methods

3.1 Sample Preparation for NAPA-LDI-MSI

1. Fill up the base molds with 2.5% CMC.
2. Detach the nodules from the plant roots in liquid nitrogen using tweezers and embed them in CMC (*see* **Note 1**).
3. Freeze the embedding medium in a cryostat microtome at −15 to −20 °C (*see* **Note 2**). Transfer the frozen medium containing the embedded sample to a specimen stage and allow the embedded sample to reach thermal equilibrium.
4. Prepare 5–10 μm tissue cryosections at −15 to −20 °C. The frozen section can be thaw mounted on a NAPA imaging chip by carefully picking it up with a fine-tipped brush, placing it on the NAPA surface, and touching the backside of the chip.
5. Dry the tissue sections in a vacuum desiccator for 15–30 min.
6. As the sections are drying, place a piece of conductive double-sided adhesive carbon tape on a MALDI sample plate.
7. After the tissue sections are dried, take the NAPA imaging chips out of the desiccator and attach them to the carbon tape.

3.2 NAPA-LDI-MSI

1. Insert the MALDI sample plate into the MALDI ion source chamber.
2. Select and scan the region of interest to generate the corresponding optical image (*see* Fig. 1). Ensure that no carbon tape is included in the region. In Fig. 1, the entire tissue section was selected for imaging.
3. Set the step size to the desired value. In our case, the step size was set to 100 μm. If oversampling is needed, the step size can be set to, for example, 50 μm, a value smaller than the laser spot dimensions (100 × 80 μm^2). Smaller step sizes will result in longer imaging periods.
4. Adjust the laser energy until optimal ion signal is obtained. In our experiments, laser fluences between 100 and 120 mJ/cm^2 provided the highest ion signal with minimal in-source fragmentation that had deleterious effects on the ion intensities for certain endogenous compounds (*see* **Note 3**).

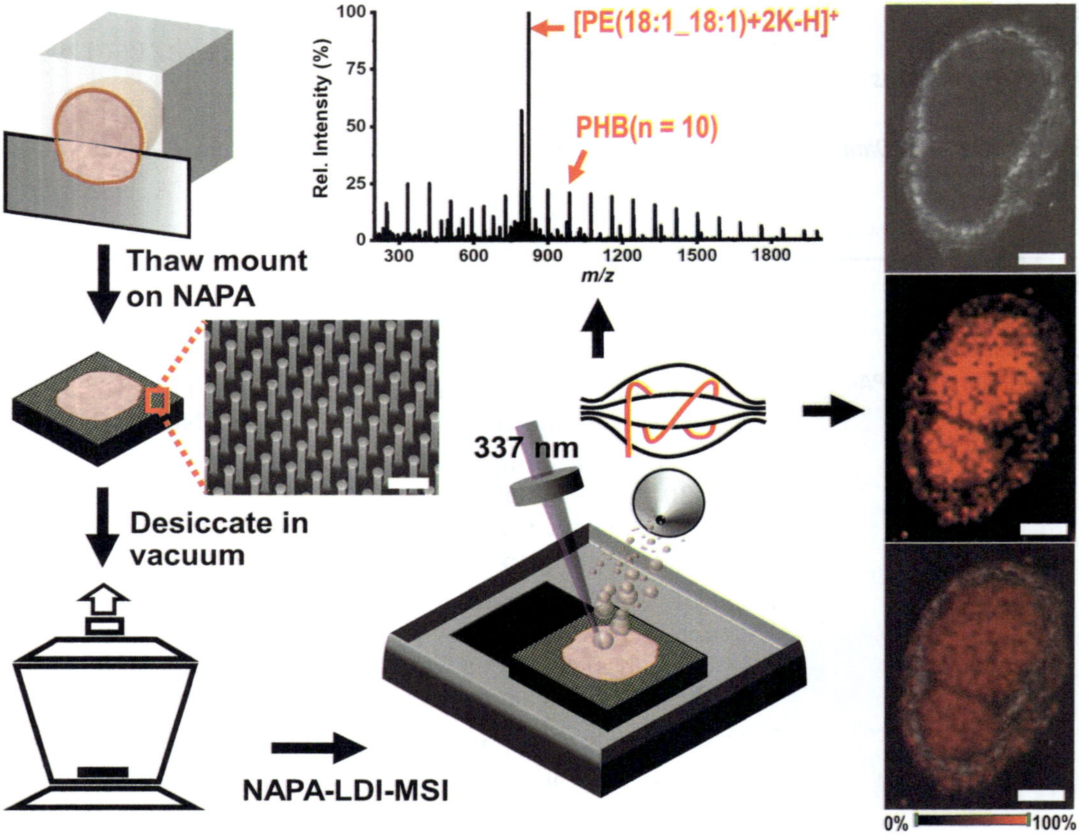

Fig. 1 Workflow of sample preparation and untargeted NAPA-LDI-MSI of biological tissues. Cryosections 5–10 μm in thickness are prepared from the tissue and then thaw-mounted on NAPA imaging chips (see scanning electron microscope image of the nanoposts in inset, scale bar = 500 nm). The region of interest, in this case, the entire sample surface (see optical image in top right panel), is interrogated with the laser beam (337 nm), and a representative mass spectrum (see top panel) is generated from averaging all the scans from the analyzed region. The ion intensities are also recorded as a function of two-dimensional spatial coordinates in the sample to produce the corresponding ion images (see example with *m/z* 792.4355 in middle right panel). Optical and chemical images (top and middle right panels, respectively) are overlaid to visualize the localization of chemical species with respect to the morphological features in the tissue section (see example in bottom right panel). Scale bars = 1 mm. (Adapted with permission [23]. Copyright 2021, John Wiley & Sons)

5. Select the number of laser shots per pixel that generates optimal ion signal. In our case, 3–10 laser shots per pixel resulted in the best ion signal.
6. Select the *m/z* acquisition range. There is a tradeoff between the *m/z* range and the instrument sensitivity. For the work presented here, the *m/z* range was set to $200 \leq m/z \leq 2000$.
7. The automatic gain control (AGC) is turned off in order to keep the number of laser shots per pixel constant.

8. Depending on the objective of the study, the mass resolving power, R, can be varied between $R = 30{,}000$ and $R = 100{,}000$. Increasing R results in increased acquisition time.
9. Begin MSI data acquisition for the region of interest with the selected instrumental parameters. The process involves automated stage movement in a rectangular raster, whereby mass spectra are recorded pixel-by-pixel for ions generated from the laser irradiation of the scanned area.

3.3 Mass Spectra and Chemical Image Analysis

One of the unique characteristics of MSI is the ability to acquire spatial distributions for molecules in an untargeted fashion. Untargeted MSI can discover the distributions for chemical species that previously had not been identified in the tissue. This can be accomplished by examining the mass spectra from the scanned region, selecting *m/z* peaks of interest, and inspecting the corresponding ion images, as described below. Depending on the sample area and the spatial resolution, images can comprise of hundreds to tens of thousands of individual mass spectra.

1. The recorded mass spectra from each pixel are saved in a ".raw" file that can be opened with the software XCalibur.
2. The files contain the ion chromatograms, which display the total ion count (TIC) and the signal intensities for selected ions as a function of time or scan number.
3. To visualize a representative mass spectrum for the scanned area, the individual scans are averaged across the entire ion chromatogram. In Fig. 1, a representative NAPA spectrum from a soybean root nodule tissue section contained *m/z* peaks that corresponded to primary and secondary metabolites, lipids, and biooligomers, for example, polyhydroxybutyric acid (PHB), with varying number of repeat units, *n*.
4. To visualize the spatial distributions for the abundances of chemical species in the tissue section, the corresponding ion intensities are represented on a false-color scale as a function of two-dimensional spatial coordinates in the scanned area using the software ImageQuest or other imaging software (*see* Figs. 1, 2 and 3).
5. The spatial distributions can be investigated in a targeted (preselected ions) and untargeted approach. For targeted analysis, find the *m/z* values for the ions of interest from the averaged mass spectra across the scanned region (see mass spectra in Figs. 1 and 2) and insert them into the "mass range" search box in ImageQuest. For example, previous work revealed the presence of heme b and soyasaponins in soybean root nodules infected with wild-type *B. japonicum* [28, 29]. Accordingly, the

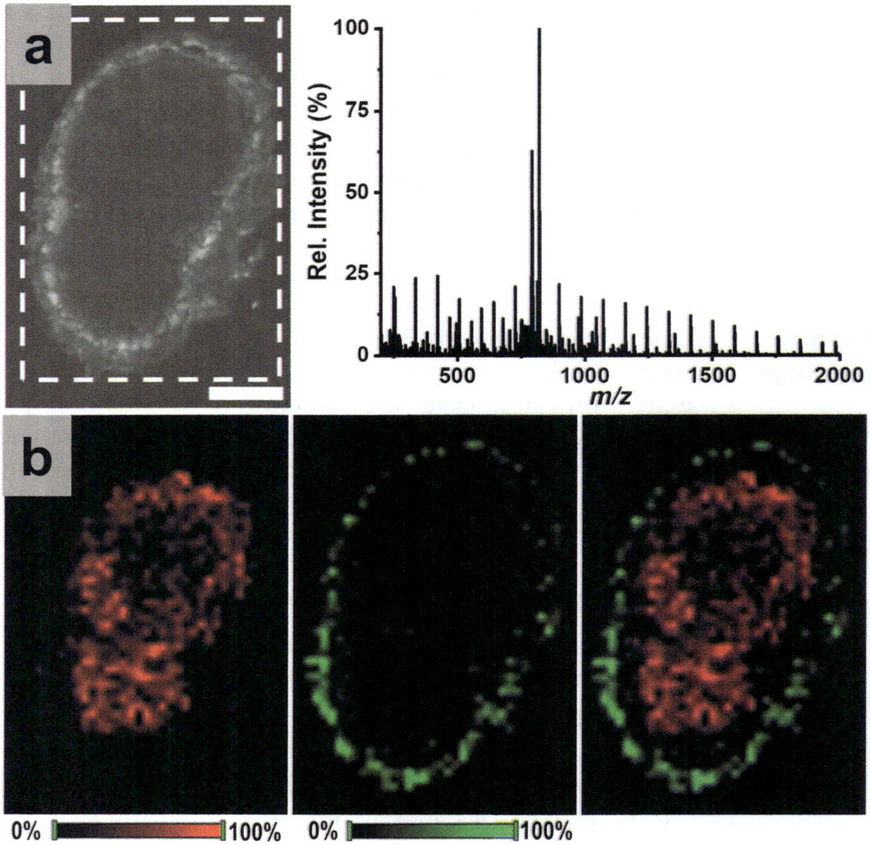

Fig. 2 Example of targeted imaging of two different chemical species by NAPA-LDI-MSI. (**a**) Optical image for a 10-μm thick root nodule tissue section before NAPA-LDI-MSI. Scans from the entire tissue section outlined in the white dashed box are averaged to yield the mass spectrum in the top right panel. (**b**) Spatial distributions for targeted chemical species, such as [heme b]$^+$ (*m/z* 616.1752, red) (left), [soyasaponin II + 2K-H]$^+$ (*m/z* 989.4273, green) (middle), and the two ions combined (right). Scale bar = 1 mm. (Adapted with permission [23]. Copyright 2021, John Wiley & Sons)

spatial distribution for these chemical species in the tissue analyzed by NAPA-LDI-MSI can be visualized by searching for the intensities of the corresponding ions in the scanned region (*see* Fig. 2).

6. To visualize the colocalization of different ions in the tissue region, the individual chemical images for the ions of interest can be combined and represented by different color schemes (*see* Fig. 2).

7. For untargeted analysis, average the individual scans within a selected area of the tissue. The areas may be selected based on morphological features. For example, in a soybean root nodule tissue section, the outer cortex and infection zone are made up of uninfected and infected plant cells, respectively, that can be distinguished under a light microscope (*see* Fig. 3).

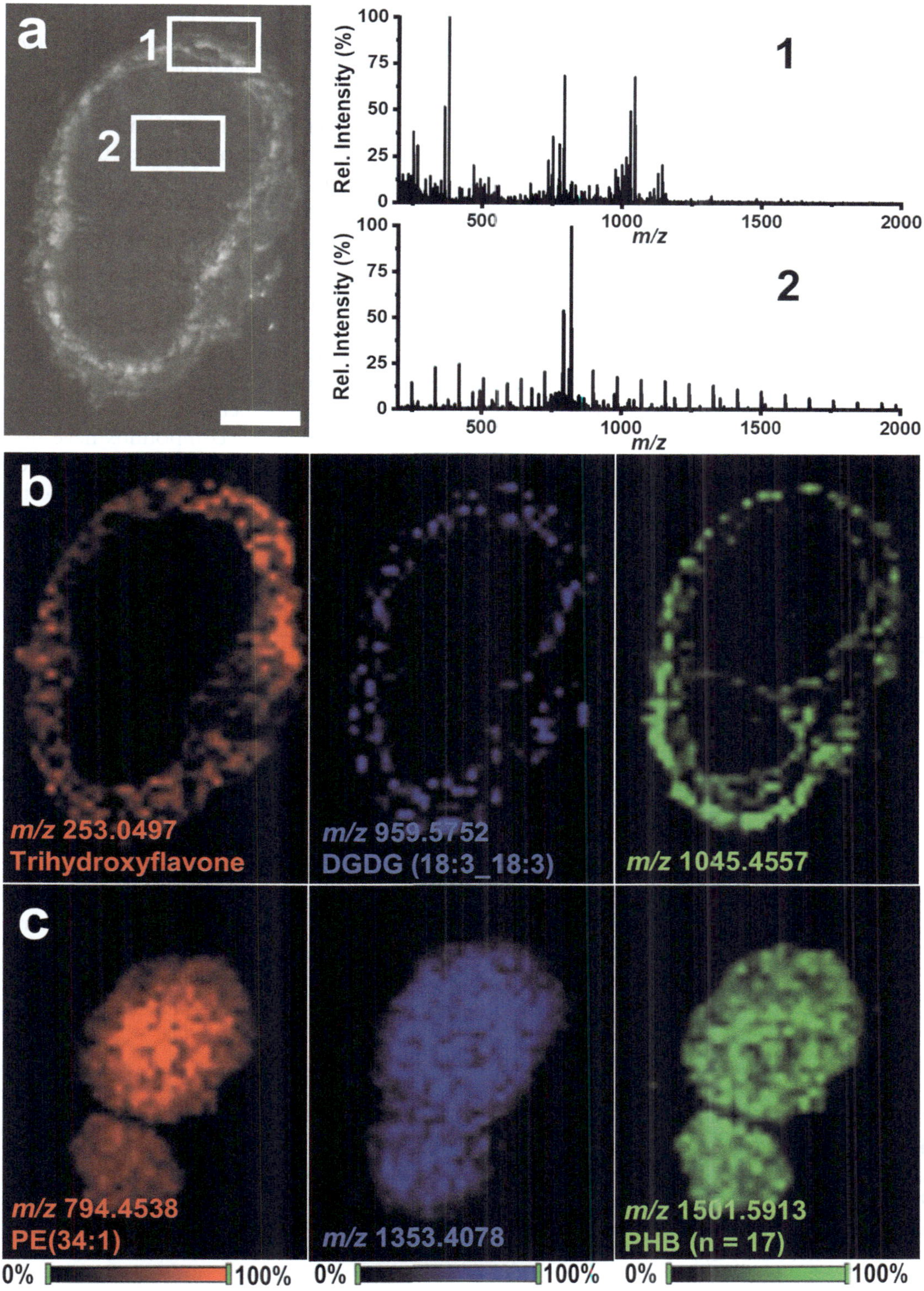

Fig. 3 Untargeted imaging of several chemical species by NAPA-LDI-MSI. (**a**) Optical image for a 10-μm thick tissue section from root nodule before NAPA-LDI-MSI. Scans from different regions, such as the outer cortex

8. Inspect the resulting mass spectrum and generate chemical images for ions that were detected in the selected region (*see* Fig. 3). This approach is particularly useful for low-abundance ions for which the *m/z* peaks are not obvious in the representative mass spectrum from the entire tissue region. For example, the relative signal intensity for the ion at *m/z* 1045.4557 appears significantly higher in the mass spectrum obtained from scanning area "1" (*see* Fig. 3) compared to the one produced from averaging the scans across the entire tissue section (*see* Fig. 2).
9. Untargeted MSI can reveal the spatial distributions for known chemical species that have not been localized before in the tissue, and, in some cases, for ions that have not been identified. For example, PHB and some digalactosyldiacylglycerols (DGDGs) have been previously detected in soybean root nodules, but chemical images for these compounds have not been reported [30, 31]. In contrast, other chemical species could not be identified from previous work and require additional steps, for example, tandem mass spectrometry, for their identification (e.g., the chemical images for unknown ions at *m/z* 1045.4557 and *m/z* 1353.4048 in Fig. 3).

4 Notes

1. Tissue embedding in CMC is not always required for preparing thin cryosections. When feasible, frozen tissue samples can be directly affixed on a specimen stage without embedding.
2. The cryosectioning temperature may vary depending on the tissue type and section thickness.
3. In addition to the laser pulse energy, the angle of incidence for the laser radiation and the polarization angle are other factors that affect the ion yield in NAPA-LDI-MS [13, 15]. For example, changing the angle of incidence from 45° to normal incidence and the polarization plane from p- to s-polarized light both results in significant drop in ion yields [13, 15].

Fig. 3 (continued) (rectangle 1) and the infection zone (rectangle 2), are averaged separately to generate the corresponding mass spectra (top right panel). Peaks for particular *m/z* in these spectra are then selected, and the corresponding ion images in the outer cortex (**b**) and infection zone (**c**) are produced for identified and unknown (only *m/z* values are shown) ions. Scale bar = 1 mm. (Adapted with permission [23]. Copyright 2021, John Wiley & Sons)

Acknowledgments

This work is supported by the US National Science Foundation, Plant Genome Program under grant no. IoS-1734145.

References

1. Samarah LZ, Vertes A (2020) Mass spectrometry imaging based on laser desorption ionization from inorganic and nanophotonic platforms. VIEW 1(4):20200063. https://doi.org/10.1002/VIW.20200063
2. Buchberger AR, DeLaney K, Johnson J, Li LJ (2018) Mass spectrometry imaging: a review of emerging advancements and future insights. Anal Chem 90(1):240–265. https://doi.org/10.1021/acs.analchem.7b04733
3. Morris NJ, Anderson H, Thibeault B, Vertes A, Powell MJ, Razunguzwa TT (2015) Laser desorption ionization (LDI) silicon nanopost array chips fabricated using deep UV projection lithography and deep reactive ion etching. RSC Adv 5(88):72051–72057. https://doi.org/10.1039/c5ra11875a
4. Shanta SR, Kim TY, Hong JH, Lee JH, Shin CY, Kim KH, Kim YH, Kim SK, Kim KP (2012) A new combination MALDI matrix for small molecule analysis: application to imaging mass spectrometry for drugs and metabolites. Analyst 137(24):5757–5762. https://doi.org/10.1039/c2an35782h
5. Dreisewerd K (2003) The desorption process in MALDI. Chem Rev 103(2):395–425. https://doi.org/10.1021/cr010375i
6. Garden RW, Sweedler JV (2000) Heterogeneity within MALDI samples as revealed by mass spectrometric imaging. Anal Chem 72(1):30–36. https://doi.org/10.1021/ac9908997
7. Lai YH, Cai YH, Lee H, Ou YM, Hsiao CH, Tsao CW, Chang HT, Wang YS (2016) Reducing spatial heterogeneity of MALDI samples with Marangoni flows during sample preparation. J Am Soc Mass Spectrom 27(8):1314–1321. https://doi.org/10.1007/s13361-016-1406-0
8. Kinumi T, Saisu T, Takayama M, Niwa H (2000) Matrix-assisted laser desorption/ionization time-of-flight mass spectrometry using an inorganic particle matrix for small molecule analysis. J Mass Spectrom 35(3):417–422. https://doi.org/10.1002/(sici)1096-9888(200003)35:3<417::Aid-jms952>3.0.Co;2-#
9. Northen TR, Yanes O, Northen MT, Marrinucci D, Uritboonthai W, Apon J, Golledge SL, Nordstrom A, Siuzdak G (2007) Clathrate nanostructures for mass spectrometry. Nature 449(7165):1033–1036. https://doi.org/10.1038/nature06195
10. Shen ZX, Thomas JJ, Averbuj C, Broo KM, Engelhard M, Crowell JE, Finn MG, Siuzdak G (2001) Porous silicon as a versatile platform for laser desorption/ionization mass spectrometry. Anal Chem 73(3):612–619. https://doi.org/10.1021/ac000746f
11. Wei J, Buriak JM, Siuzdak G (1999) Desorption-ionization mass spectrometry on porous silicon. Nature 399(6733):243–246. https://doi.org/10.1038/20400
12. Xu SY, Li YF, Zou HF, Qiu JS, Guo Z, Guo BC (2003) Carbon nanotubes as assisted matrix for laser desorption/ionization time-of-flight mass spectrometry. Anal Chem 75(22):6191–6195. https://doi.org/10.1021/ac0345695
13. Walker BN, Razunguzwa T, Powell M, Knochenmuss R, Vertes A (2009) Nanophotonic ion production from silicon microcolumn arrays. Angew Chem Int Ed 48(9):1669–1672. https://doi.org/10.1002/anie.200805114
14. Walker BN, Stolee JA, Vertes A (2012) Nanophotonic ionization for ultratrace and single-cell analysis by mass spectrometry. Anal Chem 84(18):7756–7762. https://doi.org/10.1021/ac301238k
15. Walker BN, Stolee JA, Pickel DL, Retterer ST, Vertes A (2010) Tailored silicon nanopost arrays for resonant nanophotonic ion production. J Phys Chem C 114(11):4835–4840. https://doi.org/10.1021/jp9110103
16. Huang L, Wan JJ, Wei X, Liu Y, Huang JY, Sun XM, Zhang R, Gurav DD, Vedarethinam V, Li Y, Chen RP, Qian K (2017) Plasmonic silver nanoshells for drug and metabolite detection. Nat Commun 8:220. https://doi.org/10.1038/s41467-017-00220-4
17. Su HY, Liu TT, Huang L, Huang JY, Cao J, Yang HQ, Ye J, Liu J, Qian K (2018) Plasmonic Janus hybrids for the detection of small metabolites. J Mater Chem B 6(44):7280–7287. https://doi.org/10.1039/c8tb01587b
18. Takats Z, Wiseman JM, Gologan B, Cooks RG (2004) Mass spectrometry sampling under ambient conditions with desorption electrospray ionization. Science

306(5695):471–473. https://doi.org/10.1126/science.1104404

19. Nemes P, Vertes A (2007) Laser ablation electrospray ionization for atmospheric pressure, in vivo, and imaging mass spectrometry. Anal Chem 79(21):8098–8106. https://doi.org/10.1021/ac071181r
20. Haapala M, Pol J, Saarela V, Arvola V, Kotiaho T, Ketola RA, Franssila S, Kauppila TJ, Kostiainen R (2007) Desorption atmospheric pressure photoionization. Anal Chem 79(20):7867–7872. https://doi.org/10.1021/ac071152g
21. Stopka SA, Rong C, Korte AR, Yadavilli S, Nazarian J, Razunguzwa TT, Morris NJ, Vertes A (2016) Molecular imaging of biological samples on nanophotonic laser desorption ionization platforms. Angew Chem Int Ed 55(14):4482–4486. https://doi.org/10.1002/anie.201511691
22. Fincher JA, Korte AR, Dyer JE, Yadavilli S, Morris NJ, Jones DR, Shanmugam VK, Pirlo RK, Vertes A (2020) Mass spectrometry imaging of triglycerides in biological tissues by laser desorption ionization from silicon nanopost arrays. J Mass Spectrom 55(4):e4443. https://doi.org/10.1002/jms.4443
23. Samarah LZ, Tran TH, Stacey G, Vertes A (2021) Mass spectrometry imaging of biooligomer polydispersity in plant tissues by laser desorption ionization from silicon nanopost arrays. Angew Chem Int Ed 60(16):9071–9077. https://doi.org/10.1002/anie.202015251
24. Feenstra AD, O'Neill KC, Yagnik GB, Lee YJ (2016) Organic-inorganic binary mixture matrix for comprehensive laser-desorption ionization mass spectrometric analysis and imaging of medium-size molecules including phospholipids, glycerolipids, and oligosaccharides. RSC Adv 6(101):99260–99268. https://doi.org/10.1039/c6ra20469d
25. Nozaki K, Nakabayashi Y, Murakami T, Miyazato A, Osaka I (2019) Novel approach to enhance sensitivity in surface-assisted laser desorption/ionization mass spectrometry imaging using deposited organic-inorganic hybrid matrices. J Mass Spectrom 54(7):612–619. https://doi.org/10.1002/jms.4370
26. Fincher JA, Korte AR, Yadavilli S, Morris NJ, Vertes A (2020) Multimodal imaging of biological tissues using combined MALDI and NAPA-LDI mass spectrometry for enhanced molecular coverage. Analyst 145(21):6910–6918. https://doi.org/10.1039/d0an00836b
27. Samarah LZ, Tran TH, Stacey G, Vertes A (2020) In vivo chemical analysis of plant sap from the xylem and single parenchymal cells by capillary microsampling electrospray ionization mass spectrometry. Anal Chem 92(10):7299–7306. https://doi.org/10.1021/acs.analchem.0c00939
28. Stopka SA, Agtuca BJ, Koppenaal DW, Pasa-Tolic L, Stacey G, Vertes A, Anderton CR (2017) Laser-ablation electrospray ionization mass spectrometry with ion mobility separation reveals metabolites in the symbiotic interactions of soybean roots and rhizobia. Plant J 91(2):340–354. https://doi.org/10.1111/tpj.13569
29. Agtuca BJ, Stopka SA, Evans S, Samarah L, Liu Y, Xu D, Stacey MG, Koppenaal DW, Pasa-Tolic L, Anderton CR, Vertes A, Stacey G (2020) Metabolomic profiling of wild-type and mutant soybean root nodules using laser-ablation electrospray ionization mass spectrometry reveals altered metabolism. Plant J 103(5):1937–1958. https://doi.org/10.1111/tpj.14815
30. Essigmann B, Guler S, Narang RA, Linke D, Benning C (1998) Phosphate availability affects the thylakoid lipid composition and the expression of SQD1, a gene required for sulfolipid biosynthesis in Arabidopsis thaliana. Proc Natl Acad Sci U S A 95(4):1950–1955. https://doi.org/10.1073/pnas.95.4.1950
31. Alves LPS, do Amaral FP, Kim D, Bom MT, Gavidia MP, Teixeira CS, Holthman F, Pedrosa FD, de Souza EM, Chubatsu LS, Muller-Santos M, Stacey G (2019) Importance of poly-3-hydroxybutyrate metabolism to the ability of Herbaspirillum seropedicae to promote plant growth. Appl Environ Microbiol 85(6):e02586–e02518. https://doi.org/10.1128/aem.02586-18

Chapter 7

Metal-Assisted Laser Desorption Ionization Imaging Mass Spectrometry

Frédéric Fournelle and Pierre Chaurand

Abstract

Matrix-assisted laser desorption ionization (MALDI) remains the reference method to generate molecular images of proteins and lipids within thin tissue sections. However, traditional MALDI imaging mass spectrometry (IMS) suffers from low matrix homogeneity and high signal background in low mass range caused by matrix signals. To overcome these issues, alternative workflow and methods have been developed. Of these, metal-assisted laser desorption ionization (LDI) has become a reference technique to ionize low molecular weight compounds while allowing IMS at very high spatial resolutions with very low background signal in the low mass range. Silver and gold remain the two most used metals for the detection of neutral lipids including cholesterol, free fatty acids, and triglycerides. In this chapter, we demonstrate the potential of metal-assisted LDI IMS through the analysis of spinal cord and kidney thin tissue sections after silver and gold metal deposition. We also detail typical step-by-step workflows and discuss the strength of the methods.

Key words Imaging MS, Laser desorption ionization, Neutral lipids, Cholesterol, Silver, Gold, SALDI, Spinal cord, Metal sputter

1 Introduction

Over the past decades, imaging mass spectrometry (IMS) has been under constant evolution and new developments have mainly benefited biological applications [1]. Out of all available technique, matrix-assisted laser desorption ionization (MALDI) IMS remains the most prominent technology to analyze biomolecular spatial distribution within thin biological tissue sections [2, 3]. MALDI IMS allows the acquisition of both qualitative [3–5] and quantitative [6–8] information for a large variety of biomolecules such as metabolites [9–11], lipids [12–17], proteins [3, 18–20], and carbohydrates [11, 21, 22] in a wide range of species. A typical MALDI IMS workflow includes tissue sectioning and thaw-

Young-Jin Lee (ed.), *Mass Spectrometry Imaging of Small Molecules: Methods and Protocols*,
Methods in Molecular Biology, vol. 2437, https://doi.org/10.1007/978-1-0716-2030-4_7,
© The Author(s), under exclusive license to Springer Science+Business Media, LLC, part of Springer Nature 2022

mounting on an analytical slide, matrix applications, and data acquisition [1]. Sample preparation is of premier importance and any flaws in protocol or execution may greatly impact the quality of the results. One critical step involved in sample preparation is the deposition of a highly homogenous matrix layer of small matrix crystals over the sample surface with any defects in matrix homogeneity reflecting in analytical quality [23].

Traditional MALDI matrices generally suffer from a common issue which goes back to the intrinsic properties of organic matrices and their low molecular weights. These organic matrices are known to generate intense background signals, typically below 500 Da, which limits the detection (and imaging) of low molecular weight compounds such as metabolites [24, 25]. To overcome these issues, new organic matrices with very high molecular weight such as meso-Tetrakis(pentafluorophenyl)porphyrin have been developed, which allow the detection of small molecules with limited background signals below 1000 Da [26] .

Another strategy has been to replace the organic matrix by a metallic layer. Metallic matrices have become a very interesting alternative to classical organic matrices and like those, they can strongly absorb in the UV range of the lasers traditionally used for MALDI [13, 14]. Moreover, metallic layers have shown to produce a highly structured and homogenous surface, which helps reduce shot to shot variability [13]. Lastly, metallic depositions are extremely resistant to high vacuum which is known to be a problem for some low vapor pressure organic matrices such as 2,5-dihydroxyacetophenone (2,5-DHA) [25]. In 2013, Dufresne et al. demonstrated the use of a thin layer of silver to selectively ionize cholesterol and unsaturated fatty acids by laser desorption ionization (LDI) [14]. The high specificity of silver towards carbon-carbon double bonds [27, 28] allowed the detection and imaging of unsaturated species such as cholesterol and fatty acids in mouse brain and kidney tissue sections. In 2015, Lauzon et al. adapted this approach to analyze and image endogenous and exogenous substances in latent fingermarks [29]. In 2016, Dufresne et al. demonstrated that a combination of sodium salt and sputtered gold on tissue section improved by 30 folds the detection of triacylglycerols (TAGs) compared to 2,5-dihydroxybenzoic acid (DHB) [13]. This method has been adapted to investigate the molecular distribution of TAGs in liver [30] and kidney [31] tissue sections.

In this chapter, we detail the latest methods and protocols for silver and gold-assisted LDI IMS and highlight their potentials through the analysis of spinal cord and kidney tissue sections.

2 Materials

2.1 Tissue Section Preparation

1. Cryomicrotome with a chamber temperature set at −15 °C.
2. Standard microscope glass (25 × 75 mm) and indium tin oxide (ITO) coated glass (25 × 75 mm) slides.
3. Spinal cord or liver tissue sample (or other tissue sample).
4. Regular stainless-steel cryostat cutting blades.
5. Optimal cutting temperature (OCT) compound.
6. Carboxymethylcellulose (CMC).
7. Peel-away embedding molds.
8. Regular small paint brushes.
9. Razor blades.
10. Glass desiccator connected to an in-house vacuum system.

2.2 Salt Pneumatic Spray Deposition

1. Automated pneumatic spray system. (TM sprayer, HTX Technologies, LLC, Chapel Hill, NC, USA).
2. Compressed air up to 20 psi.
3. 20 mL glass vials
4. Acetone HPLC grade.
5. Cleaning solution: 0.1% trifluoroacetic acid (TFA) in 1:1 H_2O/ACN (v/v).
6. Analytical balance.
7. A carbonate buffer solution (CBS) containing: sodium bicarbonate ($NaHCO_3$) 6.93 mg/mL, sodium carbonate (Na_2CO_3) 8.01 mg/mL, and sodium acetate (NaAC) 20.43 mg/mL, in ACN–H_2O (3:7).
8. Syringe pump allowing a 0.225 mL/min flow rate and 5 mL syringe.
9. 5 mL Luer-lock syringe
10. Luer-lock connector to connect a 5 mL Luer-lock syringe to the capillary of the spray system.

2.3 Metal Sputter Deposition

1. Sputter coater system with rotational slide holders (Cressington 308R, Ted Pella Inc., Redding, CA).
2. Ultra-high purity Ag and Au metallic sputter targets (99.999% purity).
3. Tank of high purity compressed argon and nitrogen gas (99.9%).

2.4 LDI IMS Data Acquisition

1. A document scanner.
2. A MALDI mass spectrometer (ultrafleXtreme, Bruker Daltonics, Billerica, MA).

3 Methods

3.1 Tissue Sectioning

This section describes how previously frozen small tissue samples likes spinal cord are embedded in CMC solution using a cryomold and subsequently sectioned with a cryostat. Further IMS analyses are performed on a MALDI-TOF MS instrument, which requires the use of conductive slides. ITO-coated glass slides are our first choice for thaw-mounting all sections (*see* **Note 2**).

1. Prepare a 1% CMC solution in water at 60 °C with agitation until completely dissolved. Store at 4 °C between usage.
2. Break the tip of a disposable glass pipette ~5 cm above the tip. This will help aspiration of the cold CMC solution.
3. Fill half of a cryomold with 1% CMC. Let the CMC cool in the cryostat main chamber until it starts to harden.
4. Place the spinal cord with the desired orientation inside the mold and fill the rest of the cryomold with CMC (*see* **Note 3**).
5. Place the cryomold on dry ice back in the −80 °C freezer until completely solid (*see* **Note 4**).
6. Place the mold inside the precooled cryostat chamber 30 min prior the sectioning. This allows for sample temperature equilibration. All cryostat surfaces should be thoroughly cleaned with ethanol prior to usage. The cryostat chamber temperature should be set at −15 °C.
7. Remove the sample from the mold and roughly trim the excess CMC. More precise removal of excess CMC can be done before or after trimming.
8. For spinal cord, mount the CMC block on a cryostat holder inside the main cryostat chamber using OCT compound as 'glue' with the desired sectioning orientation. For spinal cord, we typically cut transversal sections. Wait until OCT has fully solidified before removing excess CMC using a razor blade. This is performed to minimize excess CMC surrounding the embedded tissue sample.
9. For liver, directly mount a fresh-frozen liver sample on a cryostat holder inside the main cryostat chamber using OCT compound as 'glue'. The sample needs to be properly anchored to the holder but does not need to be fully embedded in OCT. In this case, the cryostat blade will never cut through the OCT. Excess OCT may need to be removed using a razor blade prior to sectioning.

10. To begin the sectioning process, mount a new cryostat blade in the appropriate cryostat holder according to the manufacturer instruction. To limit any contamination, we recommend rinsing the blade with organic solvents (acetone or methanol) prior to use. Note that these blades are extremely sharp and should be handled with care. Start roughly trimming the CMC block at 70 μm thickness using the 'Trim' function of the cryostat until the sample is visible within the CMC block. Trimmed sections should be discarded to keep the sectioning environment clean.
11. For thin sectioning, if possible, slide the cryostat blade inside the blade holder to use a new part of the blade (*see* **Note 5**). A new blade may also be used.
12. Exit the 'Trim' function and section the tissue bloc at the desired thickness. For spinal cord and liver, we normally cut 12-μm thick sections. The first few sections typically need to be discarded.
13. To thaw-mount tissue sections on the ITO-coated slides, the slides should initially be cold and placed inside the main cryostat chamber at least 5 min prior to sectioning.
14. Delicately manipulate thin tissue sections with the brushes avoiding contact with the metallic part of the brush (*see* **Note 6**). This may melt the sections.
15. Place or drag the section onto a cold ITO-coated glass slide. Once the section is placed as desired, while keeping the slide in the cryostat chamber, warm the bottom of the slide on one's hand. Because of heat transfer, the section should now thaw-mount on the slide. Maintain heat contact until the slide is mostly dry. This takes about 30 s and the section should now be white opaque. The slide should be kept cold if multiple sections are to be collected on the same slide.
16. Collect serial sections on the microscope glass slides for H&E staining and further histological comparisons (*see* **Note** 7).
17. Repeat **steps 15** and **16** until the desired number of sections has been collected.
18. When sectioning is finished, the slide can be removed from the cryostat. To avoid moisture condensation on the slide, we recommend to again warm the slides on one's hand inside the cryostat chamber for ~30 s as described in **step 15** before removal from the cryostat chamber.

3.2 Tissue Sections Use and Preservation

This section briefly describes preprocessing or preservation prior to IMS.

1. If the section is to be immediately analyzed by IMS, we recommend placing the slide inside of a (glass or plastic) vacuum desiccator for 30 min to fully dehydrate the section.

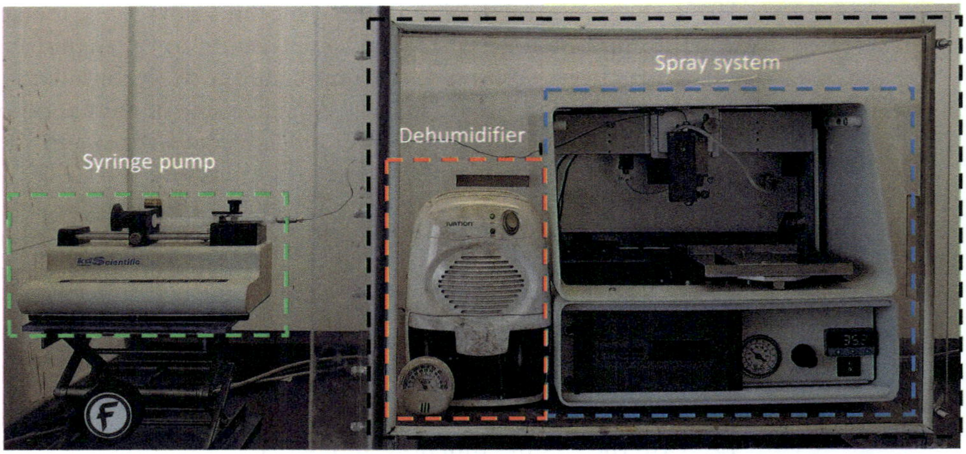

Fig. 1 Photomicrograph of the spray deposition setup. Green dotted lines include syringe pump and Luer-lock syringe, black dotted lines outline the Plexiglass box, red dotted lines include the small dehumidifier and hygrometer and blue dotted lines outline the HTX TM sprayer unit. Electric wiring and gas lines are fit in a small hole at the back of the Plexiglass box

2. Slide storage for a few days is also possible at −80 °C. After **step 1**, place the slide inside a small slide box within a vacuum bag. Purge the vacuum bag with nitrogen three times prior to vacuum sealing (*see* **Note 9**). This will limit molecular degradation [32, 33].
3. When the section is to be analyzed, remove the bag from the freezer and let it reequilibrate at room temperature.
4. When room temperature is reached, the slide should be desiccated for 30 min.

3.3 Salt Spray Deposition

This section describes salt spray deposition on liver tissue sections. Salt-spray deposition is only performed prior to gold sputter deposition to enhance the ionization efficiency of TAGs [13]. Our lab is using a first generation TM sprayer which is placed inside a Plexiglass box equipped with a small dehumidifier system (Fig. 1). For safety concerns, the Plexiglass box is installed inside a fume hood.

1. Start the dehumidifier at least 30 min to 1 h prior to salt deposition (*see* **Note 10**).
2. Turn on both the TM sprayer instrument and software.
3. Open compressed air valves and adjust pressure to 20 psi. Set the spray temperature to 75 °C and allow the system to stabilize for 5 min.
4. Flush and clean thoroughly the system's solvent lines with 5 mL of cleaning solution.
5. Set the syringe pump flow rate to 0.225 mL/min and create or load a TM sprayer method with the following parameters (*see*

Note 11): 15 passes at 7 mm of track spacing with 3.5 mm offset for odd number passages with 3485 mm/min velocity and 6 s of drying time after each passage. Nozzle is positioned 6 cm above the slide. For spray dimensions, make sure to spray at least 0.5 cm wider on both sides of the slide to ensure homogeneous salt coverage on the slide.

6. Connect the syringe to the TM sprayer line and start the syringe pump. Wait until the spray is fully homogenous and continuous (*see* **Note 12**).
7. Place the slide at the desire position on the instrument tray and stabilize it with adhesive tape.
8. Close the Plexiglass door and start the method on the TM-sprayer software.
9. When the spray deposition is complete, remove the slide from the tray. Salt deposition should be visible observing a faint white homogenous film throughout the sprayed area.
10. Thoroughly clean the TM sprayer system with the cleaning solution. This includes the spray head, slide holder and spray line.
11. Turn off the instrument software and air supply.
12. Leave on the dehumidifier and close the Plexiglass door.

3.4 Metal Sputter Deposition

This section describes the use of a sputter coater instrument for tissue sample preparation. Sputter deposition allows for very thin homogeneous metal layer deposition on tissue sections toward LDI IMS analysis (*see* **Note 13**). Different groups have demonstrated the use of sputter deposition for IMS including gold, silver, platinum, chrome, and titanium [34]. Silver and gold remain the most used metals for their high sensitivity toward distinct classes of biomolecules [34]. The sputter system we use generally produces highly homogeneous metal layers on single sections, but a gradient in metal thickness is often observed across the slide. More information to minimize this gradient can be found in Subheading 4. The protocol described here was optimized for a Cressington 308R sputter coater and can be adapted to other sputter systems.

1. Open the argon and nitrogen gas tanks to 40 psi and 80 psi, respectively.
2. Turn on and vent the sputter system. Remove the lid to gain access to the interior of the sputter system chamber (Fig. 2) (*see* **Note 14**).
3. Place the slide on the sample holder (Fig. 2).
4. Inside the chamber, install the slide holder on the rotational platform of the sputter system and set rotation speed around 20 rpm (*see* **Note 15**). Make sure the plate holder is properly mounted or it may dislodge during the deposition.

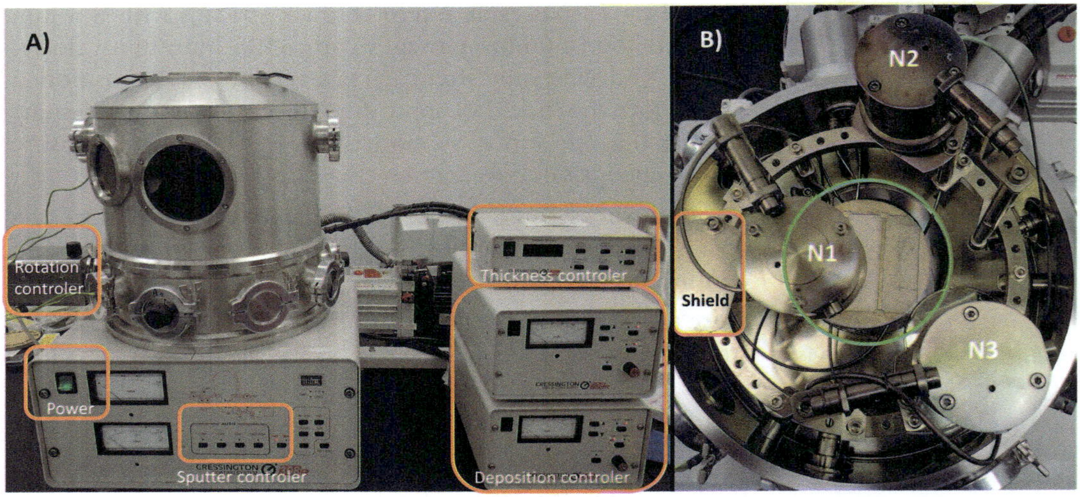

Fig. 2 Cressington 308R sputter coater. Photomicrograph of (**a**) the system with controllers, and (**b**) inside of the sputter coater chamber showing the head N1 in the ideal position for metal sputter deposition, and heads N2 and N3 in standby position facing their respective shields. Green circle enlightens a custom-built slides holder allowing to mount up to three slides

5. Using the rotational arm, position the desire target head (silver or gold) over the slide. Make sure the metallic head is properly positioned directly over the center rotational platform (*see* **Note 16**). When this is done, use a permanent ink pen to mark the position on the rotational arm. Place the head back in the standby position facing the shield (Fig. 2).
6. Clean the sputter chamber O-rings with Kimwipes and close the lid.
7. Pump the system down to 10^{-5} mbar. This step takes about 15 min.
8. Select the desired metal head (silver or gold) and set the appropriate current value using the up/down arrows on the deposition controller unit (Fig. 2a). We normally use 40 mA for gold and 80 mA for silver.
9. Push the Leak switches and wait until the argon pressure inside the chamber is stable.
10. Push the Start switch to perform a 5–10 s metal deposition on the shield to ensure a complete metal oxide layer removal on the selected high purity metal target head (*see* **Note 17**).
11. Use the rotational arm to perfectly position the chosen head (silver or gold) above the slide holder using the mark made on the rotational arm in **step 5**.
12. Activate rotation of the slide holder and confirm rotation visually using the glass portals of the chamber.

13. Push the 'Start' switch and perform deposition (*see* **Note 18**). We normally perform deposition times of 40 s for gold on liver sections and 30 s for silver on spinal cord sections (*see* **Note 19**).
14. When deposition time is over, push the 'Start' switch again to stop the deposition process.
15. Push the 'Vent' switch to vent the system with nitrogen and wait until the system reaches atmospheric pressure. This step takes about 2 min.
16. Remove the slide holder and liberate the slide.
17. To shut down the system, clean the lid with a Kimwipe, put the lid back on the chamber and push the 'Pump' switch. Wait until the system is back at 10^{-5} mbar and press shutdown. Wait until shutdown is complete and turn off all controllers, sputter system and close the argon and nitrogen gas tanks.

3.5 Metal-Assisted LDI IMS Data Acquisition

This section describes how metal-assisted LDI IMS data is acquired by TOF MS on a ultrafleXtreme MALDI TOF/TOF instrument (Bruker Daltonics, Billerica, MA) after metal sputter deposition. Silver-assisted LDI MS (AgLDI MS) allows the easy detection of silver (Ag^+) and silver cluster (Ag_2^+, Ag_3^+, Ag_5^+, and Ag_7^+) ions (Fig. 3), which can be used for spectral pre- and postacquisition calibration. Gold-assisted LDI MS (AuLDI IMS) allows for the detection of Au^+ ions as well as the Au_2^+ and Au_3^+ cluster ions. The gold clusters are however of limited use of mass calibration in the TAG MW range.

1. Once metal sputter deposition is complete, scan an image of the slide. We use a conventional document scanner at 3200 dpi resolution.
2. Insert the slide in the mass spectrometer using a slide holder target plate and load the image of the slide in the flexImaging software to perform teach point calibration, select the region (s) of interest of the tissue section to be analyzed, define an IMS spatial resolution, and load the MS acquisition method.
3. The TOF MS ion source should be cleaned prior to IMS data acquisition. For cleaning protocols, refer to the manufacturer manual.
4. Select the desired MS acquisition method in flexControl and if necessary, optimize laser spot size on target, laser power (*see* **Note 20**), delayed extraction parameters, detector gain, number of laser shots per position, random walk, and mass calibration.
5. Confirm MS acquisition method validity on all selected tissue regions (*see* **Note 21**).

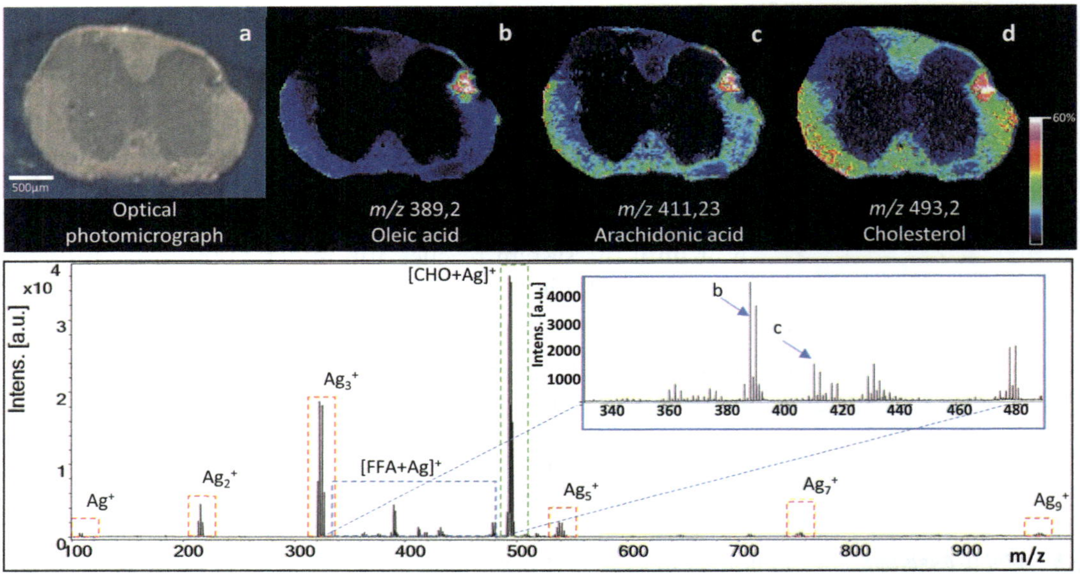

Fig. 3 Positive ion AgLDI MS and IMS of a spinal cord thin tissue section acquired at 20 μm of spatial resolution. (**a**) Photomicrograph of the section after silver sputter deposition, (**b–d**) $[M+Ag]^+$ ion images for oleic acid (**b**), arachidonic acid (**c**) and cholesterol (**d**). Bottom inset shows a representative AgLDI MS spectrum with an emphasis on free fatty acid (FFA) region. The silver cluster ions have also been outlined in the red dotted boxes

6. Confirm metal signals are intense and present throughout the slide.
7. In flexImaging, save the auto execute sequence.
8. Load the saved 'auto execute' method in Batch Runner and launch the data acquisition sequence.

3.6 Analysis of AgLDI and AuLDI IMS Datasets

This section describes important steps and particularities regarding data analysis of AgLDI and AuLDI IMS datasets. As stated previously, for AgLDI IMS, the average spectrum will be dominated with silver clusters. Silver has two major stable isotopes Ag_{107} (exact mass: 106.905095 u) and Ag_{109} (exact mass: 108.904754 u) with approximately a 51:49 abundance ratio. Silver has a predominant affinity for olefins and particularly for cholesterol (CHO) and its derivatives. In the positive ion mode, AgLDI will predominantly favor the formation of silver adduct molecular ions. Cholesterol (isotopic MW 386.35 u) is therefore detected as $[CHO+Ag]^+$ generating a signal doublet at m/z 493.23 and 495.23 when considering both silver isotopes. For AuLDI, the average mass spectrum is dominated by TAGs and most species will be detected as sodiated adducts, $[TAG+Na]^+$, due to the sodium salt deposition prior to gold sputtering and the high affinity of TAGs for sodium (Fig. 4).

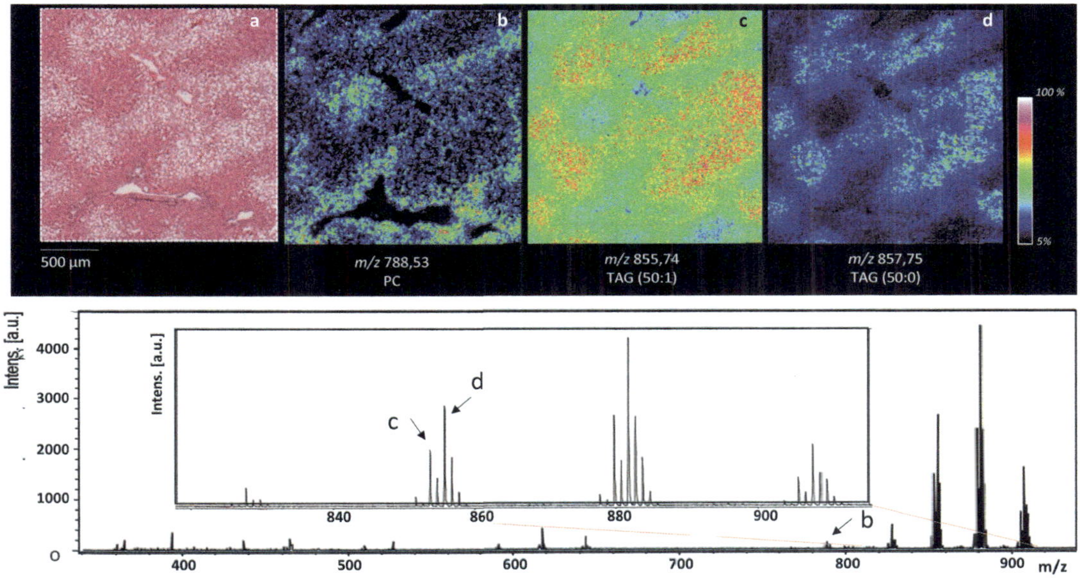

Fig. 4 Positive ion AuLDI MS and IMS acquired from a liver section afflicted by nonalcoholic fatty liver disease (NAFLD) at 10 μm of spatial resolution. (**a**) Photomicrograph of a serial section after H&E staining. (**b**–**d**) $[M+Na]^+$ ion images of a PC (**b**), TAG(50:1) (**c**), TAG(50:0) (**d**). Bottom inset shows a representative AuLDI MS spectrum with an emphasis on the TAGs region between 820 and 930 Da

1. Open the selected dataset in flexImaging or any compatible IMS analysis software.
2. Confirm the presence and experimental mass of all detected Ag clusters. If the mass error is over your instrumental limitation, internal mass recalibration may be required (*see* **Note 22**). With the Bruker suite of software, this task can be performed in flexAnalysis.
3. For AgLDI IMS, generate images of the most predominant silver clusters and confirm homogeneity on and off samples throughout the slide.
4. Focus your attention between 300 and 500 Da where most of signals of interest (free fatty acids and cholesterol) are located (Fig. 3). Look for any peak doublets separate by 2 Da with approximately 51:49 ratio, which indicates the presence of silver adducts (*see* **Note 23**).
5. For AuLDI IMS, generate images for Au^+, Au_2^+ and Au_3^+ which should be present and homogenously distributed throughout the sample.
6. Focus your attention between 500 and 950 Da where diacylglycerols (DAGs) and triacyclglycerols (TAGs) are detected (Fig. 4).

7. When possible, perform on-tissue MS/MS to confirm the identity of compounds of interest.
8. If desired, export your dataset in imzML format for further quantitative comparative analyses. For this task, we use the Cardinal software package [35] (*see* **Note 24**).

4 Notes

1. To prepare a CBS solution, weight and dissolve all salts in water using a sonic bath prior to addition of ACN. From experience, this will greatly reduce dissolving times.
2. ITO coating is only on one side of the slides and sections need to be mounted on the ITO coated side. The ITO-coated side can be determined by measuring resistivity on both sides using a standard voltmeter. Once Identified, we recommend referencing with a permanent marker.
3. Be mindful of the orientation and placement of the spinal cord sample inside the cryomold. This will greatly help to obtain perfect transversal sections. We normally mark the sample orientation on the cryomold and transfer all the marks on the CMC block right after demolding using a permanent marker.
4. We normally wait overnight before processing the sample. CMC should now be solid and opaque.
5. CMC can heavily damage cryostat blades and poor-quality sections will be obtained if the same blade edge is used for trimming and sectioning. Be therefore mindful not to touch the edge of the blade with razor blades and brushes during the sectioning process.
6. To manipulate the sections, make sure every object including glass slides, ITO-coated slide and brushes that will touch the sections are at cryostat temperature. Otherwise, the section will immediately melt.
7. To date, we have not been able to remove the metallic layers without major degradation of the tissue sections. This prevents any histological staining post-IMS. Ideally, strictly serial sections should be collected to perform histological staining. This will allow IMS and histology alignments without major distortion. For most tissues, a classical hematoxylin and eosin coloration is performed.
8. Make sure to leave all object including brushes and slides inside the cryostat while sectioning to prevent water condensation. Partially close the cryostat slide glass door to prevent water condensation inside the cryostat chamber and on the slides.

When excess condensation is observed (ice formation), the cryostat needs to be shut down, defrosted, and thoroughly cleaned with ethanol (see manufacturer's operating manual).

9. When purging the vacuum bag for long-term section storage, make sure that the bag in fully inflated and deflated during the three cycles of nitrogen purge. Make a first seal along the edge of the bag and leave a 3 cm opening on one side. After the purges, inflate the bag fully with nitrogen and vacuum seal the small opening. The vacuum sealed bags allow for long-term storage and better freezer organization compared to inflated Ziploc bags.
10. The dehumidifier installed within the Plexiglass spray box (Fig. 1) helps to maintain a constant low humidity inside the spray chamber and allows for more reproducible day to day spraying conditions. We normally never turn it off and we keep the Plexiglass door shut when the TM-sprayer is not in use.
11. Make sure the dry time between each CBS spray passage is long enough to allow for a complete drying of the slide. Small method optimization including flow rate and velocity adjustments may be required when adapting the spray deposition to other or newer spray systems.
12. We normally use Kimwipes or water-sensitive paper to confirm the consistency and stability of the CBS spray. Time required to reach stability will differ according to the solvent line length and flow rate. If the spray is not consistent, we normally use a soap/water solution to check for leaks on all connections.
13. Metal sputter deposition has shown to render sufficiently conductive nonconductive surfaces (such as paper, cardboard or plastic surfaces) to obtain high quality IMS data on TOF MS instruments without noticeable surface charging effects [36]. ITO coated slides are therefore not absolutely required if the metallic layer is thick and homogenous enough throughout the slide and contacts the slide holder ensuring electric conductivity.
14. Make sure to wear clean gloves prior to manipulating any parts inserted inside the sputter chamber. Any particle left inside will be sucked into the vacuum and will increase vacuum pump time. Make sure all surfaces are clean and dry. Otherwise, clean these with methanol and ensure proper drying before use.
15. We use a custom build slide holder which mounts on the rotational platform of the sputter instrument (*see* Fig. 2). This holder allows to prepare three slides simultaneously and ensures reproducibility in slide positioning from deposition to the deposition.

16. Properly positioning the chosen head directly over the center of slide is critical as the rate of deposition is faster in the middle of the metallic target compare to the edges. Any misposition will result in a lack of homogeneity in metal thickness throughout the slide and will result in variation in MS signal intensity.
17. On the sputter system we use, the sputter heads can be manipulated from outside the chamber. During initial oxide layer removal on the metal targets, make sure you place the metallic head over the shield to prevent any metal deposition on the slides. Otherwise, this will produce an uneven metal deposition on the slide.
18. During the sputter deposition process, plasma discharge should be visible throughout the chamber window. If this is not the case, this indicates a problem with the system that should be addressed prior to the deposition. We have experienced some issues with metal head wire connectors, and it should be the first thing to check.
19. Metallic layer thickness should be understood and optimized like a matrix layer thickness is normally optimized for each tissue sample type and target molecules. This optimization must be performed to ensure reproducibility and maximum efficiency.
20. When optimizing a method for metal-assisted LDI IMS, be sure to appropriately adjust the laser energy correctly to obtain high signal intensity throughout the m/z range of interest while minimizing in-source fragmentation. Metallic layers do not protect the analytes as well as traditional organic matrices. The transition between maximum signal intensity and in-source fragmentation is very sharp and should be evaluated and properly set for every metal and for every experiment. This evaluation can be performed on a reference tissue section placed on the slide.
21. For method validation, we typically use mouse brain homogenate sections [6] for AgLDI IMS and mouse kidney sections for AuLDI IMS. For AgLDI IMS, the $[CHO+Ag]^+$ signal should be the most intense peak detected. The Ag_3^+ and Ag_7^+ peaks should also be detected easily. For AuLDI IMS, when targeting renal pelvis and regions with visceral fat, sodiated triacylglycerol's should be the most intense peaks while background noise below 400 Da should be minimal. If not, this might be an indicator of excessive in-source fragmentation cause by too much laser power.
22. To perform post-acquisition mass recalibration, make sure you first make a copy of your entire dataset. Post-acquisition mass recalibration may require custom scripts and will differ between data formats. Please contact your instrument manufacturer for assistance.

23. Be aware of triplet peaks separate by 2 Da which may indicate Ag adducts on fatty acids differing by one unsaturation. For example, oleic acid (MW 282.25 u) stearic acid (MW 284.27 u) will have a common ion at 391.16 Da corresponding to respectively ^{109}Ag and ^{107}Ag adducts. To only visualize one specie, either select the first or third ion of the triplet. For fatty acids, beyond the isotopic silver ion distribution, signals will generally have a 0.10 to 0.25 Da mass defect, which can help secure identification.
24. Software like the Cardinal package allows for internal signal normalization using single features. Since silver and gold ions (or clusters) are homogeneously detected throughout tissue sections, these ions can be used to normalize your IMS dataset. This can help improve signal homogeneity by minimizing the effects of 'hot spots' and empty pixels.

Acknowledgments

The authors would like to acknowledge Pr. Jean-François Masson (Dept of Chemistry, Université de Montréal) for unlimited access to the metal sputtering system, Pr. Alan C. Peterson (Dept of Human Genetics, McGill University) for providing the spinal cord samples, and Dr. Nidia Lauzon for acquiring the spinal cord IMS data.

References

1. Gessel MM, Norris JL, Caprioli RM (2014) MALDI imaging mass spectrometry: spatial molecular analysis to enable a new age of discovery. J Proteome 107:71–82. https://doi.org/10.1016/j.jprot.2014.03.021
2. Stoeckli M, Chaurand P, Hallahan DE, Caprioli RM (2001) Imaging mass spectrometry: a new technology for the analysis of protein expression in mammalian tissues. Nat Med 7 (4):493–496. https://doi.org/10.1038/86573
3. Chaurand P, Schwartz SA, Billheimer D, Xu BJ, Crecelius A, Caprioli RM (2004) Integrating histology and imaging mass spectrometry. Anal Chem 76(4):1145–1155. https://doi.org/10.1021/ac0351264
4. Patterson NH, Tuck M, Lewis A, Kaushansky A, Norris JL, Van de Plas R, Caprioli RM (2018) Next generation histology-directed imaging mass spectrometry driven by autofluorescence microscopy. Anal Chem 90(21):12404–12413. https://doi.org/10.1021/acs.analchem.8b02885
5. Schöne C, Höfler H, Walch A (2013) MALDI imaging mass spectrometry in cancer research: combining proteomic profiling and histological evaluation. Clin Biochem 46(6):539–545. https://doi.org/10.1016/j.clinbiochem.2013.01.018
6. Groseclose MR, Castellino S (2013) A mimetic tissue model for the quantification of drug distributions by MALDI imaging mass spectrometry. Anal Chem 85(21):10099–10106. https://doi.org/10.1021/ac400892z
7. Takai N, Tanaka Y, Saji H (2014) Quantification of small molecule drugs in biological tissue sections by imaging mass spectrometry using surrogate tissue-based calibration standards. Mass Spectrom 3(1):A0025. https://doi.org/10.5702/massspectrometry.A0025
8. Rzagalinski I, Volmer DA (2017) Quantification of low molecular weight compounds by MALDI imaging mass spectrometry – a tutorial review. Biochim Biophys Acta 1865 (7):726–739. https://doi.org/10.1016/j.bbapap.2016.12.011

9. Shrivas K, Hayasaka T, Sugiura Y, Setou M (2011) Method for simultaneous imaging of endogenous low molecular weight metabolites in mouse brain using TiO_2 nanoparticles in nanoparticle-assisted laser desorption/ionization-imaging mass spectrometry. Anal Chem 83(19):7283–7289. https://doi.org/10.1021/ac201602s
10. Yang FY, Chen JH, Ruan QQ, Saqib HSA, He WY, You MS (2020) Mass spectrometry imaging: an emerging technology for the analysis of metabolites in insects. Arch Insect Biochem Physiol 103(4):11. https://doi.org/10.1002/arch.21643
11. Wang J, Yang E, Chaurand P, Raghavan V (2021) Visualizing the distribution of strawberry plant metabolites at different maturity stages by MALDI-TOF imaging mass spectrometry. Food Chem 345:128838. https://doi.org/10.1016/j.foodchem.2020.128838
12. Thomas A, Charbonneau JL, Fournaise E, Chaurand P (2012) Sublimation of new matrix candidates for high spatial resolution imaging mass spectrometry of lipids: enhanced information in both positive and negative polarities after 1,5-Diaminonapthalene deposition. Anal Chem 84(4):2048–2054. https://doi.org/10.1021/ac2033547
13. Dufresne M, Masson J-F, Chaurand P (2016) Sodium-doped gold-assisted laser desorption ionization for enhanced imaging mass spectrometry of Triacylglycerols from thin tissue sections. Anal Chem 88(11):6018–6025. https://doi.org/10.1021/acs.analchem.6b01141
14. Dufresne M, Thomas A, Breault-Turcot J, Masson J-F, Chaurand P (2013) Silver-assisted laser desorption ionization for high spatial resolution imaging mass spectrometry of olefins from thin tissue sections. Anal Chem 85:3318–3324. https://doi.org/10.1021/ac3037415
15. Yang E, Dufresne M, Chaurand P (2017) Enhancing ganglioside species detection for MALDI-TOF imaging mass spectrometry in negative Reflectron mode. Int J Mass Spectrom 437:3–9. https://doi.org/10.1016/j.ijms.2017.09.011
16. Dufresne M, Patterson NH, Norris JL, Caprioli RM (2019) Combining salt doping and matrix sublimation for high spatial resolution MALDI imaging mass spectrometry of neutral lipids. Anal Chem 91(20):12928–12934. https://doi.org/10.1021/acs.analchem.9b02974
17. Perry WJ, Patterson NH, Prentice BM, Neumann EK, Caprioli RM, Spraggins JM (2020) Uncovering matrix effects on lipid analyses in MALDI imaging mass spectrometry experiments. J Mass Spectrom 55(4):e4491. https://doi.org/10.1002/jms.4491
18. Thomas A, Patterson NH, Laveaux Charbonneau J, Chaurand P (2013) Orthogonal organic and aqueous-based washes of tissue sections to enhance protein sensitivity by MALDI imaging mass spectrometry. J Mass Spectrom 48(1):42–48. https://doi.org/10.1002/jms.3114
19. Yang J, Caprioli RM (2011) Matrix sublimation/recrystallization for imaging proteins by mass spectrometry at high spatial resolution. Anal Chem 83(14):5728–5734. https://doi.org/10.1021/ac200998a
20. Caprioli RM, Farmer TB, Gile J (1997) Molecular imaging of biological samples: localization of peptides and proteins using MALDI-TOF MS. Anal Chem 69(23):4751–4760. https://doi.org/10.1021/ac970888i
21. Enomoto H, Sato K, Miyamoto K, Ohtsuka A, Yamane H (2018) Distribution analysis of anthocyanins, sugars, and organic acids in strawberry fruits using matrix-assisted laser desorption/ionization-imaging mass spectrometry. J Agric Food Chem 66(19):4958–4965. https://doi.org/10.1021/acs.jafc.8b00853
22. Horikawa K, Hirama T, Shimura H, Jitsuyama Y, Suzuki T (2019) Visualization of soluble carbohydrate distribution in apple fruit flesh utilizing MALDI–TOF MS imaging. Plant Sci 278:107–112. https://doi.org/10.1016/j.plantsci.2018.08.014
23. Li S, Zhang Y, Ja L, Han J, Guan M, Yang H, Lin Y, Xiong S, Zhao Z (2016) Electrospray deposition device used to precisely control the matrix crystal to improve the performance of MALDI MSI. Sci Rep 6:37903. https://doi.org/10.1038/srep37903
24. Molin L, Seraglia R, Dani F, Prof G, Traldi P (2011) The double nature of 1,5-diaminonaphthalene as matrix-assisted laser desorption/ionization matrix: some experimental evidence of the protonation and reduction mechanisms. Rapid Commun Mass Spectrom 25:3091–3096. https://doi.org/10.1002/rcm.5201
25. Leopold J, Popkova Y, Engel KM, Schiller J (2018) Recent developments of useful MALDI matrices for the mass spectrometric characterization of lipids. Biomol Ther 8(4):25. https://doi.org/10.3390/biom8040173
26. Ayorinde FO, Hambright P, Porter TN, Keith QL Jr (1999) Use of meso- tetrakis (pentafluorophenyl)porphyrin as a matrix for low molecular weight alkylphenol ethoxylates

in laser desorption/ionization time-of-flight mass spectrometry. Rapid Commun Mass Spectrom 13(24):2474–2479

27. Nikolova-Damyanova B (2009) Retention of lipids in silver ion high-performance liquid chromatography: facts and assumptions. J Chromatogr A 1216(10):1815–1824. https://doi.org/10.1016/j.chroma.2008.10.097
28. Jackson AU, Shum T, Sokol E, Dill A, Cooks RG (2011) Enhanced detection of olefins using ambient ionization mass spectrometry: ag+ adducts of biologically relevant alkenes. Anal Bioanal Chem 399(1):367–376. https://doi.org/10.1007/s00216-010-4349-5
29. Lauzon N, Dufresne M, Chauhan V, Chaurand P (2015) Development of laser desorption imaging mass spectrometry methods to investigate the molecular composition of latent fingermarks. J Am Soc Mass Spectrom 26 (6):878–886. https://doi.org/10.1007/s13361-015-1123-0
30. Alamri H, Patterson NH, Yang E, Zoroquiain P, Lazaris A, Chaurand P, Metrakos P (2019) Mapping the triglyceride distribution in NAFLD human liver by MALDI imaging mass spectrometry reveals molecular differences in micro and macro steatosis. Anal Bioanal Chem 411(4):885–894. https://doi.org/10.1007/s00216-018-1506-8
31. Fournelle F, Yang E, Dufresne M, Chaurand P (2020) Minimizing visceral fat delocalization on tissue sections with porous aluminum oxide slides for imaging mass spectrometry. Anal Chem 92(7):5158–5167. https://doi.org/10.1021/acs.analchem.9b05665
32. Patterson NH, Thomas A, Chaurand P (2014) Monitoring time-dependent degradation of phospholipids in sectioned tissues by MALDI imaging mass spectrometry. J Mass Spectrom 49(7):622–627. https://doi.org/10.1002/jms.3382
33. Lukowski JK, Pamreddy A, Velickovic D, Zhang G, Pasa-Tolic L, Alexandrov T, Sharma K, Anderton CR (2020) Storage conditions of human kidney tissue sections affect spatial Lipidomics analysis reproducibility. J Am Soc Mass Spectrom 31(12):2538–2546. https://doi.org/10.1021/jasms.0c00256
34. Hansen RL, Dueñas ME, Lee YJ (2019) Sputter-coated metal screening for small molecule analysis and high-spatial resolution imaging in laser desorption ionization mass spectrometry. J Am Soc Mass Spectrom 30 (2):299–308. https://doi.org/10.1007/s13361-018-2081-0
35. Bemis KD, Harry A, Eberlin LS, Ferreira C, van de Ven SM, Mallick P, Stolowitz M, Vitek O (2015) Cardinal: an R package for statistical analysis of mass spectrometry-based imaging experiments. Bioinformatics 31 (14):2418–2420. https://doi.org/10.1093/bioinformatics/btv146
36. Lauzon N, Dufresne M, Beaudoin A, Chaurand P (2017) Forensic analysis of latent fingermarks by silver-assisted LDI imaging MS on nonconductive surfaces. J Mass Spectrom 52 (6):397–404. https://doi.org/10.1002/jms.3938

Chapter 8

Mass Spectrometry Imaging of Metabolites by Nanostructure Initiator Mass Spectrometry with Fluorinated Gold Nanoparticles

Amelia Palermo

Abstract

Nanostructure initiator mass spectrometry (NIMS) with fluorinated gold nanoparticles (f-AuNPs) enables the detection and spatial localization of a breath of polar metabolites and lipids with high spatial resolution and ultrasensitivity. Here we describe the methods and procedures for the synthesis and application of f-AuNPs for NIMS of small molecule metabolites and lipids in biological tissues, encompassing sample preparation, mass spectrometric detection, and data analysis and interpretation.

Key words Nanostructure initiator mass spectrometry, Metabolomics, Mass spectrometry imaging, Metabolism heterogeneity, Gold nanoparticles, Metabolite imaging

1 Introduction

Mass spectrometry imaging on nanostructured surfaces aims at detecting, quantifying and localizing analytes with high sensitivity and spatial resolution [1–5]. This approach leverages the unique properties of nanostructured surfaces to facilitate the desorption and ionization of analytes for MSI detection. For instance, MSI workflows based on metallic nanoparticles (e.g., silver, platinum, gold), carbon nanotubes, nanostructured semiconductors, and silica-based nanowires, [1, 6–11] greatly contributed to the breath of mass spectrometry (MS)-based strategies available for charting metabolites and lipids over space (i.e., spatial metabolomics) [12]. Among these, NIMS on porous surfaces (e.g., silicon 2D surfaces, gold nanoparticles i.e., AuNPs) enables the direct in situ detection of analytes through a straightforward sample preparation approach, and a cohesive analytical workflow [13–15]. More specifically, NIMS with functionalized AuNPs (f-AuNPs) employs gold nanoparticles nanostructured with self-assembled monolayers

Young-Jin Lee (ed.), *Mass Spectrometry Imaging of Small Molecules: Methods and Protocols*,
Methods in Molecular Biology, vol. 2437, https://doi.org/10.1007/973-1-0716-2030-4_8,
© The Author(s), under exclusive license to Springer Science+Business Media, LLC, part of Springer Nature 2022

of desorption/ionization molecules. In its most common setup, gold nanoparticles are functionalized with perfluorinated hydrocarbon chains through a facile and safe synthetic protocol, and can be solubilized in perfluorohexane (PFH) for the even distribution on the surface of the sample. This represents a significant advantage as opposed to previous protocols for the synthesis of perfluorinated NIMS surfaces, which included the use of the hazardous chemical hydrofluoric acid (HF) [16]. Analytes are ionized and detected using a matrix assisted laser desorption ionization (MALDI) source coupled with high-resolution, high-accuracy mass spectrometry (MS) (e.g., Waters SYNAPT G2-Si, Bruker MALDI-2 timsTOF flex). This enables the targeted and untargeted, simultaneous detection of a variety of polar and nonpolar, small molecule metabolites and lipids, that can be used as a direct proxy/functional readout for the activity of metabolic pathways in biological systems [13].

Here we present the workflow and methods for the application of NIMS with f-AuNPs for charting metabolites and lipids in biological tissues, for the systems-level exploration of metabolism in space. As opposed to classic MALDI matrixes, desorption/ionization on f-AuNPs is softer and allows for the use of lower energies of the incident laser. This translates in lower in source molecular fragmentation for reduced background noise in the low m/z range, and decreased diameter of the incident laser, which enables high lateral resolution. Moreover, the Teflon-like nature of the surface area of f-AuNPs allows the unbiased desorption/ionization of molecules of diverse chemical nature (sugars, amino acids, lipids, and steroids, among others), and prevents the solubilization and spatial dislocation of polar metabolites, which can limit the spatial resolution achieved by classic hydrophilic MALDI matrixes. NISM with f-AuNPs has been demonstrated amenable to the high-throughput untargeted analysis of more than 3000 metabolic features over space, and for mapping the distribution of targeted metabolites and small molecule drugs at ultra-high sensitivity [13, 15, 17]. All in all, NIMS with f-AuNPs unlocks the exploration of metabolism in space, from whole biological systems to single cells and subcellular biosynthetic architectures [14].

2 Materials

All reagents should be of analytical grade and stored at room temperature, with the exception of octanethiol functionalized gold nanoparticles. All solutions should be prepared using ultrapure water. Review your waste disposal regulations and guidelines when disposing waste materials.

2.1 f-AuNPs Preparation

1. 2% (w/v) octanethiol functionalized gold nanoparticles (2–4 nm diameter, 20 mg/mL in toluene). Store at 2–8 °C.
2. Fluorinated ligands: 1H,1H,2H,2H-perfluorooctanethiol or analogue (e.g., 1H,1H,2H,2H-perfluorodecanethiol).
3. LC grade solvents: ethanol, acetone, hexane, methanol, tetradecafluorohexane.
4. Ultrahigh purity nitrogen.
5. Temperature controlled oven (up to 80 °C).
6. 30 mL clear glass vials with cap.

2.2 Silicon Plate Preparation

1. Silicon Wafer (P-type Silicon Wafer, thickness 0.5 mm).
2. Flat plastic container.
3. Piranha solution (concentrated sulfuric acid/hydrogen peroxide, 3/1 in v/v) (**Warning!** *See* **Note 1**).
4. Methanol.
5. Cutter with diamond blade for sawing silicon wafers.

2.3 Sample Preparation

1. Optimal cutting temperature compound (OCT) embedding medium.
2. Liquid nitrogen.
3. Isopentane.
4. Petri dish.
5. Disposable histology molds.
6. Metal beaker.
7. Twelve inch forceps.
8. 1/4 in., #2 brush with angled tip.
9. Standard compound solutions of internal lock masses (e.g., tryptophan).
10. Desktop digital scanner.
11. −80 °C cold storage.
12. Glass vacuum desiccator.

2.4 Instrumentation

1. Cryostat equipped with microtome.
2. MALDI mass spectrometer (high-resolution, high-accuracy MS, e.g., Waters SYNAPT HDMS G2-Si qTOF, Bruker MALDI-2 timsTOF flex).

2.5 Software and Spectral Databases

1. MassLynx™ software (tuning and calibration for Waters SYNAPT HDMS G2-Si qTOF, or analogous vendor software depending on MS system).

2. High Definition Imaging (HDI) 1.4 software with Region of Interest functionality (Waters), or analogous software for raw MSI data acquisition, processing, visualization, and analysis.
3. R/Python/Matlab (either of them) for extensive data analysis and modeling in batch autonomous metabolite putative annotation.
4. iPhat3 (https://pathways.embl.de), MetaboAnalyst 5.0 (https://www.metaboanalyst.ca) for metabolic pathway analysis [18, 19].
5. METLIN (https://metlin.scripps.edu/), LIPID MAPS (https://www.lipidmaps.org), HMDB (https://hmdb.ca) for metabolite annotation and ID [20–22].
6. METASPACE data repository and annotation (https://metaspace2020.eu) for data sharing and annotation [23].

3 Methods

All procedures should be carried out at room temperature, unless otherwise stated. Work under fume hood when appropriate.

3.1 f-AuNPs Synthesis

1. Place 1 mL of 2% (w/v) octanethiol functionalized gold nanoparticles in a 30 mL clear glass vial, add 20 mL of neat ethanol.
2. Cap the vial and sonicate for 1 min, centrifuge at 12,000 rcf, 4 °C for 20 min.
3. Remove ethanol and add 20 mL of acetone to the precipitate.
4. Sonicate for 1 min, centrifuge at 12,000 rcf, 4 °C for 20 min.
5. Remove acetone and repeat the acetone wash for a total of three times.
6. Remove acetone and add 1 mL of hexane to the precipitate (Fig. 1a).
7. Test the solution of gold nanoparticles in hexane for the presence of tetraoctyl ammonium ion at m/z 466.53 ($[M+H]^+$) by MALDI-MS (this is a common contaminant in commercial octanethiol functionalized gold nanoparticles). If tetraoctyl ammonium is detected (>5000 counts), repeat washing protocol (**steps 1–6**). If no contaminant is detected, proceed to the next step.
8. Dry the precipitate under N_2 flow and weigh the gold nanoparticles.
9. Resuspend 20 mg of nanoparticles in 1 mL hexane.
10. Add 200 μL of fluorinated ligand (1H,1H,2H,2H-perfluorodecanethiol or 1H,1H,2H,2H-perfluorooctanethiol) under N_2 flow, and stir the solution for 72 h using a magnetic stir bar in a capped 30 mL glass vial.

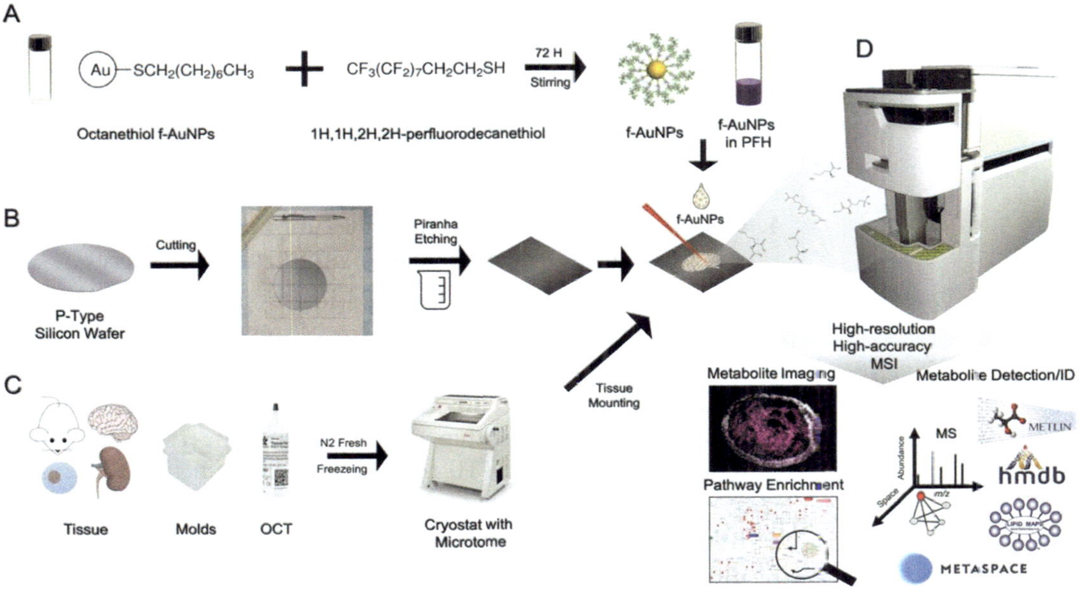

Fig. 1 Workflow for NIMS of metabolites and lipids in biological tissues using f-AuNPs. (**a**) f-AuNPs synthesis; (**b**) Silicon plate preparation; (**c**) Sample preparation; (**d**) Untargeted/targeted MSI detection, data processing and interpretation

11. Evaporate the solvent under N_2 flow. Wash the f-AuNPs with hexane (×3 times), acetone (×3 times), methanol/acetone/water (1/1/1) (×2 times). Between each wash, sonicate the solution (1 min) and centrifugate at 12,000 rcf, 4 °C for 20 min.
12. Remove supernatant and dry the f-AuNPs at 90 °C in the oven without cap.
13. Prepare a solution of f-AuNPs in PFH at 3 mg/mL concentration. Test the solution for contaminants by MALDI-MS and store at room temperature until use.

3.2 Silicon Plate Preparation

1. Place one P-type silicon wafer on an even surface and cut the wafer according to the dimension of the MALDI plate used as support (Fig. 1b).
2. Place the wafers in the Piranha solution for 2 h.
3. Wash the wafers in methanol and dry them under N_2 flow. Store the silicon wafers in a clean plastic container at room temperature until use (Fig. 1b).

3.3 Sample Preparation

1. Separate the tissue/organ/cells of interest from the original biological context and cover it with OCT for few minutes in a petri dish (this acclimates tissue to OCT).

2. Label a disposable histology mold (Cryomold). Transfer the sample in fresh OCT and orientate it in the Cryomold (*see* **Note 2**).
3. Fresh freeze the sample: prepare a metal beaker filled with isopentane (2/3 of total volume of the beaker) and place it in a Dewar of liquid nitrogen.
4. Grab the cryomold containing the sample with a 12 in. forceps and freeze it in isopentane bath, without fully submerge.
5. Transfer the sample to a covered foam cooler of dry ice and continue with the next samples (*see* **Notes 3** and **4**).
6. Sample cryosectioning: equilibrate the cryostat chamber temperature for at least 30 min before use. Select the optimal temperature for this step depending on the type of frozen tissue/sample (e.g., −10 to −15 °C for brain tissue, −25 to −50 °C for fat tissue) (*see* **Note 5**) (Fig. 1c).
7. Bring the tissue in the cryostat and wait for the tissue temperature equilibrated with the temperature of the cryostat.
8. Mount the OCT block on the embedding medium, trim it and start cryosections into 5 μm sections. Use the brush to grab and maneuver the tissue section on the stage (*see* **Note 6**).
9. Standard solution and lock mass compounds: spot standard solutions/lock mass compounds on the piranha etched p-type silicon wafer, evaporate and cover with the solution of f-AuNPs in PFH (*see* **Note 7**).
10. Thaw-mount the tissue slice on a piranha etched p-type silicon wafer.
11. Distribute the 3 mg/mL solution of f-AuNP in PFH on the surface of the sample using a glass pipette or electrospray device (*see* **Notes 8** and **9**).
12. Keep the prepared sample wafer under vacuum in the glass vacuum desiccator at room temperature until analysis.

3.4 MSI Analysis

1. Tune and calibrate the MS system before analysis (e.g., with red phosphorus clusters), for example, using the MassLynx™ Software.
2. Scan the surface of the sample on the silicon wafer using a desktop digital scanner. The acquired image is used as reference for setting the teaching points and positioning the wafer for MSI acquisition.
3. Use a MALDI-MS equipped with a Nd:YAG laser operating at 355 nm. The following condition can be used for MALDI SYNAPT G2-Si (Waters, Milford, MA): laser energy 175, sampling rate 3 s, and repetition rate of 2500 Hz. Use the High Definition Imaging (HDI) 1.4 software with Region of

Interest functionality, for setting the acquisition parameters and selecting the region of interest from the surface of the sample. Use the image acquired at **step 2** as reference.

4. Additional parameters, as for instance the raster step and image size, can be configured in the HDI software (*see* **Note 10**).
5. MSI acquisition is performed using a MALDI SYNAPT G2-Si, as for the m/z range of interest (e.g., 50–1000 Da) (*see* **Notes 11** and **12**) (Fig. 1c).

3.5 Data Processing and Interpretation

1. Perform metabolic feature detection by processing the raw MSI spectra using the High Definition Imaging (HDI) 1.4 (or other vendor specific or nonspecific software). This software allows for normalization of metabolite intensities, baseline subtraction, internal calibration with lock-mass, visualization of metabolite distribution in 2D space, statistical analysis among selected regions of interest (e.g., principal component analysis, PCA). It also provides the data matrix reporting the accurate m/z for the metabolic features detected in each pixel of the MS image, together with the measured intensity (raw or normalized). This data matrix can be exported in text format for further autonomous putative annotation and statistical analysis, for instance, in R or Matlab (Fig. 1d).
2. Perform metabolite and lipid putative annotation by accurate m/z match against metabolite/lipid databases including METLIN, LIPID MAPS, HDMB using an in-house R script or through database web access. Metabolite identity can be confirmed by MS/MS experiments, and comparison with pure standards when possible (Fig. 1d).
3. To interpret the metabolic changes at systems-level, in the broader context of the metabolome and lipidome, the detected metabolites and lipids can be uploaded as input in pathway enrichment analysis tools, such as iPath3, MetaboAnalyst 5.0 (Fig. 1d).

4 Notes

1. The Piranha solution should be prepared and handled with great care because it is highly corrosive and a powerful oxidizer. The handling of this solution requires special protection equipment including a full-face shield, heavy-duty rubber gloves, acid apron wared on top of the lab coat. Always use glass container as this solution can degrade plastic, and work under the hood. Always add hydrogen peroxide to sulfuric acid while gently stirring, not the opposite. For more information visit: https://ehs.princeton.edu/book/export/html/513.

2. Carefully avoid formation of bubbles at this stage. The sectioning surface is at the bottom of the Cryomold.
3. The block can crack at this stage. To avoid this, remove the sample when there is still a small drop of unfrozen OCT.
4. Samples can be wrapped in individually labeled foils, sealed in plastic bags, placed in a freezer box and stored at −80 °C until use.
5. Use the reference chart provided by the cryostat vendor for details [24].
6. Adjust cryostat sectioning antiroll plate to prevent frozen section from curling. Attention should be posed to avoid contact of OCT with the surface of the sample to avoid contamination.
7. MS calibrant compounds can be used during acquisition, for instance tryptophan (205.01.05 m/z) as internal lock mass.
8. f-AuNPs should be distributed evenly on the surface of the sample.
9. Alternatively, f-AuNPs can be perfused in the animal using a blood substitute f-AuNP solution (50% 20 mg mL^{-1} f-AuNP/prefluorodecilin and 50% Pluronic F-68 (0.5%) in Milli-Q water [15].
10. These are usually selected based on the size of the region of interest, and the desired spatial resolution, and on the type of MALDI source used. Commercial instruments currently allow laser step down to 10–5 μm.
11. Depending on the targeted metabolites and lipids, the desorption/ionization can be performed in negative or positive mode.
12. The MALDI SYNAPT G2-Si can be interfaced with ion mobility, which can be used for resolving isobaric compounds.

References

1. Wel J, Buriak JM, Siuzdak G (1999) Desorption-ionization mass spectrometry on porous silicon. Nature 399(6733):243–246. https://doi.org/10.1038/20400
2. Kurczy ME, Northen TR, Trauger SA, Siuzdak G (2015) Nanostructure imaging mass spectrometry: the role of fluorocarbons in metabolite analysis and yoctomole level sensitivity. Methods Mol Biol 1203:141–149. https://doi.org/10.1007/978-1-4939-1357-2_14
3. Calavia R, Annanouch FE, Correig X, Yanes O (2012) Nanostructure initiator mass spectrometry for tissue imaging in metabolomics: future prospects and perspectives. J Proteome 75(16):5061–5068
4. Greving MP, Patti GJ, Siuzdak G (2011) Nanostructure-initiator mass spectrometry metabolite analysis and imaging. Anal Chem 83:2–7. https://doi.org/10.1021/ac101565f
5. Moening TN, Brown VL, He L (2014) Nanostructure-initiator mass spectrometry (NIMS) for molecular mapping of animal tissues. In: Mass spectrometry imaging of small molecules. Methods in molecular biology (methods and protocols), vol 1203. Humana Press, New York, NY, pp 151–157
6. Tang HW, Ng KM, Lu W, Che CM (2009) Ion desorption efficiency and internal energy transfer in carbon-based surface-assisted laser desorption/ionization mass spectrometry:

desorption mechanism(s) and the design of SALDI substrates. Anal Chem 81(12): 4720–4729. https://doi.org/10.1021/ac8026367

7. Zhu Q, Wang Z, Wang Y et al (2020) Investigation of surface morphology on ion desorption in SALDI-MS on tailored silicon Nanopillar arrays. J Phys Chem C 124(4): 2450–2457. https://doi.org/10.1021/acs.jpcc.9b09520
8. Go EP, Apon JV, Luo G et al (2005) Desorption/ionization on silicon nanowires. Anal Chem 77(6):1641–1646. https://doi.org/10.1021/ac048460o
9. Walker BN, Razunguzwa T, Powell M et al (2009) Nanophotonic ion production from silicon microcolumn arrays. Angew Chem Int Ed Engl 48(9):1669–1672. https://doi.org/10.1002/anie.200805114
10. Han M, Sunner J (2000) An activated carbon substrate surface for laser desorption mass spectrometry. J Am Soc Mass Spectrom 11(7): 644–649. https://doi.org/10.1016/S1044-0305(00)00129-X
11. Northen TR, Yanes O, Northen MT et al (2007) Clathrate nanostructures for mass spectrometry. Nature 449:1033–1036. https://doi.org/10.1038/nature06195
12. Wang T, Cheng X, Xu H et al (2019) Perspective on advances in laser-based high-resolution mass spectrometry imaging. Anal Chem 92(1): 543–553. https://doi.org/10.1021/acs.analchem.9b04067
13. Palermo A, Forsberg EM, Warth B et al (2018) Fluorinated gold nanoparticles for nanostructure imaging mass spectrometry. ACS Nano 12(7):6938–6948. https://doi.org/10.1021/acsnano.8b02376
14. Palermo A (2020) Charting metabolism heterogeneity by nanostructure imaging mass spectrometry: from biological systems to subcellular functions. J Am Soc Mass Spectrom 31(12):2392–2400. https://doi.org/10.1021/jasms.0c00204
15. Kurczy ME, Zhu ZJ, Ivanisevic J et al (2015) Comprehensive bioimaging with fluorinated nanoparticles using breathable liquids. Nat Commun 6:5998. https://doi.org/10.1038/ncomms6998
16. Woo HK, Northen TR, Yanes O, Siuzdak G (2008) Nanostructure-initiator mass spectrometry: a protocol for preparing and applying NIMS surfaces for high-sensitivity mass analysis. Nat Protoc 3(8):1341–1349. https://doi.org/10.1038/nprot.2008.110
17. O'Brien PJ, Lee M, Spilker ME et al (2013) Monitoring metabolic responses to chemotherapy in single cells and tumors using nanostructure-initiator mass spectrometry (NIMS) imaging. Cancer Metab 1(1):4. https://doi.org/10.1186/2049-3002-1-4
18. Darzi Y, Letunic I, Bork P, Yamada T (2018) IPath3.0: interactive pathways explorer v3. Nucleic Acids Res 46(W1):W510–W513. https://doi.org/10.1093/nar/gky299
19. Xia J, Psychogios N, Young N, Wishart DS (2009) MetaboAnalyst: a web server for metabolomic data analysis and interpretation. Nucleic Acids Res 37:W652–W660. https://doi.org/10.1093/nar/gkp356
20. Guijas C, Montenegro-Burke JR, Domingo-Almenara X et al (2018) Comment on METLIN: a technology platform for identifying knowns and unknowns. Anal Chem 90: 13126–13127. https://doi.org/10.1021/acs.analchem.8b03613
21. O'Donnell VB, Dennis EA, Wakelam MJO, Subramaniam S (2019) LIPID MAPS: serving the next generation of lipid researchers with tools, resources, data, and training. Sci Signal 12(563):eaaw2964. https://doi.org/10.1126/scisignal.aaw2964
22. Wishart DS, Feunang YD, Marcu A et al (2018) HMDB 4.0: the human metabolome database for 2018. Nucleic Acids Res 46(D1): D608–D617. https://doi.org/10.1093/nar/gkx1089
23. Alexandrov T, Ovchinnikova K, Palmer A, et al (2019) METASPACE: A community-populated knowledge base of spatial metabolomes in health and disease. bioRxiv. https://doi.org/10.1101/539478
24. Shi SR, Liu C, Pootrakul L et al (2008) Evaluation of the value of frozen tissue section used as "gold standard" for immunohistochemistry. Am J Clin Pathol 129(3):358–366. https://doi.org/10.1309/7CXUYXT23E5AL8KQ

Chapter 9

Imaging of Plant Hormones with Nanoparticle-Assisted Laser Desorption/Ionization Mass Spectrometry

Shu Taira and Katsuhiro Shiono

Abstract

Plant hormones can act in synergistic and antagonistic ways in response to biotic and abiotic stresses and during plant growth and development. Thus, a technique is needed to simultaneously determine the distribution and concentration of several plant hormones. A relatively new technology, mass spectrometry imaging (MSI), enables the direct mapping and imaging of biomolecules on tissue sections. MSI permits simultaneous detection of multiple analytes on a single section of plant tissue, even in the absence of target-specific markers such as antibodies. Recently, MSI has been used to localize multiple, small molecule ($m/z < 500$) plant hormones by the nanoparticle-assisted laser desorption/ionization (Nano-PALDI) mass spectrometry (MS) method. Here, we illustrate a technology for multiple-hormone imaging using Nano-PALDI MSI and discuss its potential in investigating the role of hormone signaling in plant development and stress responses.

Key words Visualization, High-resolution, Nanoparticle, Hormone, Plant

1 Introduction

Environmental responses, development, and morphogenesis in plants are determined by the coactivation and reciprocal actions of multiple plant hormones. Most of the major plant hormones are small molecules with m/z less than 500 (Table 1). In addition, the size of plant tissues is often small. Thus, in this chapter, we describe mass spectrometry imaging (MSI) procedures and introduce the nanoparticle-assisted laser desorption/ionization (Nano-PALDI) mass spectrometry (MS) method as a suitable method for small molecule imaging.

In the case of soft ionization, which results in avoidance of fragmentation of molecules, such as the matrix-assisted laser desorption/ionization (MALDI) method, the mass of the intact target molecules can be measured. The MALDI method, which uses an organic matrix (i.e., α-cyano-4-hydroxycinnamic acid,

Young-Jin Lee (ed.), *Mass Spectrometry Imaging of Small Molecules: Methods and Protocols*,
Methods in Molecular Biology, vol. 2437, https://doi.org/10.1007/978-1-0716-2030-4_9,
© The Author(s), under exclusive license to Springer Science+Business Media, LLC, part of Springer Nature 2022

Table 1
Plant hormones and associated compounds used in the present study

Compound	Abbreviation	Compound used as standard	MW	Solvent
Abscisic acid	ABA	(±)-Abscisic acid	264.32	MeOH
Auxin	IAA	3-Indoleacetic acid	175.19	MeOH
Brassinosteroid	Br	Brassinolide	480.68	EtOH
Cytokinin	*tZ*	*trans*-Zeatin	219.25	DMSO
Cytokinin	iP	6-(γ,γ-Dimethylallylamino) purine	203.24	DMSO
Jasmonic acid	JA	(±)-Jasmonic acid	210.27	DMSO
Salicylic acid	SA	2-Hydroxybenzoic acid	138.12	Water
Ethylene precursor	ACC	1-Aminocyclopropane-1-carboxylic acid	101.10	Water
Heavy hydrogen-labeled ABA	D_6-ABA	$[^2H_6]$(+)-*cis, trans*-Abscisic acid	270.36	MeOH

CHCA; 2,5-dihydroxybenzoic acid, DHB; sinapic acid, SA) for efficient ionization of target materials, is the most popular ionization method. The target is co-crystalized with a large excess of organic matrix and subsequently ionized by laser energy. One of the primary merits of MALDI-MS is its ability to semi-quantitate multiple targets in a single measurement.

MSI is an expansion of MS techniques in which ion images are reconstructed from tens of thousands of MS spectra with x and y positions to reveal the target molecule's distribution in the absence of specific markers such as antibodies and fluorescent dyes. Immunostaining, histochemical assays and gene expression induction methods, known as label-based methods, are often used to detect and image the distribution of target molecules [1]. MSI has accumulated nearly 20 years of practical knowledge of its application, mainly in the medical field [2, 3]. In recent years, MSI has been increasingly applied in the field of plant science [4–6] and food science [7, 8]. Because this method is based on mass spectrometry, multiple molecules of different masses can be imaged simultaneously in a single measurement and on a single section without the use of antibody and fluorescent dye. The MSI experimental procedure can be divided into four main steps: (1) preparation of the sample section, (2) acquisition of an optical image, (3) matrix application and (4) two-dimensional measurement with a mass spectrometer.

Although live imaging is not possible, the method can be applied to any plant species in any environment, providing that tissue sections can be prepared. In the MALDI-MSI procedure, an organic matrix is typically sprayed on the surface of the sections to assist the desorption and ionization of target molecules. It is a common concern that MALDI-based MSI is difficult to use for the imaging of low molecular weight targets because of the high background interferences of organic matrices. Since MSI is based on two-dimensional mass measurements on a section and the image is constructed after the measurement, the image resolution depends on (1) the laser irradiation diameter, (2) the irradiation step interval, and (3) the size of the organic matrix crystal on a section (*see* **Note 1**).

As research using MSI increases, there is an increasing demand for the acquisition of high-resolution images and measurements covering wide mass ranges. Thus, the development of a new assisting ionization reagent and improved instrument specifications are required. Recently, a nanomaterial-based ionization method has attracted the attention of the MS field for ionization of small molecules, in which the background signal is reduced in the low mass region ($m/z < 500$) [9]. We developed Nano-PALDI MS for use in ionizing both small molecules without background signal in the low mass region [10–14] (Fig. 1). This involves the use of a metal oxide nanoparticle surrounded by amorphous SiO_2 with a modified functional group (diameter (d) = 3.6 nm) (Fig. 2a). The core of the particles absorbs the laser radiation, and the silicate network and amino groups covering the particles transfer some of

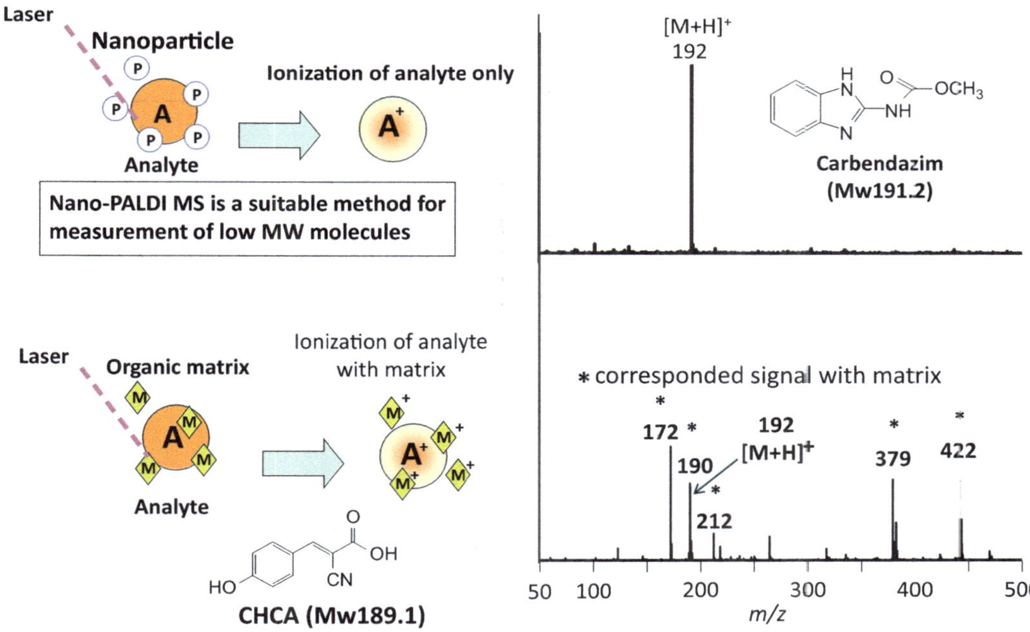

Fig. 1 Comparison of Nano-PALDI (top) vs. MALDI-MS (bottom) for small molecule analysis. Several matrix-derived signals (such as proton addition, dehydrated ion, Na-added ion, and dimer ion) are detected in MALDI-MS, making it difficult to detect analyte signals

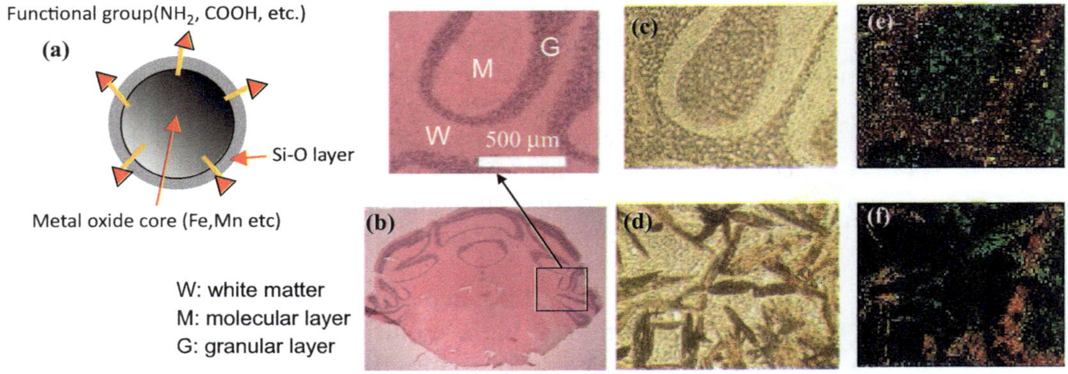

Fig. 2 Schematic illustration of the nanoparticle used for the Nano-PALDI method (**a**). Optical images of rat cerebellum tissues after (**b**) staining with hematoxylin and eosin (HE), (**c**) spray-coated with nanoparticles or (**d**) 2,5-dihydroxybenzoic acid (DHB) as an organic matrix. MS images reconstructed for (**e**) nanoparticle or (**f**) DHB-coated tissue section. Distinct lipid distributions are observed for *m*/*z* 820.7 (phosphatidylcholine: PC (1-alk 38:3), green) localized at the molecular layer (M) and *m*/*z* 850.8 (glycolipid: GalCer(C24h:0), red) at the white matter (W) with Nano-PALDI. (Modified with permission from American Chemical Society [14])

the energy to the analyte. Nanoparticles (NPs) are only physically adsorbed on the sample surface, so even if the particles aggregate to form a secondary particle size, the size is limited to about 100 nm in diameter [14]. Thus, Nano-PALDI MSI can easily acquire high-resolution images due to the absence of crystallization effects observed in the MALDI method (Fig. 2b–f). In this chapter, Nano-PALDI based MSI for plant imaging is introduced.

2 Materials

Prepare solutions for plant samples using deionized water and high purity reagent.

2.1 Materials for Plant Growth

1. Rice seed (*Oryza sativa* L. cv. Nipponbare).
2. 0.6% (w/v) sodium hypochlorite
3. Mesh float holed with Styrofoam for floating on the nutrient solution (mesh size: 5 mm).
4. Plastic container.
5. Air pump.
6. Growth chamber 1.5 × 1.5-cm mesh sheet (mesh size: 5 mm).
7. 2-mL plastic tubes.
8. 10 mM stock solution of nine hormones and associated compounds: the powders of ABA [(±)-ABA], D_6-ABA ([$^2H_6$](+)-cis, trans-ABA) and IAA (3-indoleacetic acid) are dissolved in methanol, the powders of *tZ*, iP, JA [(±)-JA] are dissolved in

DMSO, the powder of Br (brassinolide) are dissolved in ethanol and the powder of ACC (1-aminocyclopropanecarboxylic acid) and SA (2-hydroxybenzoic acid) were dissolved in water.

9. 100 μM hormone mix: All 10 mM hormone stocks are mixed and diluted to 100 μM each with water.
10. Water without hormone mix: aqueous solution containing 12% (v/v) methanol, 2% (v/v) DMSO, and 1% (v/v) ethanol as a negative control.
11. Full-strength nutrient solution: 3.75 mM KNO_3, 1.5 mM $CaSO_4$, 0.6 mM NH_4NO_3, 0.4 mM $MgSO_4$, 0.2 mM KH_2PO_4, 0.1 mM Na_2SiO_3, 50 μM Fe(III)-EDTA, 50 μM KCl, 25 μM H_3BO_3, 2 μM $MnSO_4$, 2 μM/$ZnSO_4$, 1 μM $NiSO_4$, 0.5 μM $CuSO_4$, and 0.5 μM Na_2MoO_4. The solution also contains 2.5 mM MES and is pH-adjusted to 6.5 using KOH.
12. Razor.

2.2 Nanoparticle Synthesis

1. $FeCl_2 \cdot 4H_2O$ (40 mM).
2. (3-Aminopropyl)triethoxysilane (γ-APTES).
3. Ultrapure water (≥18.2 Ω).
4. Fourier transform infrared spectrophotometer (FT-IR).
5. Transmission electron microscope (TEM).
6. Adsorption spectrometer (ultra-violet visible (UV-VIS) spectrometry).
7. Sinaptic acid (SA) and α-cyano-4-hydroxycinnamic acid (CHCA) (10 mM, respectively).

2.3 Mass Spectrometry Imaging

1. Mass spectrometer with imaging capability.
2. Cryostat.
3. Super cryoembedding medium (SCEM, SECTION-LAB Co. Ltd.).
4. Plastic cryomold.
5. Glass slides coated with indium tin oxide (ITO).
6. Automated pneumatic sprayer (TM-Sprayer, HTX Technologies., LLC).

3 Methods

3.1 Plant Growth

1. Sterilize rice seeds (*Oryza sativa* L. cv. Nipponbare) for 30 min in 0.6% (w/v) sodium hypochlorite, and wash thoroughly with deionized water.

2. For imbibition, place the seeds in Petri dishes containing about 5 mL of deionized water for 1 day in darkness at 28 °C.
3. One day after imbibition, place the seeds on the mesh floating on one quarter-strength nutrient solution in a plastic container with air bubbling using an air pump. These plants grow in an growth chamber at 28 °C under light conditions (photosynthetic photon flux density of 249 μmol m^{-2} s^{-1}, 24 h light).
4. Three days after imbibition, fill 2-mL tubes with 100 μM hormone mix or water without hormone mix (control). Then, make it sure the roots are immersed in the solution and their shoots are held on the mesh sheet. Incubate plants in the growth chamber for 2 h.
5. After 2 h incubation, collect roots and prepare samples as in Subheading 3.5 for Nano-PALDI-MSI.

3.2 Preparation and Characterization of NPs

1. Add 2.5 mL ferrous chloride solution to 2.5 mL γ-aminopropyltriethoxysilane (γ-APTES) solution (undiluted solution) (*see* **Note 2**).
2. Stir for 1 h at room temperature and subsequently centrifuge the solution for 1 h at 22,000 × *g* at 4 °C.
3. Wash the precipitate three times with ultrapure water, suspend the precipitate in methanol, and store at 4 °C.
4. Confirm the presence of amino and hydroxy groups, the Si-O layer, and the light adsorption pattern by FT-IR attenuated total-reflectance spectroscopy and UV-Vis spectrometry (Fig. 3a, b).
5. Examine the morphology of the NPs using a TEM (Fig. 3c).

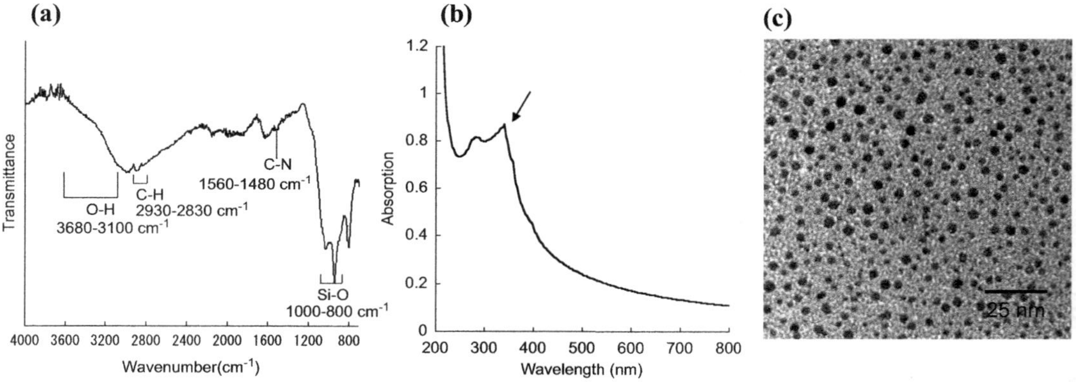

Fig. 3 Characterization of synthesized nanoparticles: (**a**) FT-IR absorption spectrum, (**b**) UV-VIS adsorption spectrum, and (**c**) TEM image. NPs have the size of 3.6 ± 0.03 nm (mean ± SEM; $n = 10^3$) and do not aggregate. (Modified with permission from American Chemical Society [14])

3.3 Synthesis of Organic Matrix-Modified NPs

When measuring small and high MW samples using NPs at single measurement, the synaptic acid (SA)-modified NPs is recommended [11]. The surface of metal oxide NPs could be modified through covalent bonding to amino groups on SA. Because the SA-modified-NPs did not undergo self-ionization, no background noise was produced (unlike SA only, which produced a noise signal in the low molecular weight region). The SA-NPs acted as an ionization-assisting reagent for both small and large molecules.

1. Dissolve organic matrix (SA or CHCA) and carbonyldiimidazole (CDI) in 5 mL DMF to a concentration of 0.1 M and 0.2 M, respectively.
2. Add the solution to a 5 mL NP suspension and heat to 35 °C.
3. After stirring for 20 h, centrifuge the solution at 22,000 × *g* and 4 °C for 1 h.
4. Wash the precipitate three times with DMF, suspend in methanol, and store at 4 °C.

3.4 Test MS Measurement of Standard Samples

Synthesized NPs should be validated for their ability to assist the desorption and ionization of analytes of interest using a MALDI-TOFMS.

1. Pipet 1 μL of suspended NPs onto a MALD-MS target plate.
2. After the NP is dried, drop 1 μL standard hormone mix (ABA, IAA, Br, *tZ*, iP, JA, SA, ACC, and D_6-ABA; 100 μM each) on top of the NP spot. The number of hormones on spot is 100 pmol.
3. Acquire MALDI-MS spectrum of the spot to confirm the hormone ion signals (Fig. 4).
4. Obtain tandem MS spectra using collision-induced dissociation method. (Table 2) (*see* **Note 3**).
5. Calculate the limit of detection (*LOD*) and coefficients of determination (R^2s) from the semilog plots for the signal intensity versus sample amount. *LOD* is calculated as $3.3 \times (s/a)$, where s is the standard deviation of a signal-to-noise ratio and a is the slope of the calibration curve (Table 3) (*see* **Note 4**).

3.5 Tissue Sample Preparation

1. Cut roots with a razor to collect the root portions at the elongation zone (region from 2.5 to 10 mm behind the root apex).
2. Embed the portions with the super-embedding medium in the plastic cryomold and fresh freeze with liquid nitrogen.
3. Cut the specimen block into 8–20 μm sections using a cryostat with the chamber set to −23 °C and the object holder set to −25 °C.

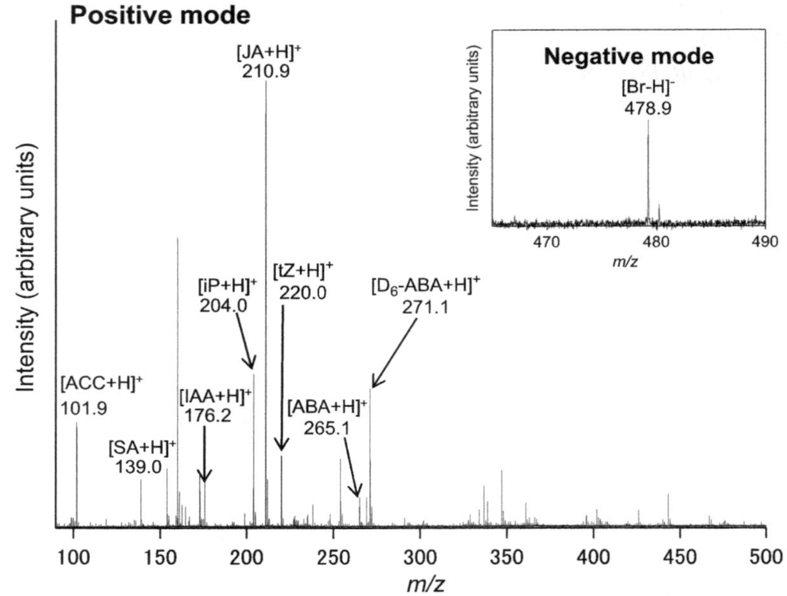

Fig. 4 Microscopic images of rice root cross-section (**a**) with and (**b**) without staining. MS images of abscisic acid (ABA) with a spatial resolution of (**c**) 5 and (**d**) 20 μm. Combination images of MS and optical image for (**e**) 5 μm and (**f**) 20 μm resolution

Table 2
Comparison of limits of detection and coefficients of determination

	Nano-PALDI		MALDI	
Compound	**LOD (fmol)**	**R^2**	**LOD (fmol)**	**R^2**
ABA	1.3	0.902	2.1	0.888
IAA	1.6	0.998	N.D.	N.D.
Br	1.1	0.913	N.D.	N.D.
tZ	1.9	0.997	8.0	0.969
iP	1.6	0.982	52	0.987
JA	2.5	0.998	23	0.998
SA	7.0	0.983	21	0.947
ACC	0.5	0.859	5.2	0.997
D_6-ABA	0.8	0.895	N.D.	N.D.

R^2 (coefficient of determination) was derived from semilog plots of signal intensity versus amount of each compounds. *LOD* Limit of detection, *N.D.* not detected

Reproduced with permission from American Chemical Society [10]

Table 3
Comparison of standard hormones and detected molecules for tandem MS

Compound	Precursor ion (*m/z*)	Product ion (*m/z*) of standard	Product ion (*m/z*) on the root section
ABA	265.1	247.0, 219.4	247.0
IAA	327.2	131.0	131.1
Br	478.9	363.1, 349.1, 333.0, 129.1	349.1, 129.1
*t*Z	220.0	202.0, 147.7, 135.7	202.1, 147.7, 135.7
iP	204.0	147.9, 135.9	147.9, 135.9
JA	210.9	166.0, 150.1, 57.1	150.1, 57.1
SA	139.0	121.0	121.1
ACC	101.9	84.0	84.0
D_6-ABA	271.1	253.0, 225.4	253.1

Reproduced with permission from American Chemical Society [10]

4. Gently thaw mount the sections onto slides coated with ITO coated glass.
5. If pathological analysis is required, prepare an additional section for sample staining (i.e., hematoxylin-eosin and immunostaining).
6. Obtain optical images of the sections using a scanner before MSI analysis, as is typically required. Use an image resolution of 4800–6400 dpi.
7. If analysis of high-resolution sample images is required, obtain microscopic images (Fig. 4) (*see* **Note 5**).

3.6 Application of Nanoparticles or Organic Matrix on Sections (Fig. 5)

1. For Nano-PALDI MSI, suspend 1 mg NPs in 1 mL methanol and spray on the plant section mounted on ITO-coated glass slides using an artistic airbrush.
2. Maintain a distance of 4 cm between the airbrush and the section to avoid migration of the target by wetting the NP suspension (*see* **Note 6**).
3. For one of examples of MALDI MSI, spray the sections with the matrix solution (e.g., 10 mM CHCA in 1:1 acetonitrile–water (v/v) using an automated pneumatic sprayer.
4. Spray ten passes under the following conditions: flow rate, 120 μL/min; air flow, 10 psi; nozzle speed, 1100 mm/min, at 75 °C.

3.7 MS Imaging of Plant Root

1. Before measurement, set each instrument parameters for a suitable imaging experiment (i.e., *m/z* range, laser power, gain, irradiated laser width) depending on the instrument.

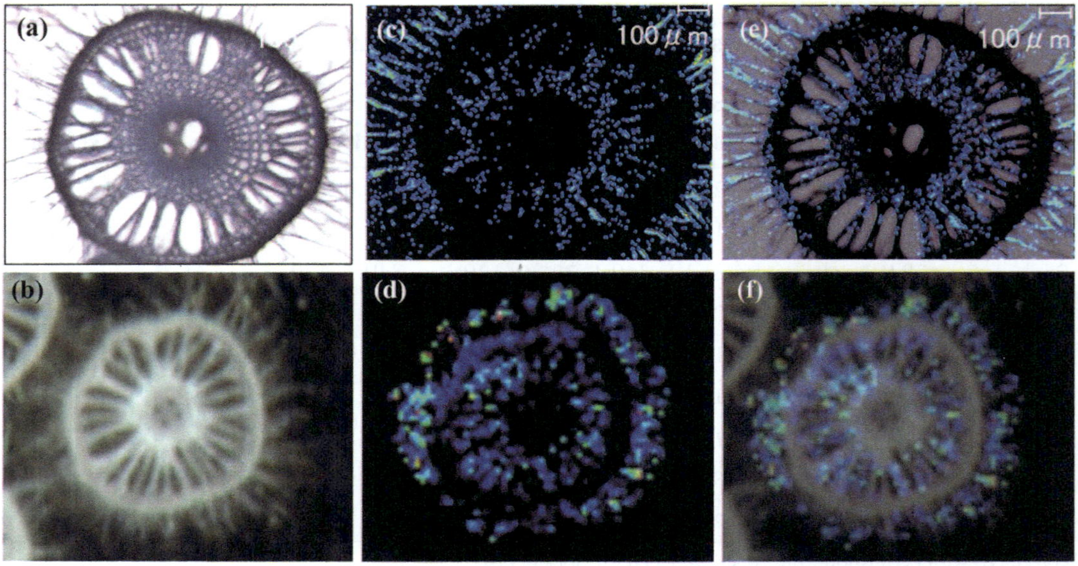

Fig. 5 Mass spectra of a standard mixture of plant hormone-related analytes obtained by Nano-PALDI-MS. Br is detected in negative ion mode and all others are detected in positive ion mode. (Reproduced with permission from American Chemical Society [10])

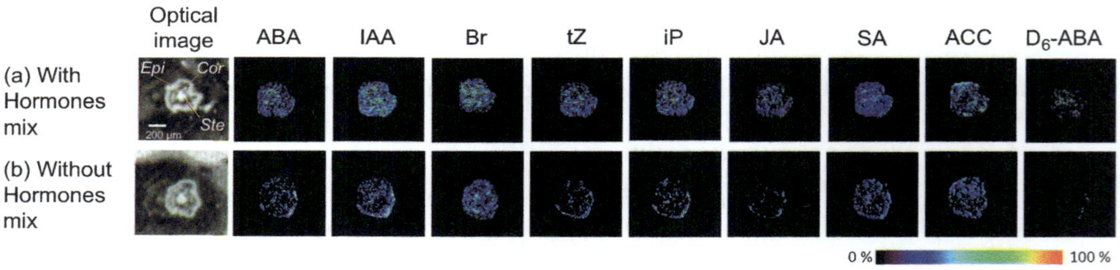

Fig. 6 Nano-PALDI-MS imaging of plant hormones and associated compounds in rice root sections after incubating in a water (**a**) with or (**b**) without plant hormone mix for 2 h. (Reproduced with permission from American Chemical Society [10])

2. Obtain images by irradiating each data point on the section with 50–200 laser shots in the positive or negative ion detection mode of the mass spectrometer in conjunction with the desired ion species.
3. Approximately 1000 to 200,000 data points with 5–500 μm raster step are required for each root of oryza section (in Fig. 6, we adopted 15 μm spacing between spots) (*see* **Note** 7).
4. Use imaging software to reconstruct the MS image from the obtained MS spectra with a mass bin width of $m/z \pm 0.01$ from the exact mass.
5. Obtain the relative intensities of the epidermis, cortex and stele regions by discarding the background peaks.

6. Use the accurate mass of the ions for image generation; Confirm the mass accuracy and root-mean-square error to avoid false-positive signals (*see* **Note 8**).

4 Notes

1. There is a trade-off between ion signals and image resolution. The laser spot size and the raster step interval are the most determining factors of the spatial resolution. However, the smaller the laser irradiation diameter, the smaller the number of ions obtained in a single spot, and thus the worse the S/N. The size of the organic matrix crystal also dictates the maximum spatial resolution. Thus, it is necessary to make the co-crystal as small as possible. Note that the results of MSI using organic matrices are affected not only by the crystal size but also by the uniformity of the matrix. In the case of Nano-PALDI MSI, the spatial resolution is not limited by the crystal size of the matrix and thus higher resolution analysis is possible [14].
2. The order in which reagents are added is important in preparing monodispersed NPs. The $FeCl_2{\cdot}4H_2O$ solution should be added to the γ-APTES solution.
3. All the product ions on the root sections should match with the product ions of the hormone standard mix by tandem MS.
4. In Nano-PALDI, the limits of detection (*LODs*) range from 0.5 to 7.0 fmol, while those for MALDI are higher (Table 3). Additionally, the coefficients of determination (R^2s) of semilog plots of the signal intensity versus quantity are relatively higher for Nano-PALDI.
5. A virtual slide that scans glass slides at high speed and converts them into high-resolution digital data scanner is recommended for MSI. Alternatively, there is an instrument that incorporates a built-in microscope as original equipment.
6. Do not use pneumatic sprayer to apply NPs on section to avoid getting clogged in the flow path.
7. Measurement time depends on the pulse laser frequency, the number of laser shots and the imaging region. For example, a measurement area of about 4 mm^2 is produced with laser spot areas (100 shots) that are detected with a spot-to-spot center distance of 15 μm in each direction on the section. When a 10 kHz laser is used, the measurement time is estimated using the following equation:

$$(2000\ [\mu m]/15\ [\mu m])^2 \times (100/10,000\ [s^{-1}]) = 177\ [s]$$

Newer instruments can achieve rapid measurement and high spatial resolution imaging; however, it should be noted that the first trial should be conducted using a low spatial resolution to roughly estimate the localization of the target; then one can determine whether high-resolution imaging is achieved.

8. It should be noted that comparison of MS images is semiquantitative because each signal value is derived from the relative intensity normalized to the highest intensity spot on the slide. It is recommended that an internal standard signal be used for normalization, if possible.

Acknowledgments

This work is partly supported by the Japan Science and Technology (JST) Grants-in-A-STEP (VP30118067678 to S.T.).

References

1. Isoda R, Yoshinari A, Ishikawa Y, Sadoine M, Simon R, Frommer WB, Nakamura M (2021) Sensors for the quantification, localization and analysis of the dynamics of plant hormones. Plant J 105(2):542–557. https://doi.org/10.1111/tpj.15096
2. Shariatgorji M, Nilsson A, Fridjonsdottir E, Vallianatou T, Källback P, Katan L, Sävmarker J, Mantas I, Zhang X, Bezard E, Svenningsson P, Odell LR, Andrén PE (2019) Comprehensive mapping of neurotransmitter networks by MALDI–MS imaging. Nat Methods 16(10):1021–1028. https://doi.org/10.1038/s41592-019-0551-3
3. Sugiura Y, Katsumata Y, Sano M, Honda K, Kajimura M, Fukuda K, Suematsu M (2016) Visualization of in vivo metabolic flows reveals accelerated utilization of glucose and lactate in penumbra of ischemic heart. Sci Rep 6:32361. https://doi.org/10.1038/Srep32361
4. Gayral M, Fanuel M, Rogniaux H, Dalgalarrondo M, Elmorjani K, Bakan B, Marion D (2018) The spatiotemporal deposition of Lysophosphatidylcholine within starch granules of maize endosperm and its relationships to the expression of genes involved in endoplasmic reticulum–Amyloplast lipid trafficking and Galactolipid synthesis. Plant Cell Physiol 60(1):139–151. https://doi.org/10.1093/pcp/pcy198
5. Li B, Bhandari DR, Römpp A, Spengler B (2016) High-resolution MALDI mass spectrometry imaging of gallotannins and monoterpene glucosides in the root of Paeonia lactiflora. Sci Rep 6(1):36074. https://doi.org/10.1038/srep36074
6. Horn PJ, Korte AR, Neogi PB, Love E, Fuchs J, Strupat K, Borisjuk L, Shulaev V, Lee Y-J, Chapman KD (2012) Spatial mapping of lipids at cellular resolution in embryos of cotton. Plant Cell 24(2):622–636
7. Shiota M, Shimomura Y, Kotera M, Taira S (2018) Mass spectrometric imaging of localization of fat molecules in water-in-oil emulsions containing semi-solid fat. Food Chem 245:1218–1223. https://doi.org/10.1016/j.foodchem.2017.11.009
8. Tatsuta Y, Kasai K, Maruyama C, Hamano Y, Matsuo K, Katano H, Taira S (2017) Imaging mass spectrometry analysis of ubiquinol localization in the mouse brain following short-term administration. Sci Rep 7(1):12990. https://doi.org/10.1038/s41598-017-13257-8
9. Kawasaki H, Yonezawa T, Watanabe T, Arakawa R (2007) Platinum Nanoflowers for surface-assisted laser desorption/ionization mass spectrometry of biomolecules. J Phys Chem C 111(44):16278–16283. https://doi.org/10.1021/jp075159d
10. Shiono K, Taira S (2020) Imaging of multiple plant hormones in roots of Rice (Oryza sativa) using nanoparticle-assisted laser desorption/ionization mass spectrometry. J Agric Food Chem 68(24):6770–6775. https://doi.org/10.1021/acs.jafc.0c00749

11. Komori H, Hashizaki R, Osaka I, Hibi T, Katano H, Taira S (2015) Nanoparticle-assisted laser desorption/ionization using sinapic acid-modified iron oxide nanoparticles for mass spectrometry analysis. Analyst 140 (24):8134–8137. https://doi.org/10.1039/C5AN02081F
12. Taira S, Taguchi H, Fukuda R, Uematsu K, Ichiyanagi Y, Tanaka Y, Fujii Y, Katano H (2014) Silver oxide based nanoparticle assisted laser desorption/ionization mass spectrometry for the detection of low molecular weight compounds. Mass Spectrom 3:S0025. https://doi.org/10.5702/massspectrometry.S0025
13. Taira S, Osaka I, Shimma S, Kaneko D, Hiroki T, Kawamura-Konishi Y, Ichiyanagi Y (2012) Oligonucleotide analysis by nanoparticle-assisted laser desorption/ionization mass spectrometry. Analyst 137 (9):2006–2010
14. Taira S, Sugiura Y, Moritake S, Shimma S, Ichiyanagi Y, Setou M (2008) Nanoparticle-assisted laser desorption/ionization based mass imaging with cellular resolution. Anal Chem 80(12):4761–4766. https://doi.org/10.1021/ac800081z

Part III

Quantitative Mass Spectrometry Imaging

Chapter 10

Nanosecond Photochemical Reaction (nsPCR) for Enhanced Mass Spectrometric Identification, Quantification, and Visualization of Metabolites and Neuropeptides

Gongyu Li, Yuan Liu, and Lingjun Li

Abstract

Small-molecule (e.g., metabolite) and low-abundance neuropeptide analyses by mass spectrometry (MS) represent important research directions and have witnessed tremendous growth and developments during past decades. With innate advantages of MS and gentle nature of soft ionization techniques including electrospray ionization (ESI) and matrix-assisted laser desorption/ionization (MALDI), profiling and visualization of these bioactive metabolites and neuropeptides have undergone technological advancements that can be applied to real biological systems, although numerous challenges still exist. We herein present a rapid and efficient strategy to improve both metabolite and neuropeptide analysis, the nanosecond photochemical reaction (nsPCR)-enabled fast chemical derivatization. Amine-directed chemoselectivity facilitates the rapid tagging on amine-containing metabolites and neuropeptides, resulting in improved detection sensitivity. Additionally, the nsPCR generates a localized pH jump zone and enables localized thermophoresis at nanosecond timescale which benefits on-demand matrix removal during MALDI-MS identification and visualization of low-abundance biomolecules. A step-by-step nsPCR experimental protocol is introduced in detail herein for both spot analysis and imaging analysis, followed by suggestions for data analysis to ensure successful application of the nsPCR strategy.

Key words MALDI, Mass spectrometry imaging, Amine-containing metabolites, On-tissue derivatization, Photochemical reaction, nsPCR

1 Introduction

Amongst the diverse small molecule space, one of the top representative classes of interest to the biological community is metabolites which include neurotransmitters, amino acid analogs, and lipids. Qualitative identification and quantitative analysis of small-molecule metabolites are essential to improve our understanding of pathogenic pathways, enable detecting many diseases at early stages, and to offer choices of personalized therapeutic strategies [1]. This area of research has been fueled by powerful mass spectrometry (MS) and nuclear magnetic resonance (NMR)

Young-Jin Lee (ed.), *Mass Spectrometry Imaging of Small Molecules: Methods and Protocols*,
Methods in Molecular Biology, vol. 2437, https://doi.org/10.1007/978-1-0716-2030-4_10,
© The Author(s), under exclusive license to Springer Science+Business Media, LLC, part of Springer Nature 2022

techniques, experiencing rapid advances and exponential growth during the past two decades, including improved sample preparation, instrumentation developments, and enhanced data analysis tools [2]. NMR is a noninvasive method capable of direct metabolite analysis from real biological environment such as cells and fresh tissues with careful sample preparation. While NMR can provide unambiguous structural information, it is often limited by the amount of sample required for comprehensive metabolite profiling (typically less than 100 metabolites can be assessed in single run).

Liquid chromatography in conjunction with MS, LC-MS, has grown to be the method of choice with high specificity, sensitivity (3 orders of magnitude higher than NMR) and throughput. Nonetheless, several analytical challenges for LC-MS-based metabolite analysis include: (a) matrix effect that needs to be evaluated and corrected carefully, (b) compound-dependent response factor, (c) need for authentic standard (can be expensive and/or unattainable for some cases) required for quantification, and (d) adequate sample preparation critical to the success of LC-MS-based metabolite analysis. Alternatively, matrix-assisted laser desorption/ionization (MALDI)-MS has increasingly been employed as a complementary fast tool for metabolite analysis with improved versatility for spatial distribution information and sample preparation requirement [3]. Despite these merits, challenges associated with MALDI-MS–based metabolite analysis include in-source fragmentation, isobaric interference, poor ionization efficiency, and ion suppression by MALDI matrix [4]. Our previous study highlighted the importance of the optimization of matrix application method [5]. More recently, improved small molecule imaging analysis has been achieved via a laser-based tissue handling method with reduced thickness and chemical derivatization time [6]. In addition, matrix functionalization also seems to be a promising strategy to achieve high-resolution visualization of low-abundance metabolites with delicate chemical design and multistep synthesis procedures [7].

Notably, chemical derivatization has drawn much attention for small-molecule metabolite analysis [4, 8]. Previous studies have demonstrated the chemical derivatization–enabled better separation, higher ionization efficiency and enhanced MS^2 fragmentation [8, 9]. A recent chemical derivatization method has shown great potential in improving imaging sensitivity that enabled more metabolites/neurotransmitters to be identified in crab tissues [10]. Most derivatization strategies can be used for both LC-MS and MALDI-MS imaging analyses. For example, carboxyl-containing metabolites (fatty acids and more) derivatization strategy for imaging analysis [11] and LC-MS analysis [12], fatty acid derivatization strategy for imaging analysis [13, 14] and LC-MS analysis [15], and lipid charge derivatization strategy for LC-MS [16] and imaging analysis [14], have been reported. While a cysteine charge tagging strategy enables direct analysis [17], more efforts have been devoted to improve the direct visualization analysis, including boronic acid-based catecholamine permanent charge tagging [18], amine metabolite derivatization using aldehyde for improved imaging sensitivity [4], and Girard's reagent–based glycan imaging analysis [19]. A previous review summarized a series of amine derivatization strategies for analytical use [9]. Shotgun lipidome analysis is also benefited with enhanced ionization, separation, and quantification by sequential chemical derivatization [20]. However, most of the traditional chemical derivatization strategies require either addition of substantial amount of extraneous chemicals that may perturb the native biological environment of metabolites of interest, or a certain incubation time and reaction condition that ensures the derivatization efficiency. These requirements discourage time-sensitive (unstable/dynamic) metabolite identification and quantification as such types of metabolites can undergo fast metabolomic flux/pathway that leads to the discrepancy in both identity and concentration level changes between originally sampled metabolites and finally detected ones under LC-MS or MALDI-MS analytical workflow.

To address these remaining challenges, we aim to develop fast chemical derivatization method for improved metabolite analysis. Inspired by previous in-source tagging chemistry [21, 22] and fast click photochemical strategy [23–25], we reported a versatile nanosecond photochemical reaction (nsPCR)-based ambient MS probe capable of selective and on-demand tagging amine groups at nanosecond time scale and a few tens of micrometer scale (Fig. 1) [26]. While the previous report focused on rapid structural probing and enhanced identifications of proteins, we herein aim to discuss the potential of the nsPCR-enabled MS strategy for enhanced molecular identification and spatial visualization of metabolites and neuropeptides, through both fast amine-directed chemical derivatization and in situ alleviation of matrix suppression effect as a result of the local pH jump and associated microthermophoresis. Detailed step-by-step experimental procedures for both spot analysis and tissue imaging analysis are presented along with corresponding data analysis pipelines. Benefiting from the nsPCR design, the detection sensitivity for some low-mass metabolites has been improved by up to ~5 fold in terms of limits of detection and response slope. And in some cases, small metabolites (e.g., glycine and alanine) that cannot be readily detected due to severe matrix suppression during MALDI-MS analysis have been successfully identified with nsPCR at a concentration level not detectable without nsPCR. As a proof-of-concept demonstration, improved identifications and enhanced MS imaging visualization of low-abundance neuropeptides directly from a mouse brain tissue section have also been shown with key experimental steps.

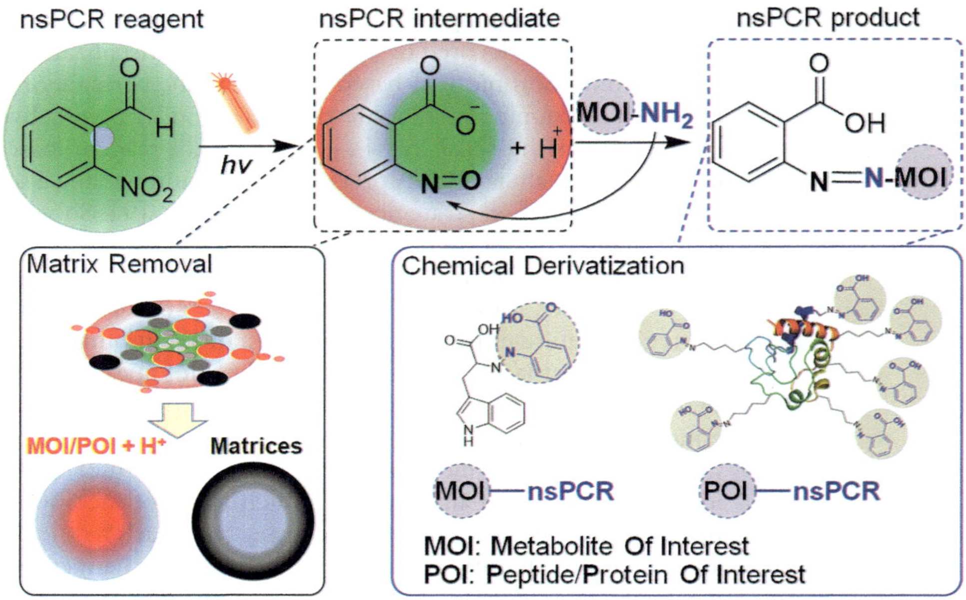

Fig. 1 The design and working principle of the nsPCR strategy for on-demand matrix removal and chemical derivatization, facilitating the molecular identification and MS visualization of a wide range of biomolecules spanning from small molecule metabolites to large peptides and proteins directly from tissues and other complex biological matrices

2 Materials

2.1 The nsPCR Reagents

1. 2-Nitrobenzaldehyde (NBA).
2. 2-Hydroxy-5-nitrobenzaldehyde (2-NBA).
3. 5-Hydroxy-2-nitrobenzaldehyde (5-NBA).

2.2 Other Reagents and Analytes

1. Dulbecco's Modified Eagle Medium (DMEM, high glucose, GlutaMAX(TM), pyruvate).
2. Amino acid stock solution: glycine, L-alanine, L-valine, L-leucine, L-glutamic acid, dopamine, L-histidine, L-phenylalanine, L-arginine, L-tryptophan are dissolved in water at 25 mM or 50 mM as stock solution. Store at −20 °C (*see* **Note 1**).
3. Matrix: α-cyano-4-hydroxycinnamic acid (CHCA).
4. Solvent for CHCA matrix: 5 ml water, 5 ml acetonitrile (ACN), and 10 μl trifluoroacetic acid (TFA) to a clean glass bottle (i.e., ACN–H_2O–TFA = 50:50:1, v/v/v). Spin the mixture gently to mix them well.
5. Solvent for nsPCR derivatization reagents: 8.4 ml ACN, 1.3 ml ethanol (EtOH), 290 μl water, and 10 μl TFA (or 10 μl FA) to a clean glass bottle (i.e., ACN–EtOH–H_2O–TFA = 84:13:2.9: 0.1, v/v/v/v). Spin the mixture gently to mix them well.
6. ITO glass slides and microscope glass slides.

2.3 Apparatus and Instrumentation

1. TM sprayer.
2. Cryostat.
3. Microscope.
4. Mass spectrometer: MALDI-LTQ-Orbitrap XL mass spectrometer, Q Exactive HF hybrid quadrupole-Orbitrap mass spectrometer (QE HF) (*see* **Note 2**).
5. SubAP MALDI source: subatmospheric pressure-MALDI (SubAP-MALDI) system coupled to a Q Exactive HF hybrid quadrupole-Orbitrap mass spectrometer (QE HF). This SubAP MALDI source is featured with a 355 nm high resolution Nd:YAG laser module along with an integrated ion funnel for improved ion collection and transmission.

3 Methods

Carry out all procedures at room temperature, unless otherwise specified.

3.1 Sample Preparation for Spot Analysis (Steps 1 and 2, Fig. 2)

1. MALDI matrix preparation. Weigh 10 mg CHCA matrix to 1.5 ml microcentrifuge tube and add 1 ml solvent for matrix. Sonicate to facilitate the dissolvement of CHCA and the final concentration of CHCA is 10 mg/ml. CHCA should be freshly prepared for the experiments.

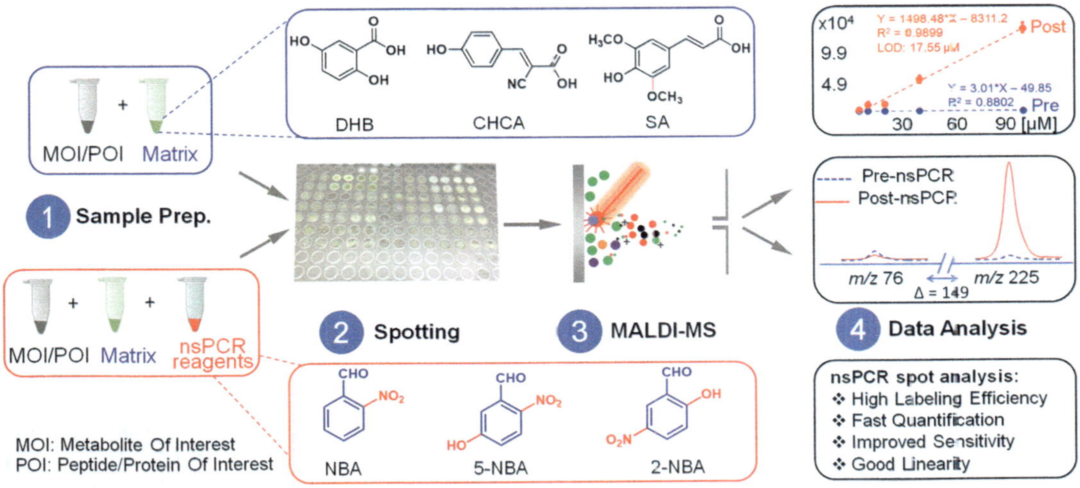

Fig. 2 Generally applicable nsPCR-enhanced MALDI-MS spot analysis workflow with three matrices (DHB, CHCA and SA) and three nsPCR reagents (NBA, 5-NBA, and 2-NBA) of choices. Key steps involve sample preparation of both MALDI chemicals and nsPCR reagents, spotting on target plate, MALDI-MS operation and post-acquisition data analysis. Accurate mass shifts for NBA, 5-NBA, and 2-NBA are 133.0116 Da, 149.0066 Da, 149.0066 Da per modification, respectively. Data shown here indicates the significantly improved identification and quantification of low-response glycine with nsPCR (data are from 5-NBA modification, 50 mM)

2. nsPCR reagent preparation. Dissolve the nsPCR reagents in the solvent for nsPCR derivatization reagents, at a concentration of 100 mM (*see* **Note 3**).
3. Serial dilution of metabolite. Serial dilution method is used to make metabolite solutions with various concentrations (e.g., 0 μM, 5 μM, 10 μM, 20 μM, 40 μM) for calibration purpose.
4. Spotting. Add 4 μl metabolite compound prepared in last step, 4 μl CHCA matrix solution and 4 μl nsPCR reagent solution to microcentrifuge tube sequentially and mix them well. Spot 2 μl of mixture on MALDI plate, at least 3 replicates for each sample. Dry down the samples for 20 min in fume hood in dark. When the samples are completely dried down, they are ready for MALDI-Orbitrap data acquisition.

3.2 Sample Preparation for Imaging Analysis (Steps 1 and 2, Fig. 3)

3.2.1 nsPCR Reagent Preparation for Imaging Analysis

1. Dissolve nsPCR reagents in the solvent for nsPCR reagent. Unless stated otherwise, 5 mg/ml of nsPCR reagents is used for imaging analysis.

3.2.2 Tissue Sample Preparation

1. Conduct animal experiments following institutional guidelines. Mice are anesthetized, perfused with chilled phosphate buffered saline, and decapitated, and the brains are removed. The brains are then cut along midline.
2. Embed each hemisphere in gelatin (100 mg/ml in water). Snap freeze in dry ice and store at −80 °C.

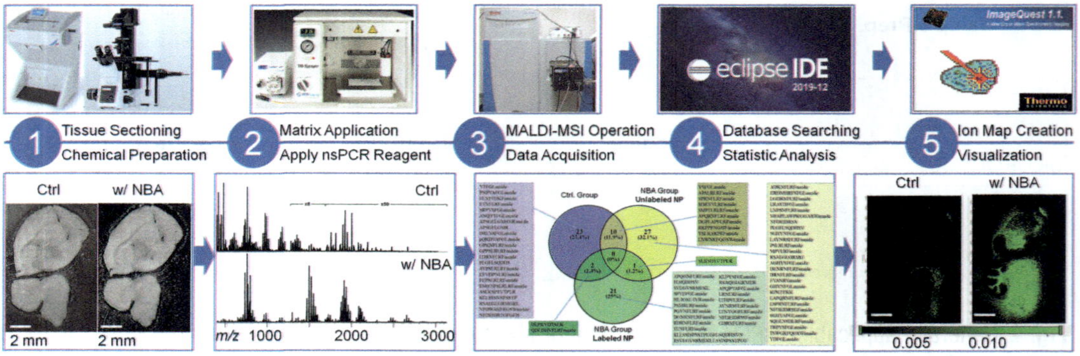

Fig. 3 Generally applicable nsPCR-enhanced MALDI-MS imaging workflow with NBA as an example nsPCR reagent. This workflow involves five-step procedure, **step 1**: tissue sectioning and chemical preparation; **step 2**: matrix and nsPCR application to tissue; **step 3**: MALDI-MS operation and data acquisition; **step 4**: database searching and statistical analysis; and **step 5**: ion map creation and visualization

3. Cryosection the gelatin-embedded tissue into 12 μm thickness on a cryostat at −20 °C and thaw-mount onto an ITO-coated slide or a microscope glass slide.
4. Acquire the optical images of tissue sections with bright field illumination.

3.2.3 Matrix and nsPCR Reagent Application to Tissues

1. Dry the prepared tissue slides for 30 min in a desiccator at room temperature right before matrix and nsPCR reagent application (*see* **Note 4**).
2. Deposit 5 mg/ml CHCA matrix using TM sprayer with the following condition: nozzle temperature of 80 °C, gas pressure of 10 psi, eight passes, moving velocity of 1100 mm/min, drying time of 30 s and flow rate of 0.2 ml/min (*see* **Note 5**).
3. Deposit nsPCR reagent with the gentle condition of 30 °C and gas pressure of 10 psi. The TM sprayer is tuned to deposit almost 1:1 (w/w) for CHCA: NBA to the tissue.
4. Put the sprayed slides in a desiccator for drying at room temperature for 30 min. At this point, the tissue slides are ready to be examined with MALDI-MSI experiments (**Step 3**, Fig. 3). Notably, the only difference between nsPCR-MALDI and regular MALDI is the supplementary application of nsPCR reagents (*see* **Note 6**).

3.3 nsPCR MS Operations

3.3.1 MS Operation for nsPCR-Enabled Spot Analysis (Step 3, Fig. 2)

Here, although the method applies to all types of MALDI instruments, we use MALDI-LTQ-Orbitrap XL mass spectrometer for spot analysis as an example, due to its high mass resolution and high mass accuracy, which enables confident peak assignment by using accurate mass matching for small molecules. Meanwhile, the LTQ Tune software enables setting up of sample sequences, permitting data acquisition for at most 96 spots in one sequence automatically. This vacuum MALDI source is featured with 60 Hz 337 nm N_2 laser.

1. MALDI-LTQ-Orbitrap XL general setting. A software tool (LTQ Tune) is used to control the XY stage motion in the ion source. For MALDI setting, plate motion is set as survey CPS, ASF is set as off and scan rate is set as 2 microscans/step. For laser setting, laser energy is set as 12 μJ, AGC is set as off and number of laser shot is set as 2.
2. Sample plate loading. Before sample loading, (1) check gas tanks level (helium: 40 psi, and nitrogen: 100 psi); (2) check temperature of water cooler (20 °C); and (3) check instrument status (everything should have a green check sign except for lower chamber pressure and laser should be a solid red). Then place the sample plate onto sample holder and insert the plate into the mass spectrometer. When the vacuum is built up, a green check will show for the lower chamber pressure.

3. MALDI-LTQ-Orbitrap XL mass spectrometer operation. The MS was operated under positive polarity mode, with mass range of *m/z* 70–400 (mass range should cover the mass of underivatized and derivatized metabolites) and mass resolution of 60,000 unless otherwise specified. The number of experiments to acquire is set as 50 for each sample spot.

3.3.2 MS Operation for nsPCR-Enabled Imaging Analysis (Step 3, Fig. 3)

As shown in our recent publications [26–28], for imaging analysis, we mainly focused on the use of a subatmospheric pressure-MALDI (SubAP-MALDI) system coupled to a Q Exactive HF hybrid quadrupole-Orbitrap mass spectrometer (QE HF), due to its improved spatial resolution, high mass resolution and high mass accuracy that permit us to confidently assign peaks by using accurate mass matching when tandem MS experiments are not achievable. Notably, this SubAP MALDI source is featured with a 355 nm high resolution Nd:YAG laser module along with an integrated ion funnel for improved ion collection and transmission.

1. SubAP-MALDI general settings. While we use iFunnel software (MassTech, Columbia, MD) to control the ion funnel operation, a software tool (Target-ng, MassTech, Columbia, MD) is used to control the XY stage motion in the ion source. Constant speed raster motion is used for MSI.
2. Set the starting laser parameters (unless otherwise specified) as laser energy of 18.9% and repetition rate of 1000 Hz. These parameters may vary based on different matrices and different types of compounds to be targeted. Set the RF voltage as 250 V (peak-to-peak).
3. Set the voltage applied to the MALDI plate as 250–300 V, and set the voltage applied to the first electrode in the 15-mm long accelerating stack located 6 mm away from the MALDI plate as 180–200 V. Set the electric potential across the first funnel (60 mm long) as 80–100 V, with the voltage drop across the second funnel (110 mm long) being 10–20 V. Set the voltage drop between the exit from the first and the entrance to the second funnel as 50–70 V.
4. Sample plate loading. Use the oil pump to make the operation source pressure of 3–5 Torr. To load the sample plate into a SubAP-MALDI source, it requires a two-step process, vacuum breakage, and vacuum build-up [27].
5. Vacuum breakage: (a) With the ion funnel set to Standby mode, eject the plate holder; (b) Turn the knob (on left side of source) clockwise until the vacuum gate valve is totally closed and the pressure gauge starts to show pressure increase. Then release the metal buckle (on right side of source) and wait ~2 min until the source automatically opens; (c) Carefully place the sample plate onto sample holder and ensure plate

does not wiggle. Standard microscopic slides (typically 75 mm × 25 mm, 1 mm thick) need to be cut into a proper length (~50 mm) prior to loading so it stretches out the slide holder for less than 5 mm for each end or the glass might be crushed during the source initiation procedure. One is advised to use 25 mm × 50 mm slides to avoid the slide shortening.

6. Vacuum built-up: (a) Close the source with buckle closed, (b) Turn on small roughing pump (first, make sure the handle on the pump hose is perpendicular to the hose) then flip handle to make it parallel to the hose to allow gas evacuation from the source. (c) When the vacuum gauge shows that the gas pressure drops below 40 Torr, start turning the knob counterclockwise and make 4 turns. (d) Turn handle 90° on the hose to disconnect small vacuum pump from SubAP MALDI source then turn off the small pump. (e) When the pressure in the source drops below 10 Torr, fully open the gate valve by turning the knob counterclockwise until its full range is achieved. The pressure shown by gas gauge should drop again. The working pressure in the MALDI source should be between 3 and 5 Torr. Make sure that the vacuum pressure in the intermediate section shown in the Tune program is below 1.9 mbar.
7. QE HF mass spectrometer operation. The MS is operated under positive polarity mode, with mass range of *m/z* 500–6000 and mass resolution of 60,000 unless otherwise specified. To ensure consistent scan time at each pixel, the maximum injection time is set to 300 ms and automatic gain control target is tuned to 3×10^6.

3.4 nsPCR Data Analysis

3.4.1 Data Analysis for nsPCR-Enabled Spot Analysis (Step 4, Fig. 2)

1. Data overview and general processing. Xcalibur (Thermo Scientific, Bremen, Germany) is used to process spectra. (a) Display noise and exact mass with 4 decimals then show spectrum list. (b) Select all the scans and view spectrum list. (c) In the display options, display all peaks. Precision is set as 4 decimals. (d) Then copy the spectrum list to excel for further data analysis.
2. Quantification of metabolites. (a) First, calculate the theoretical mass of each metabolite and derivatized metabolite through ChemCalc [29]. For example, glycine is derivatized with 5-NBA, adding Δ *m/z* 149.0113 to protonated glycine ($[M+H]^+ = 76.0399$), resulting a *m/z* 225.0511 for derivatized glycine. (b) In the excel files with exported spectrum list, summarize the signal intensities of protonated glycine and derivatized glycine based on accurate mass matching (±5 ppm). (c) Then plot signal intensity—concentration calibration curve for each metabolite. Table 1 shows some representative data for glycine, L-alanine, L-valine, L-leucine.

Table 1
Fast, internal standard-free and incubation-free quantification of low-mass metabolites improved by the nsPCR strategy[a]

Name	Mass	Labeling efficiency	Pre-nsPCR quantification	Post-nsPCR quantification	Comparison (k_2/k_1)
Glycine	75.0320	98.69% ± 1.85%	Not quantifiable	$k_2 = 2143.7$ $R^2 = 0.9251$	∞
Alanine	89.0477	97.88% ± 1.65%	Not quantifiable	$k_2 = 2810.3$ $R^2 = 0.9549$	∞
Valine	117.0790	97.34% ± 1.84%	$k_1 = 1552.8$ $R^2 = 0.9984$	$k_2 = 7198.4$ $R^2 = 0.9940$	4.64
Leucine	131.0946	99.32% ± 0.48%	$k_1 = 5864.8$ $R^2 = 0.9745$	$k_2 = 20{,}488$ $R^2 = 0.9806$	3.49

[a]Note, ∞ means significant enhancement in quantification capability. Data in this table are derived from 5-NBA (100 mM)-based nsPCR. The calculation of labeling efficiency is based on the results from data point of metabolite concentration of 20 μM. k_1/k_2, the slope values for quantification linear curves of regular MALDI-MS and nsPCR-MALDI-MS method, respectively

*3.4.2 Data Analysis for nsPCR-Enabled Imaging Analysis (**Steps 4** and **5**, Fig. 3)*

1. Data overview and general processing. Xcalibur (Thermo Scientific, Bremen, Germany) is used to process spectra. ImageQuest (Thermo Scientific, Bremen, Germany) is used for image processing, including intensity normalization and ion image generation. While MSiReader software [30] is also compatible with SubAP-MALDI imaging data, data formats need to be converted beforehand with certain tools such as imzml converter as instructed by the software.
2. Database searching. (a) To prepare database searching, first we need to create a mass list. To do that, we make Xcalibur to show mass list with noise and then export the top 1000 ions with exact mass list. (b) Next step, copy the list into an excel file and calculate the signal to noise ratio, which is used further for final database searching based on signal to noise ratio (S/N > 10 is used here, Fig. 4a). (c) Sweden peptide database (SwePep, Uppsala University, Uppsala, Sweden) and an in-house crustacean neuropeptide database (Li Lab, University of Wisconsin-Madison, Madison, WI) are used for accurate mass matching of neuropeptides with 10 ppm tolerance. Peaks of protonated ion $[M+H]^+$, corresponding sodiated ion $[M+Na]^+$, and potassium adduct $[M+K]^+$ would be accounted for the same peptide. Based on these criteria, as shown in Fig. 4a, in control group, only 35 neuropeptides are identified from accurate mass matching. Interestingly, in nsPCR group (NBA exemplified), we successfully identified 38 unlabeled neuropeptides and 24 labeled neuropeptides, representing approximately 58% improvement of neuropeptide coverage.

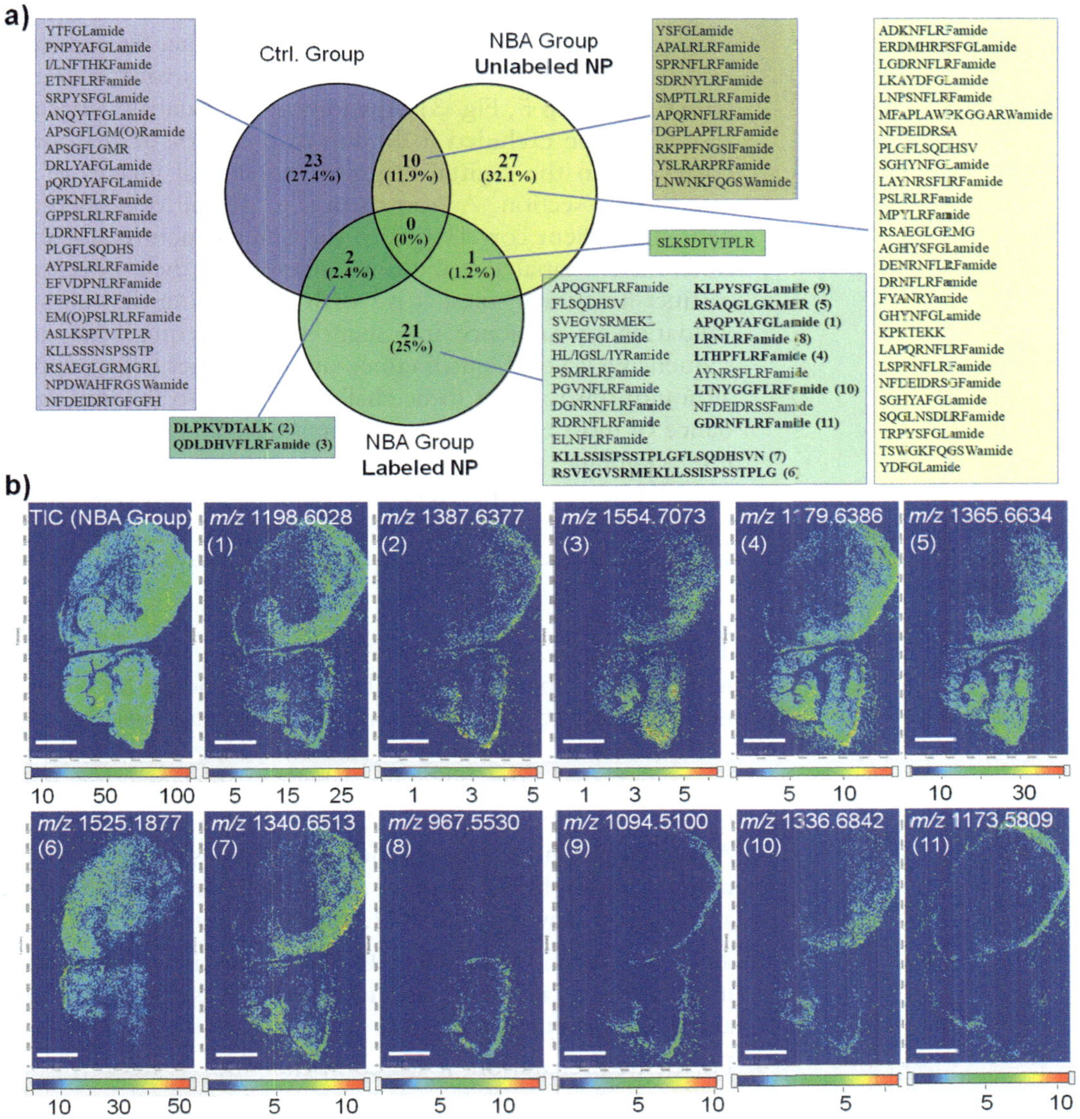

Fig. 4 Neuropeptide visualization as enhanced by the nsPCR strategy. (**a**) Identified neuropeptide lists for control mouse brain group and NBA-treated group. Venn diagram amongst control group, NBA group of unlabeled neuropeptides and NBA group of labeled neuropeptides indicates the substantially increased ID coverage (>58% deeper) with nsPCR strategy. (**b**) Representative maps of TIC and individual ions for peptide 1 to peptide 11 as visualized with nsPCR-MALDI-MSI with a step size of 50 μm. Peptide sequence information for these ions is highlighted in (**a**). All images are extracted against TIC and obtained with following settings in ImageQuest software: Scaling—Linear; Normalize—Raw Data; Histogram—Raw Data; 2D smooth—Spline; Tolerance—10 ppm. Scale bars, 2 mm

3. Ion map creation and visualization. When database searching is finished, a full list of identifications can be further filtered based on colocalization of their spatial distributions with biological structures (**Step 5**, Fig. 3). Any ions in the database searching results will be excluded if the ion map cannot be reasonably correlated to the sagittal diagram and optical images of the same tissue section. As shown in Fig. 4b, all neuropeptides exhibit excellent correlations to the brain structure and show differential spatial distributions, suggesting their potentially distinct functional roles. It is worth mentioning that ultimate validation of sequence information often requires de novo sequencing via on-tissue fragmentation of target neuropeptide, although this remains to be challenging due to the low abundance of most endogenous neuropeptides and/or inadequate ion sampling efficiency for most commercially available MALDI mass spectrometers while running an MSI experiment.

4 Notes

1. The nsPCR strategy is designed for signal enhancement of low-response species, especially for small molecules that are not amenable to fragmentation in normal mass spectrometers with limited low mass range (e.g., mostly $m/z > 50$). It should be noted that, the apparent labeling efficiency for high-response species will not be as good as low-response ones.
2. Although we only exemplify with certain types of mass spectrometers in this book chapter, the nsPCR strategy should be applicable to any types of MALDI instruments in principle.
3. The concentrations of the nsPCR reagents need to be optimized case by case. We found that 100 mM is good for 2NBA and 5NBA in most cases.
4. For tissue preparation, if stored in the −80 °C freezer, move section to dry box at room temperature to allow tissue to dry out prior to spraying with matrix. If sample is wet when sprayed, the matrix will not crystalize uniformly. If freshly sectioned, store in a dry box prior to spraying with matrix to ensure no water condensation and minimize any delocalization of analyte of interest.
5. The concentration of nsPCR reagents and its dissolving solvents should be carefully optimized to avoid any potential contamination and/or clogging of TM sprayer nozzle.
6. The chemical environment can affect the nsPCR labeling efficiency. For example, while we observe high labeling efficiency (e.g., >97% in Table 1) for those low-response small-molecule metabolites and amino acids, there are still a certain portion of

neuropeptides that are not effectively nsPCR-tagged under the current conditions. This could be due to the changed chemical environment of reactive groups, and/or too low density of these neuropeptides to be tagged at nanosecond time scale.

Acknowledgments

This work was funded in part by NIH (R56MH110215, R01DK071801, and P01CA250972) and NSF (CHE-1710140). The MALDI Orbitrap instrument and Q Exactive Orbitrap instrument were purchased through the support of an NIH shared instrument grant S10RR029531. LL acknowledges a Vilas Distinguished Achievement Professorship and the Charles Melbourne Johnson Distinguished Chair Professorship with funding provided by the Wisconsin Alumni Research Foundation and University of Wisconsin-Madison School of Pharmacy. GL thanks the funding support by the Fundamental Research Funds for the Central Universities (Nankai University, 020/63213057) and by the National Natural Science Foundation of China (22104064). The authors would like to thank Drs. Eugene Moskovets and Vladimir Doroshenko from MassTech Inc. for access to a SubAP MALDI source and technical support.

References

1. Nicholson JK, Holmes E, Kinross JM, Darzi AW, Takats Z, Lindon JC (2012) Metabolic phenotyping in clinical and surgical environments. Nature 491(7424):384–392. https://doi.org/10.1038/nature11708
2. Gupta P, Gupta S, Pruthi V (2020) Techniques for detection and extraction of metabolites. In: Engineering of microbial biosynthetic pathways. Springer, Singapore, pp 33–51. https://doi.org/10.1007/978-981-15-2604-6_3
3. Gachumi G, Purves RW, Hopf C, El-Aneed A (2020) Fast quantification without conventional chromatography, the growing power of mass spectrometry. Anal Chem 92(13):8628–8637. https://doi.org/10.1021/acs.analchem.0c00877
4. Manier ML, Spraggins JM, Reyzer ML, Norris JL, Caprioli RM (2014) A derivatization and validation strategy for determining the spatial localization of endogenous amine metabolites in tissues using MALDI imaging mass spectrometry. J Mass Spectrom 49(8):665–673. https://doi.org/10.1002/jms.3411
5. Gemperline E, Rawson S, Li L (2014) Optimization and comparison of multiple MALDI matrix application methods for small molecule mass spectrometric imaging. Anal Chem 86(20):10030–10035. https://doi.org/10.1021/ac5028534
6. Guo S, Tang W, Hu Y, Chen Y, Gordon A, Li B, Li P (2020) Enhancement of on-tissue chemical derivatization by laser-assisted tissue transfer for MALDI MS imaging. Anal Chem 92(1):1431–1438. https://doi.org/10.1021/acs.analchem.9b04618
7. Shariatgorji M, Nilsson A, Fridjonsdottir E, Vallianatou T, Kallback P, Katan L, Savmarker J, Mantas I, Zhang X, Bezard E, Svenningsson P, Odell LR, Andren PE (2019) Comprehensive mapping of neurotransmitter networks by MALDI-MS imaging. Nat Methods 16(10):1021–1028. https://doi.org/10.1038/s41592-019-0551-3
8. Santa T (2011) Derivatization reagents in liquid chromatography/electrospray ionization tandem mass spectrometry. Biomed Chromatogr 25(1–2):1–10. https://doi.org/10.1002/bmc.1548
9. Plotka-Wasylka JM, Morrison C, Biziuk M, Namiesnik J (2015) Chemical derivatization processes applied to amine determination in samples of different matrix composition.

Chem Rev 115(11):4693–4718. https://doi.org/10.1021/cr4006999

10. Cao Q, Wang Y, Chen B, Ma F, Hao L, Li G, Ouyang C, Li L (2019) Visualization and identification of neurotransmitters in crustacean brain via multifaceted mass spectrometric approaches. ACS Chem Neurosci 10(3):1222–1229. https://doi.org/10.1021/acschemneuro.8b00730
11. Sun C, Liu W, Geng Y, Wang X (2020) On-tissue derivatization strategy for mass spectrometry imaging of carboxyl-containing metabolites in biological tissues. Anal Chem 92(18):12126–12131. https://doi.org/10.1021/acs.analchem.0c02303
12. Deng T, Wu D, Duan C, Yan X, Du Y, Zou J, Guan Y (2017) Spatial profiling of gibberellins in a single leaf based on micro-scale matrix solid-phase dispersion and pre-column derivatization coupled with ultra-performance liquid chromatography-tandem mass spectrometry. Anal Chem 89(17):9537–9543. https://doi.org/10.1021/acs.analchem.7b02589
13. Wu Q, Comi TJ, Li B, Rubakhin SS, Sweedler JV (2016) On-tissue derivatization via electrospray deposition for matrix assisted laser desorption/ionization mass spectrometry imaging of endogenous fatty acids in rat brain tissues. Anal Chem 88(11):5988–5995. https://doi.org/10.1021/acs.analchem.6b01021
14. Wang SS, Wang YJ, Zhang J, Sun TQ, Guo YL (2019) Derivatization strategy for simultaneous molecular imaging of phospholipids and low-abundance free fatty acids in thyroid cancer tissue sections. Anal Chem 91(6):4070–4076. https://doi.org/10.1021/acs.analchem.8b05680
15. Jiang R, Jiao Y, Zhang P, Liu Y, Wang X, Huang Y, Zhang Z, Xu F (2017) Twin derivatization strategy for high-coverage quantification of free fatty acids by liquid chromatography-tandem mass spectrometry. Anal Chem 89(22):12223–12230. https://doi.org/10.1021/acs.analchem.7b03020
16. Qi W, Wang Y, Cao Y, Cao Y, Guan Q, Sun T, Zhang L, Guo Y (2020) Simultaneous analysis of fatty alcohols, fatty aldehydes, and sterols in thyroid tissues by electrospray ionization-ion mobility-mass spectrometry based on charge derivatization. Anal Chem 92(13):8644–8648. https://doi.org/10.1021/acs.analchem.0c01292
17. Zhuang M, Hou Z, Chen P, Liang G, Huang G (2020) Introducing charge tag via click reaction in living cells for single cell mass spectrometry. Chem Sci 11(28):7308–7312. https://doi.org/10.1039/d0sc00259c
18. Kaya I, Brulls SM, Dunevall J, Jennische E, Lange S, Martensson J, Ewing AG, Malmberg P, Fletcher JS (2018) On-tissue chemical derivatization of Catecholamines using 4-(N-methyl) Pyridinium Boronic acid for ToF-SIMS and LDI-ToF mass spectrometry imaging. Anal Chem 90(22):13580–13590. https://doi.org/10.1021/acs.analchem.8b03746
19. Zhang H, Shi X, Vu NQ, Li G, Li Z, Shi Y, Li M, Wang B, Welham NV, Patankar MS, Weisman P, Li L (2020) On-tissue derivatization with Girard's reagent P enhances N-glycan signals for formalin-fixed paraffin-embedded tissue sections in MALDI mass spectrometry imaging. Anal Chem 92(19):13361–13368. https://doi.org/10.1021/acs.analchem.0c02704
20. Ryan E, Reid GE (2016) Chemical derivatization and ultrahigh resolution and accurate mass spectrometry strategies for "shotgun" Lipidome analysis. Acc Chem Res 49(9):1596–1604. https://doi.org/10.1021/acs.accounts.6b00030
21. Qiao L, Roussel C, Wan J, Kong J, Yang P, Girault HH, Liu B (2008) MALDI in-source photooxidation reactions for online peptide tagging. Angew Chem Int Ed 47(14):2646–2648. https://doi.org/10.1002/anie.200703876
22. Lascoux D, Paramelle D, Subra G, Heymann M, Geourjon C, Martinez J, Forest E (2007) Discrimination and selective enhancement of signals in the MALDI mass spectrum of a protein by combining a matrix-based label for lysine residues with a neutral matrix. Angew Chem Int Ed 46(29):5594–5597. https://doi.org/10.1002/anie.200700811
23. Wienken CJ, Baaske P, Rothbauer U, Braun D, Duhr S (2010) Protein-binding assays in biological liquids using microscale thermophoresis. Nat Commun 1:1–7. https://doi.org/10.1038/ncomms1093
24. Seidel SA, Wienken CJ, Geissler S, Jerabek-Willemsen M, Duhr S, Reiter A, Trauner D, Braun D, Baaske P (2012) Label-free microscale thermophoresis discriminates sites and affinity of protein-ligand binding. Angew Chem Int Ed 51(42):10656–10659. https://doi.org/10.1002/anie.201204268
25. Moller FM, Kiess M, Braun D (2016) Photochemical microscale electrophoresis allows fast quantification of biomolecule binding. J Am Chem Soc 138(16):5363–5370. https://doi.org/10.1021/jacs.6b01756
26. Li G, Ma F, Cao Q, Zheng Z, DeLaney K, Liu R, Li L (2019) Nanosecond

photochemically promoted click chemistry for enhanced neuropeptide visualization and rapid protein labeling. Nat Commun 10(1):4697. https://doi.org/10.1038/s41467-019-12548-0

27. Li G, Cao Q, Liu Y, DeLaney K, Tian Z, Moskovets E, Li L (2019) Characterizing and alleviating ion suppression effects in atmospheric pressure matrix-assisted laser desorption/ionization. Rapid Commun Mass Spectrom 33(4):327–335. https://doi.org/10.1002/rcm.8358
28. Chen B, OuYang C, Tian Z, Xu M, Li L (2018) A high resolution atmospheric pressure matrix-assisted laser desorption/ionization-quadrupole-orbitrap MS platform enables in situ analysis of biomolecules by multi-mode ionization and acquisition. Anal Chim Acta 1007: 16–25. https://doi.org/10.1016/j.aca.2017.12.045
29. Patiny L, Borel A (2013) ChemCalc: a building block for tomorrow's chemical infrastructure. J Chem Inf Model 53(5):1223–1228. https://doi.org/10.1021/ci300563h
30. Robichaud G, Garrard KP, Barry JA, Muddiman DC (2013) MSiReader: an open-source Interface to view and analyze high resolving power MS imaging files on Matlab platform. J Am Soc Mass Spectrom 24(5):718–721. https://doi.org/10.1007/s13361-013-0607-z

Chapter 11

Extraction and Upconcentration of Adsorbates from Precisely Defined Area for Quantitative MALDI Mass Spectrometry Imaging

Sanghwan Park and Chang Young Lee

Abstract

Mass spectrometry imaging (MSI) allows label-free detection of a wide range of biomolecules and simultaneously provides their spatial distributions. In particular, MSI by matrix-assisted laser desorption/ionization mass spectrometry (MALDI) has been widely used in biomolecule analysis. However, quantitation in MALDI-MSI is limited by matrix-deposition heterogeneity, analyte extraction area, and analyte–matrix cocrystallization. In this chapter, a microstructured PDMS stamp is utilized to precisely control the matrix deposition area and the analyte extraction area. We describe here simple steps—including stamp fabrication, matrix application, and data-acquisition guideline—for the quantitative analysis of adsorbed peptides on hydrophobic surfaces.

Key words Mass spectrometry imaging, Matrix-assisted laser desorption/ionization, Quantitation, Peptides, Controlled extraction

1 Introduction

With the advance of bioanalytical platforms, direct analysis of small molecules adsorbed on various functional substrates has become increasingly important [1–5]. Mass spectrometry imaging (MSI) by matrix-assisted laser desorption/ionization mass spectrometry (MALDI) has been utilized for the direct analysis of a wide range of adsorbed biomolecules; it can not only identify them but can also provide information about their spatial distribution [6, 7]. Additional advantages of MALDI-MSI include its extensive mass range, high sensitivity, high salt tolerance, and minimal fragmentation of biomolecules. Quantitative analysis of the adsorbate is important for reliable estimation of surface abundance, but heterogeneity in matrix application and the extraction area of the adsorbate make using MALDI-MSI for this purpose challenging [8]. The

Young-Jin Lee (ed.), *Mass Spectrometry Imaging of Small Molecules: Methods and Protocols*, Methods in Molecular Biology, vol. 2437, https://doi.org/10.1007/978-1-0716-2030-4_11,
© The Author(s), under exclusive license to Springer Science+Business Media, LLC, part of Springer Nature 2022

heterogeneous cocrystallization behavior of the analyte and matrix molecules causes the "hot-spot" or "sweet-spot" effect: the ion yield varies depending on the position irradiated by the laser [9]. Therefore, a new methodological approach that can control the matrix deposition and extraction area precisely should be developed for quantitative MALDI-MSI of adsorbates.

In this chapter, we present a simple approach using a polydimethylsiloxane (PDMS) stamp to apply a uniform and controlled matrix for quantitative MALDI-MSI of peptides adsorbed on a substrate. This method has been shown to allow the relative quantitation of adsorbed peptides on a hydrophobic functionalized substrate by controlling the analyte extraction area and matrix deposition [10]. Below, we specify the protocol for sample preparation using a PDMS stamp for MALDI imaging of surface-adsorbed molecules. The workflow and principle of this protocol are outlined in Fig. 1. When the PDMS stamp filled with matrix solution is brought into contact with the analyte-adsorbed substrate, the surface-bound analyte molecules are extracted from the contact area by the transferred matrix solution. When the stamp is peeled off, the matrix solution remains on the substrate. Then, the analyte and matrix molecules are concentrated and cocrystallized as the solvent evaporates. The walls of the PDMS stamp prevent cross-contamination of the extracted analyte molecules. Thus, a single analyte–matrix cocrystal contains only those analyte molecules extracted from a particular area of the microwell. The relative quantitation of the surface concentration is achieved by analyzing the matrix crystal containing the analyte molecules extracted from each specified region.

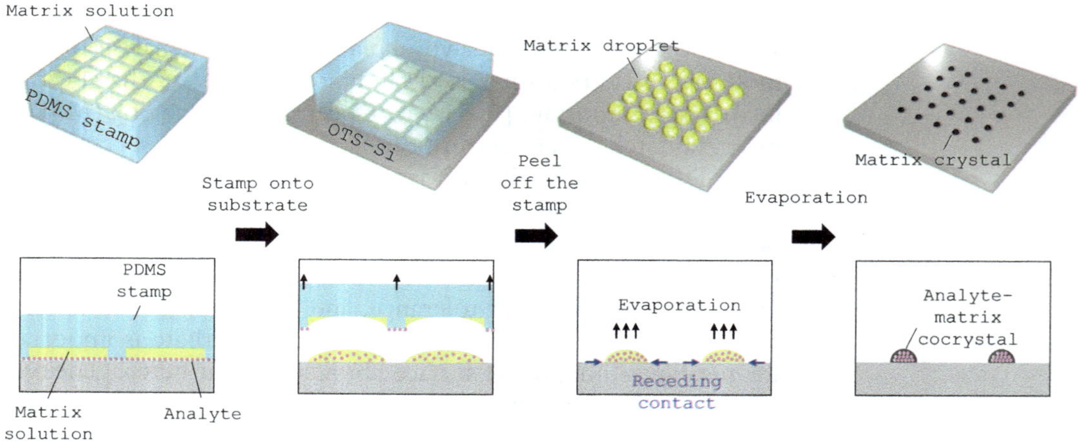

Fig. 1 Schematics and cross-sectional views illustrating the transfer of matrix solution to substrate by a microstructured PDMS stamp. Matrix solution is transferred from microwell to analyte-adsorbed substrate by PDMS stamping. Surface-bound analyte molecules extracted by the transferred matrix solution form analyte–matrix cocrystals. (Reproduced with permission from [10] Copyright (2020) American Chemical Society)

2 Materials

2.1 Preparation of Microstructured PDMS Stamp

1. Silicon wafer.
2. SU-8 photoresist (SU-8-2050, MicroChem Corp.)
3. SU-8 developer.
4. Photomask.
5. PDMS prepolymer kit (SYLGARD 184 Silicone Elastomer Kit, Dow Corning Corp.)
6. Chloromethylsilane.

2.2 Solvent and Reagents

1. Deionized water.
2. Acetonitrile.
3. Trifluoroacetic acid (TFA).
4. Octadecyltrichlorosilane (OTS).
5. Chloroform.
6. Hexane.
7. 2,5-Dihydroxybenzoic acid (DHB).

2.3 Instrumentation

1. Spin coater.
2. Hot plate.
3. Vacuum desiccator.
4. Oven.
5. Mask aligner.
6. Optical microscope.
7. Water purification system.
8. MALDI-TOF mass spectrometer.
9. Sample adaptor for MALDI-MSI.
10. Modified sample holder (25 mm × 75 mm stainless-steel plate with a height difference equal to the thickness of the sample substrate).

2.4 Software

1. Control/analysis software for MALDI-MSI.
2. OriginPro.
3. MATLAB.

3 Methods

3.1 Fabrication of Microstructured PDMS Stamp

1. Prepare a photomask for patterning the PDMS stamp having square microwells. In this protocol, the dimension of the PDMS structure is described by the width (w) and height (h)

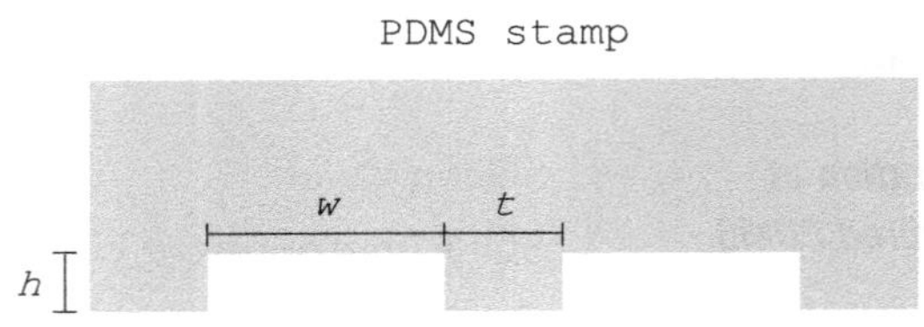

Fig. 2 Design of PDMS stamp. Dimensions of PDMS stamp are width of square microwell (*w*), wall thickness (*t*) and height (*h*)

of the square microwell and the wall thickness (t) (Fig. 2). A standard recipe provided in this section is for fabricating PDMS stamps with $w = 50$ μm, $h = 50$ μm, and $t = 20$ μm (*see* **Note 1**).

2. Pour the SU-8-2050 negative photoresist on a silicon wafer and spin-coat at 3500 rpm for 60 s (*see* **Note 2**).
3. Bake the photoresist-coated silicon wafer on a hot plate for 2 min at 65 °C and then for 7 min at 95 °C.
4. After baking, allow the wafer to cool down to room temperature (25 °C).
5. Place the wafer on a mask aligner equipped with a UV ramp (365 nm). Then expose to UV radiation at 170 mJ/cm^2 through the photomask.
6. Perform post-exposure bake of the wafer for 2 min at 65 °C followed by baking for 7 min at 95 °C (*see* **Note 3**).
7. Complete fabrication of the SU-8 master mold by soaking the wafer in SU-8 developer for 6 min to dissolve the pattern unexposed to UV.
8. Place the SU-8 master in a Petri dish and then in a vacuum desiccator with a small amount of chloromethylsilane (~0.5 mL) in a plastic weighing dish. Open the vacuum valve slightly and allow the vaporized chloromethylsilane to remain in the desiccator for 30 min (*see* **Note 4**).
9. Fill a plastic weighing dish with PDMS prepolymer (SYLGARD 184 base). Then, add the curing agent in the mass ratio of 1:10 relative to the base (*see* **Note 5**).
10. Using a plastic stirring rod, stir the mixture of SYLGARD 184 base and curing agent in a plastic weighing dish for at least 5 min. It can be observed that many air bubbles have been created within the PDMS prepolymer.
11. Pour the mixed prepolymer onto the master in the Petri dish used in **step 8** above (*see* **Note 6**).
12. Place the Petri dish in a vacuum desiccator for at least 30 min until air bubbles disappear completely from the PDMS prepolymer.

13. Place the Petri dish including both the SU-8 master and PDMS prepolymer mixture in an oven at 70 °C for 6 h (or until complete curing).
14. Detach the PDMS stamp from the SU-8 master mold using a scalpel or razor blade (*see* **Note 7**).

3.2 Analyte Extraction by Matrix Crystal Arrays

1. Prepare a solvent for matrix solubilization (0.3% TFA in 50% acetonitrile (v/v)).
2. Dissolve DHB in prepared solvent with a concentration of 60 mg/mL (*see* **Note 8**).
3. Place the PDMS stamp on a crystallizing dish with the patterned surface facing up and pour matrix solution to submerge the stamp. Air bubbles will form on the surface of the stamp and are removed in the following step.
4. Place the crystallizing dish containing the PDMS stamp into the vacuum desiccator. Use the vacuum to remove all air bubbles from the stamp, especially the ones that are trapped in the microwell structures. Air bubbles that do not detach from the surface of the stamp can be easily removed with bath sonication in less than 10 s (*see* **Note 9**).
5. Place the dish with the PDMS stamp immersed in the matrix solution beside the target sample substrate.
6. Take the PDMS stamp from the matrix solution and place it on the target substrate with the patterned surface facing down (*see* **Note 10**).
7. To extract the adsorbed analyte molecules with the matrix solution, keep the PDMS stamp in contact with the target substrate for 5 min; then, peel off the stamp (*see* **Note 11**).
8. Matrix droplets are transferred onto the substrate as the stamp is detached. An array of analyte–matrix cocrystals is formed by solvent evaporation (*see* Fig. 3 and **Note 12**).

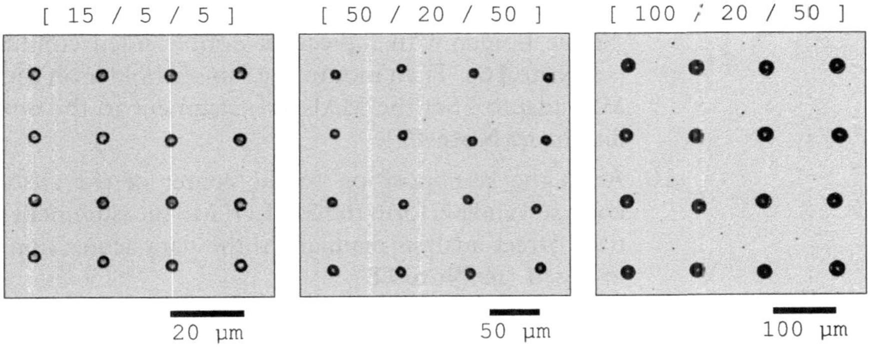

Fig. 3 Arrays of analyte–matrix cocrystals prepared by PDMS stamp. The dimension of PDMS stamps used for cocrystal array formation is indicated as [*w*/*h*/*t*]

3.3 Preparation of Calibration Standards

1. Prepare a substrate identical to the one on which the analyte molecules to be analyzed will be adsorbed. Described in this protocol is the analysis of surface-bound peptides through hydrophobic interaction using the OTS-functionalized silicon substrate (*see* **Note 13**).
2. Prepare the OTS solution with a concentration of 2.5% (v/v) in a mixture of 7:3 hexane and chloroform.
3. Immerse the pieces of silicon wafer in the OTS solution for 20 min to form a self-assembled monolayer on the silicon surface.
4. In order to stabilize the OTS self-assembled monolayer and remove excess OTS molecules, transfer the silicon substrates to a fresh chloroform and immerse them for 30 min. Repeat the procedure in another fresh chloroform bath. Then, transfer the substrates to fresh chloroform and bath-sonicate for 5 min. Finally, rinse the substrates with deionized water and dry them with nitrogen.
5. Prepare the analyte solutions at various concentrations (*see* **Note 14**).
6. Immerse the OTS-functionalized substrates into the analyte solution and incubate until adsorption–desorption equilibrium is reached (*see* **Note 15**). Repeat the process for the prepared analyte solutions to obtain the calibration standards with various surface concentrations of analytes,
7. Rinse the substrate with a deionized water to remove excess analyte that has not been adsorbed. Then, dry the substrate with nitrogen.
8. To calibrate the MALDI-MS intensity versus surface concentration of analytes, apply the matrix solution to the analyte-adsorbed substrate using the PDMS stamp as described in Subheading 3.2.
9. Mount the matrix-patterned substrates on the custom-built sample holder with a piece of double-sided conductive tape (*see* **Note 16**). Then mount the sample holder on the MALDI-MSI adaptor. Set the MALDI instrument to the optimal condition (*see* **Note 17**).
10. Align the laser position to the center of the analyte–matrix cocrystal and perform the MALDI MS measurement according to “correct analysis manual” in the data-acquisition guideline in Fig. 4 (*see* **Note 18**).
11. Obtain the calibration curve from the correlation between the surface concentration and the mass signal intensities (*see* **Note 19**).

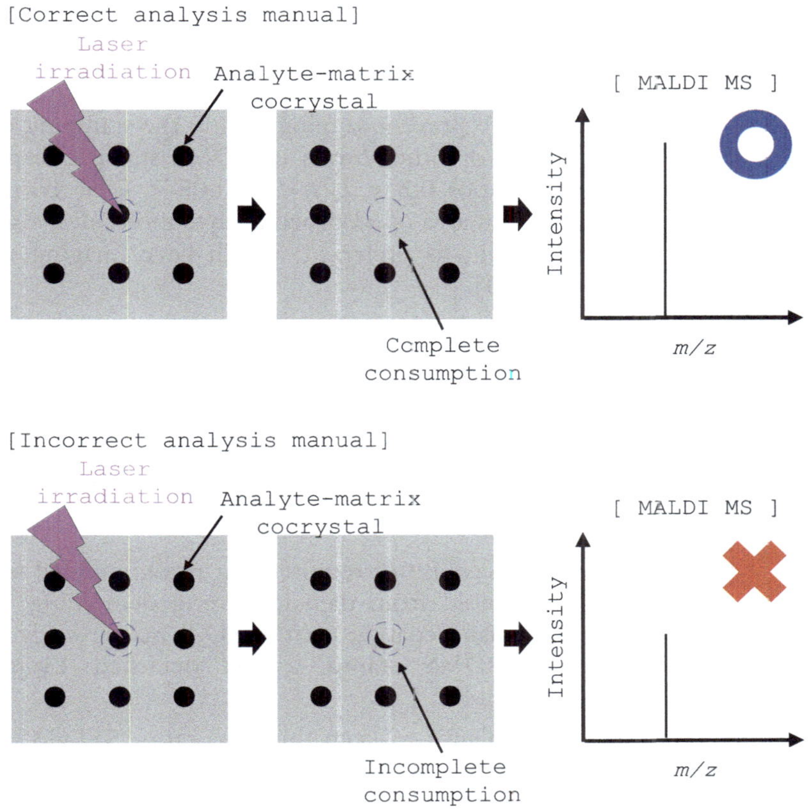

Fig. 4 Data-acquisition guideline for the quantitative MALDI mass spectrometry imaging. Analyte–matrix cocrystal should be consumed by single laser irradiation, and generated ions should be represented in the intensity of mass spectrum

3.4 MALDI-MSI and Data Analysis

1. Apply the matrix solution to the sample substrate using the PDMS stamp as described in Subheading 3.2.
2. Perform the MALDI measurement in the same way as the calibration standard measurement described in Subheading 3.3, **step 9**.
3. To obtain mass spectra, align the laser position to the center of the analyte–matrix cocrystal and irradiate the cocrystal until it is consumed completely. Please be sure to follow the Data-acquisition guideline during all data acquisition procedures.
4. Save the individual spectra with location information (*x* and *y* coordinates).
5. MALDI-MSI data can be generated with data analysis software such as OriginPro and MATLAB or with the imaging software provided by the manufacturer of MALDI-TOF mass spectrometer (*see* **Note 20**).

4 Notes

1. Consider the following to determine the dimensions of the stamp. To prevent sagging during the stamping of the matrix solution or other forms of PDMS stamp collapse, [11] use aspect ratios $0.5 < h/t < 5$, $0.05 < h/w$. To maximize the extraction area of adsorbed analyte molecules, a square micro-well design is preferred; it will have imaging resolution of $w + t$ [10].
2. All fabrication processes in this protocol are suitable for making a 50 μm-thick PDMS stamp. If different types of photoresist are used or a different microwell thickness is desired, follow the processing guidelines provided by the photoresist manufacturer for the desired stamp dimensions.
3. This baking procedure should take place directly after UV exposure.
4. This process is not required but recommended to protect the master mold from damage during demolding. The chloro-methylsilane coating on the SU-8 master surface allows the cured PDMS stamp to be detached easily from the master mold.
5. Use a balance to keep the mass ratio (SYLGARD 184 base: curing agent = 10:1). A larger amount of curing agent will result in a stiffer PDMS stamp, whereas a smaller amount in a more flexible one: either of these will affect the transfer characteristics of the matrix solution.
6. It is recommended that the thickness of the PDMS stamp be at least 5 mm.
7. When separating the PDMS stamp from the master mold, be careful not to damage the master mold. When necessary, the stamp can be separated from the master mold by breaking the Petri dish containing them.
8. Consider the type of analyte when choosing the matrix. The concentration of the matrix should be slightly lower than the solubility in the experimental environment. Avoid saturated solutions in order to prevent the formation of undissolved matrix residues due to the evaporation of small amounts of solvent in the subsequent experimental process.
9. To minimize the concentration change of matrix solution due to solvent evaporation in a vacuum desiccator, it is recommended to (1) fill the crystallizing dish with sufficient amount of matrix solution (e.g., 40 mL matrix solution for a dish 50 mm in diameter) and (2) keep the PDMS stamp under vacuum for not more than a minute.

10. When taking the PDMS stamp out from the matrix solution, hold the stamp's side with tweezers and slightly tilt it to prevent excess matrix solution from remaining on the surface. Then, remove remaining droplets of the matrix solution on the unpatterned back or side of the PDMS stamp that could flow down to the target sample during stamping by lightly touching with a laboratory wipe. In this process, be careful not to lose the matrix solution filling the microwells of the PDMS stamp due to contact with the wipe.
11. The time can be extended depending on the efficiency of the analyte extraction by the matrix. Insufficient extraction can affect the MALDI-MS signal, whereas excessive stamping time can lead to analyte loss due to analyte absorption by the PDMS stamp. It is possible to reduce the amount of loss by treating the stamp with polyethylene glycol (PEG) or 2-[methoxy(polyethyleneoxy)propyl]trimethoxysilane (OEG) [3, 12].
12. The time it takes the solvent to evaporate is affected by the volume of the matrix droplet and by the ambient temperature and humidity. Note that rapid evaporation in an extremely dry environment, such as a vacuum desiccator, cannot yield single concentrated analyte–matrix cocrystals during solvent evaporation.
13. The substrate material and functionalization molecules can be varied depending on the type of substrate and the target analyte. The process described in **steps 2–4** of the method can be modified according to the self-assembled-monolayer formation protocol. The calibration should be performed under the same molecular-adsorption conditions as the real sample system.
14. To avoid potential alteration of the adsorption and desorption equilibrium of the analytes, the analyte solution should be prepared using the same solvent as the analyte target sample solution.
15. To achieve complete adsorption–desorption equilibrium, the incubation time should be chosen according to the functional substrate and the analyte molecules. For example, adsorption of bradykinin on an OTS-functionalized silicon substrate reaches equilibrium within 1 h, and further incubation at the same peptide concentration does not affect the surface coverage of peptides.
16. A custom-built sample holder is required if the sample is not compatible with the commercial MALDI-MS sample holder. The holder should be designed so the surface of the mounted sample lies on the focal plane of the laser.

17. The operating condition for MALDI measurement, including laser intensity and detector gain, should be the same as the optimized operating condition that will be applied in real sample analysis. Before this step, perform mass calibration and optimization of the operating conditions for MALDI measurement of the target analytes.
18. When collecting MALDI mass spectra, the number of laser shots must be set large enough to consume the entire analyte–matrix cocrystal. Ions generated by consumption of an analyte–matrix cocrystal must be counted in the intensity of the single mass spectrum. It is preferable that the diameter of the laser beam be larger than that of individual analyte–matrix cocrystal.
19. When creating the calibration curve, use the average mass intensity of at least five mass spectra at each concentration to improve accuracy. When the adsorption–desorption equilibrium constant is already known for the analyte, surface concentration of the analyte can be estimated at each concentration in the solution. Otherwise, the adsorption–desorption equilibrium constant can be estimated using the correlation between the mass signal intensity and the concentration of the standard analyte solution [10].
20. Spectral information for each pixel can be obtained and easily visualized with data analysis software. When the data-acquisition guideline shown in Fig. 4 is followed for all the analyte–matrix cocrystals during MALDI-MSI, one can obtain a spatial distribution of the analytes with quantitative information.

Acknowledgments

This work was supported by the Basic Science Research Program through the National Research Foundation (NRF) of Korea (NRF-2019R1A2C1084928).

References

1. Jo K, Heien ML, Thompson LB, Zhong M, Nuzzo RG, Sweedler JV (2007) Mass spectrometric imaging of peptide release from neuronal cells within microfluidic devices. Lab Chip 7(11):1454–1460
2. Kim J, Mrksich M (2010) Profiling the selectivity of DNA ligases in an array format with mass spectrometry. Nucleic Acids Res 38(1): e2–e2
3. Zhong M, Lee CY, Croushore CA, Sweedler JV (2012) Label-free quantitation of peptide release from neurons in a microfluidic device with mass spectrometry imaging. Lab Chip 12 (11):2037–2045
4. Beloqui A, Calvo J, Serna S, Yan S, Wilson IBH, Martin-Lomas M, Reichardt NC (2013) Analysis of microarrays by MALDI-TOF MS. Angew Chem Int Ed 52(29):7477–7481

5. Yoon S, Park S, Kim MS, Lee CY (2018) Concomitant desalting and concentration of neuropeptides on a donut-shaped surface pattern for MALDI mass spectrometry. Chem Commun 54(45):5688–5691
6. McDonnell LA, Heeren RMA (2007) Imaging mass spectrometry. Mass Spectrom Rev 26 (4):606–643
7. Gessel MM, Norris JL, Caprioli RM (2014) MALDI imaging mass spectrometry: spatial molecular analysis to enable a new age of discovery. J Proteome 107:71–82
8. Ellis SR, Bruinen AL, Heeren RMA (2014) A critical evaluation of the current state-of-the-art in quantitative imaging mass spectrometry. Anal Bioanal Chem 406(5):1275–1289
9. Garden RW, Sweedler JV (2000) Heterogeneity within MALDI samples as revealed by mass spectrometric imaging. Anal Chem 72 (1):30–36
10. Park S, Yoon S, Min H, Moon SM, Choi YJ, Kim IS, Lee GH, Kim MS, Seo J, Jung W, Lee CY (2020) Compartmentalized arrays of matrix droplets for quantitative mass spectrometry imaging of adsorbed peptides. Anal Chem 92(13):8715–8721
11. Huang YY, Zhou W, Hsia KJ, Menard E, Park J-U, Rogers JA, Alleyne AG (2005) Stamp collapse in soft lithography. Langmuir 21 (17):8058–8068
12. Sui G, Wang J, Lee C-C, Lu W, Lee SP, Leyton JV, Wu AM, Tseng H-R (2006) Solution-phase surface modification in intact poly(dimethylsiloxane) microfluidic channels. Anal Chem 78 (15):5543–5551

Chapter 12

Development and Application of DropletProbe Mass Spectrometry for Examining Biodistribution of Therapeutics

Bingming Chen, Marissa Vavrek, Mark T. Cancilla, and Vilmos Kertesz

Abstract

dropletProbe mass spectrometry is a novel technique for molecular characterization of surfaces. It can be used for rapid ex vivo analysis of therapeutics from thin animal tissue sections and has been shown to improve understanding of a drug's absorption, distribution, metabolism and excretion (ADME) properties. Here, we describe the tissue distribution analysis of diclofenac from a dosed whole-body mouse thin tissue section using a dropletProbe mass spectrometry system.

Key words DropletProbe, Mass spectrometry, Diclofenac, Whole-body analysis, Tissue distribution, Metabolite identification

1 Introduction

During drug discovery, it is essential to better understand the absorption, distribution, metabolism, and excretion (ADME) properties of a drug candidate to usher it through development. Quantitative whole-body autoradiography (QWBA) has been widely utilized to understand drug distribution in preclinical animal studies [1]. However, a QWBA study requires the use of radiolabeled molecules and may not provide molecular specificity to differentiate drug from possible metabolites. Therefore, label-free mass spectrometry (MS)-based techniques are gaining in popularity throughout the pharmaceutical industry to investigate the ADME properties of drug candidates [2].

Matrix-assisted laser desorption/ionization (MALDI) MS imaging has been widely applied as a label-free technique to study ex vivo distribution of endogenous or exogeneous molecules [3–5]. MALDI-MS imaging provides molecular specific information to distinguish parent drug and metabolites. It also has appropriate

Young-Jin Lee (ed.), *Mass Spectrometry Imaging of Small Molecules: Methods and Protocols*,
Methods in Molecular Biology, vol. 2437, https://doi.org/10.1007/978-1-0716-2030-4_12,
© The Author(s), under exclusive license to Springer Science+Business Media, LLC, part of Springer Nature 2022

spatial resolution to visualize tissue morphological features. Routine analysis is acquired at 20–200 μm resolution on most commercial instruments and approximately 1 μm has been achieved with instrument modifications [3, 6]. However, MALDI can suffer from limited sensitivity and poor ionization for some drug molecules, partially due to limited analyte extraction. Moreover, it requires MALDI matrix application during sample preparation to cocrystallize with analytes and assist the ionization [3]. To overcome these limitations, other label-free MS techniques based on liquid extraction have been developed, such as liquid extraction surface analysis-MS [7], exhaustive liquid microjunction surface sampling-MS [8], dropletProbe-MS [9], laser-assisted dropletProbe-MS [10], Single-probe [11], and nanospray desorption electrospray ionization-MS [12].

In the dropletProbe system, an autosampler is employed to move a syringe to a predefined location, dispense a droplet of extraction solvent onto the sample surface, subsequently aspirate the droplet and the extracted material back into the syringe. Next, the autosampler injects the extracted material into a high-performance liquid chromatography (HPLC) mass spectrometer for analysis. Similar to MALDI-MS imaging, dropletProbe-MS can also be used to map ex vivo drug and metabolite distribution [9, 13, 14]. Unlike MALDI, the HPLC component of the dropletProbe-MS system separates analytes from interfering molecules before MS analysis, which provides better sensitivity and selectivity for tissue analysis applications [14]. dropletProbe-MS does not require additional sample preparation steps prior to MS analysis. However, the spatial resolution of dropletProbe-MS is limited by the size of the droplet and is typically around 1–2 mm. Recently, 700 μm-resolution sampling of metabolite acetaminophen sulfate from dosed rat liver was also demonstrated using this technique [15].

In this chapter, we describe a method to semiquantitatively map the spatial distribution of drugs and their metabolites using dropletProbe-MS. In the present case, we use diclofenac dosed mouse for whole-body analysis and the detailed results can be found in reference 13. The structure of diclofenac and its metabolites are in Scheme 1. The protocol can be widely applicable to detect other xenobiotics and their metabolites in various animal species.

2 Materials

2.1 Chemicals

1. Diclofenac sodium salt.
2. Labetalol.
3. Carboxymethylcellulose (CMC) sodium salt.

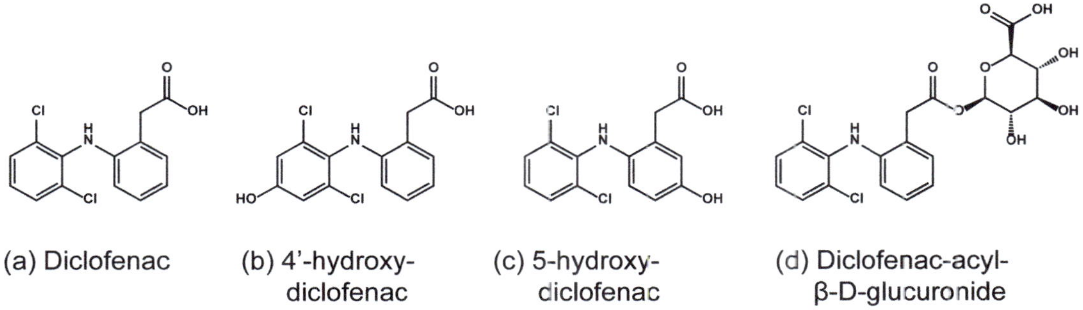

Scheme 1 Structure of diclofenac (**a**), 4′-hydroxy-diclofenac (**b**), 5-hydroxy-diclofenac (**c**), and diclofenac-acyl-β-D-glucuronide. (Reprinted with permission from reference 13. Copyright 2020 American Chemical Society)

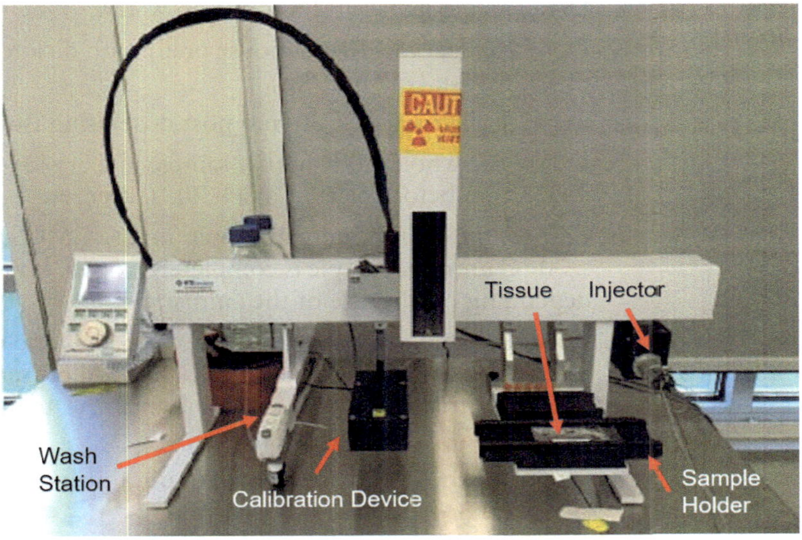

Fig. 1 Photo of the SepQuant dropletProbe system equipped with a whole-body tissue holder

4. Acetonitrile (ACN, Optima LC/MS grade).
5. Water (Optima LC/MS grade).
6. Formic acid (FA, Optima LC/MS grade).
7. Dry ice.

2.2 Animal

Female C57BL/6 mouse.

2.3 Equipment

1. SepQuant dropletProbe system (HTX Technologies) with whole-body tissue holder shown in Fig. 1 (*see* **Note 1**).
2. HPLC system.
3. High resolution accurate mass spectrometer.
4. Macrotome.

5. X-type syringe 22 (CTC Analytics AG) (*see* **Note 2**).
6. C18 column 50 × 2.1 mm (Mobile phase A as 0.1% FA in water and mobile phase B as 0.1% FA in ACN).
7. Flatbed scanner.

2.4 Miscellaneous

1. Transparent tape.
2. Double-sided tape.
3. Pipettes.

3 Methods

3.1 Animal Experiments and Whole-Body Section Preparation

1. Follow institutional guidelines for all animal experiments.
2. Dissolve diclofenac at 2 mg/mL in water.
3. Dose a female C57BL/6 mouse orally with diclofenac at a level of 15 mg/kg (7.5 mL/kg).
4. Sacrifice the mouse at 30 min post dose using isoflurane.
5. Freeze the whole-body mouse sample in a dry ice/hexane bath for 15 min before placing in a −70 °C freezer.
6. Embed the frozen animal sample in 2% CMC and freeze in −70 °C until the sample is completely frozen. It takes 12–72 h, depending on the size of the animal.
7. Place the sample into a macrotome at −20 °C and use 5% CMC to glue the tissue block onto the adaptor. Allow sample to sit in the macrotome for an hour to balance the temperature and allow the 5% CMC to solidify.
8. Section a whole-body sample at 40 μm thickness at target level (*see* **Note 3**).
9. Collect tissue section on a transparent tissue tape (*see* **Note 4**).
10. Store the tissue tape at −20 °C until analysis.

3.2 Surface Sampling

1. Prepare extraction solution of 50/50/0.05 (v/v/v) ACN/-water/FA (*see* **Note 5**).
2. Spike in labetalol at a final concentration of 1 μM as internal standard (*see* **Note 6**).
3. Take out tissue tape and store in desiccator at room temperature for approximately 15 min.
4. Mount the tissue tape onto a piece of black cardboard, then secure the black cardboard to the whole-body tissue holder of the dropletProbe platform using double-sided tape as shown in Fig. 2.
5. Open the dropletProbe instrument control software (dropletP-robe Premium) to control the instrument.

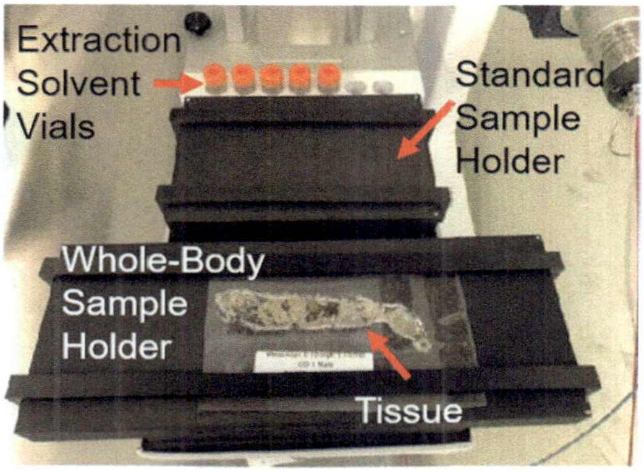

Fig. 2 Photo of a whole-body mouse section mounted onto the whole-body tissue holder of the dropletProbe system

6. Scan an optical image of the tissue mounted tray using a flatbed scanner.
7. Place the tissue mounted tray into the sample holder of the dropletProbe system.
8. Calibrate the sample holder to align the optical image with the physical coordinates of the instrument.
9. Set the extraction method on dropletProbe software: 3 extraction cycles, 0.1 mm away from sample surface, 2 μL extraction solvent, 5 μL sample volume and 5 μL injection volume (*see* **Note 7**).
10. Select the area(s) of interest to be analyzed on the scanned picture using the instrument control software (*see* **Notes 8–11**).
11. Run the sample queue in the instrument control software.

3.3 HPLC-MS Analysis

1. Connect the HPLC to the mass spectrometer for HPLC-MS analysis (*see* **Note 12**).
2. Setup a 3-min gradient at a flow rate of 0.3 mL/min as follows: (time, % mobile phase B): (0.00 min, 5%), (0.30 min, 5%), (0.35 min, 40%), (2.35 min, 90%), (2.55 min, 90%), (2.60 min, 5%), (2.98 min, 5%), (3.00 min, 50%, flow rate 0.01 mL/min) (*see* **Note 13**).
3. Setup a four-scan event method for the mass spectrometer as follows:

   Scan 1: full mass measurement at *m/z* 200–2000, positive ion FT mode, resolution of 30,000 (at *m/z* 400).

   Scan 2: targeted MS/MS for diclofenac by ion trap (*m/z* 296.02, isolation window (IW) = 3 *m/z*, collision-induced dissociation (CID), normalized collisional energy (NCE) 25).

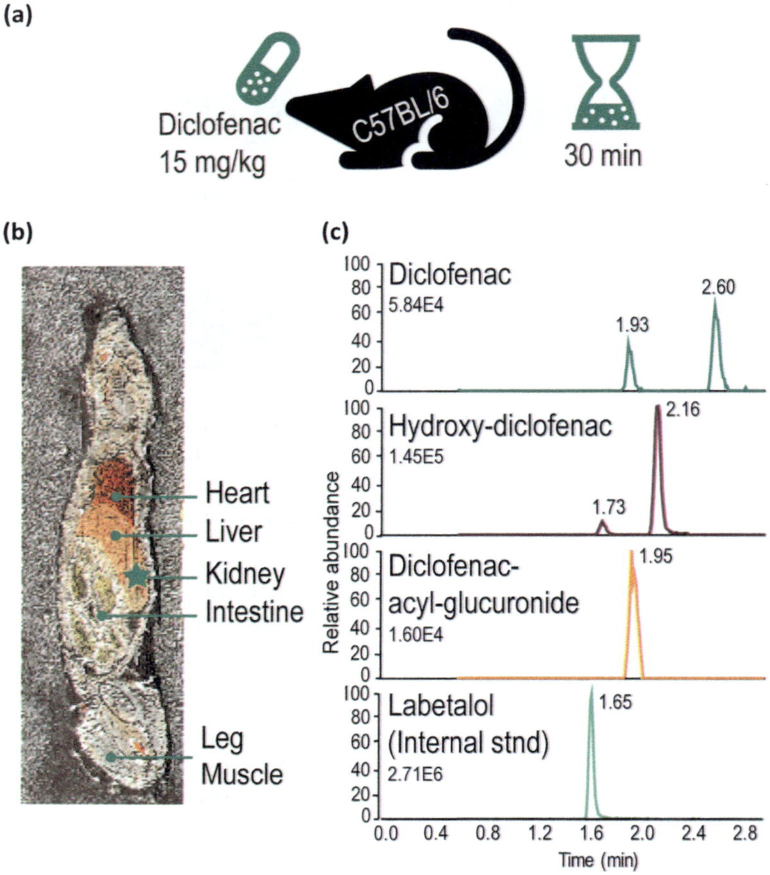

Fig. 3 Whole-body analysis of diclofenac dosed mouse on dropletProbe-MS. (**a**) Experiment setup: C57BL/6 mouse was dosed with diclofenac at 15 mg/kg and sacrificed at 30 min. (**b**) Location of extraction spots (heart, liver, kidney, intestine, and leg muscle) on the whole-body mouse section. (**c**) Representative XICs of diclofenac (*m*/*z* 296.0245), hydroxy-diclofenac (*m*/*z* 312.0194) and diclofenac acyl glucuronide metabolite (*m*/*z* 472.0566) extracted from the kidney spot and internal standard labetalol (*m*/*z* 329.1865) spiked in extraction solvent. Intensity of each peak was labeled on the XIC. (Reprinted with permission from reference 13. Copyright 2020 American Chemical Society)

Scan 3: targeted MS/MS for hydroxydiclofenac by ion trap (*m*/*z* 312.01, IW 3 *m*/*z*, CID, NCE 30).

Scan 4: targeted MS/MS for diclofenac acyl glucuronide by ion trap (*m*/*z* 472.06, IW 3 *m*/*z*, CID, NCE 25).

The dosing scheme, extraction location, example extracted ion chromatograms (XICs) of diclofenac and metabolites are shown in Fig. 3. Two peaks were observed at the *m*/*z* value of diclofenac: the peak at 1.93 min was the in-source fragmentation of the acyl-glucuronide metabolite and the

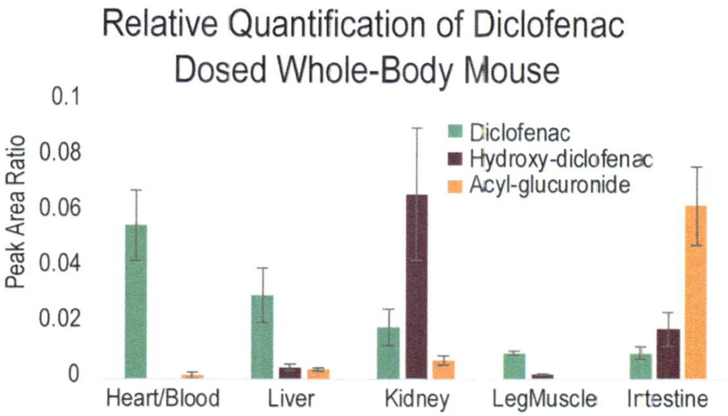

Fig. 4 Relative quantification graphs of parent drug and metabolites detected on diclofenac dosed whole-body mouse section by dropletProbe-MS. (Reprinted with permission from reference 13. Copyright 2020 American Chemical Society)

peak at 2.60 min was diclofenac. Also, two peaks were observed at the *m/z* value of hydroxy-diclofenac, which corresponded to 4′-hydroxy-diclofenac and 5-hydroxy-diclofenac.

4. Use vendor's data processing software for reporting peak areas of the different XICs.
5. Use Microsoft Excel for calculating peak area ratio and generating graphs. Resulting relative quantification of diclofenac and metabolites are included in Fig. 4 (*see* **Note 14**).

4 Notes

1. A regular tissue holder (3.5″ × 5″) is appropriate for analyzing multiwell plate, animal tissue sections and whole-body mouse section. A whole-body tissue holder (3.5″ × 9.5″) is needed for analyzing animals larger than a mouse.
2. Syringes with various needle endings are available. A taper-tip-end needle, such as 0F-BT-GT-5/0.63C (SGE Analytical Science), can be used to achieve smaller droplet size.
3. Prepare a whole-body section at a level where all organs/tissues of interest are presented in the section. Additional sections might be needed if it is impossible to align some organs together, such as kidney and spine. A few additional backup sections can be prepared in case something goes wrong in the experiment.
4. Prior to sectioning, place several transparent tissue tapes into the macrotome to chill the tape and avoid tissue section melting during collection.

5. In our previous publication [13], an extraction solvent with 50% ACN was determined to be an optimal solvent composition that works for extraction of many drugs. However, we recommend always optimizing the extraction solvent for the actual analyte of interest to improve extraction efficiency and sensitivity.
6. Spiking internal standard in the extraction solvent corrects for the sample loss during the sampling process, HPLC injection and ionization variability, but does not correct for the tissue extraction process. To correct for the extraction process, consider spraying internal standard homogeneously on the sample surface.
7. A dropletProbe sampling method has four key parameters: number of extraction cycles, amount of extraction solvent, needle-to-surface distance and sampling time. Please *see* reference 14 for detailed information. Briefly, changing these parameters can affect the formation of liquid microjunction and sensitivity.
   (a) **Liquid microjunction**: Forming an optimal liquid microjunction (shown in Fig. 5) is essential for the surface extraction process, as dropletProbe-MS relies on liquid extraction of analytes from the tissue surface. Amount of extraction solvent and needle-to-surface distance are affecting the formation of liquid microjunction: using too little extraction solvent or setting the needle to surface distance too far can fail to form a liquid microjunction. Using too much extraction solvent or setting the distance too short can result in a large droplet, which lower the spatial resolution [16].

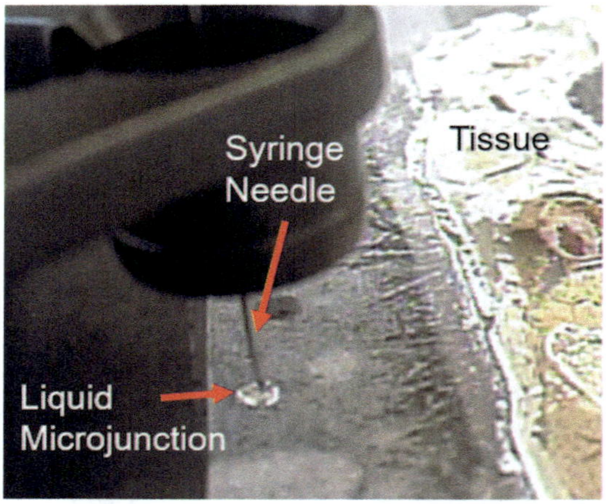

Fig. 5 Photo of the formed liquid microjunction on sample tape

(b) **Sensitivity**: increasing the amount of extraction solvent, sampling time and extraction cycles tend to increase the sensitivity, as more analytes are extracted. It is important to optimize the extraction method prior to performing the experiment.

8. Multiple areas can be selected and queued together for HPLC-MS run. Make sure to enter the correct HPLC-MS time on the dropletProbe-MS instrument control software to avoid dropletProbe-MS starting too early or too late between runs.
9. There is a built-in camera attached to the outside of the *Z*-axis of the autosampler. The user has the option to let the camera automatically taking a picture during each extraction cycle to monitor the process.
10. Droplet size can vary with different tissues. For example, a smaller droplet can be observed on brain due to the high lipid content, while a more spread droplet is observed in a hollow organ such as stomach. If spatial resolution is important in the experiment, please use the built-in camera to monitor the size of each liquid microjunction.
11. On the outside of the *Z*-axis of the autosampler there is a built-in laser displacement sensor for detecting uneven tissue surfaces and adjusting the sampling heights. Therefore, the dropletProbe system can be used to sample uneven surfaces, such as tissue biopsies or bacteria colonies.
12. In the setup shown in this study, the dropletProbe system triggers the HPLC and MS systems via contact closure. Make sure the contact closure is setup correctly.
13. The HPLC-MS method depends on the specific instrument platform connected to the dropletProbe system. Please refer to the manufacturer's manual for method setup. The method also needs to be customized to the analyte of interest.
14. The user can also use the dropletProbe Premium software to calculate, and to report peak and heatmap of selected ions.

Acknowledgments

The work of V.K. was supported by the US Department of Energy, Office of Science, Biological and Environmental Research, Bioimaging Science Program. The authors acknowledge Alain Creissen and Sloan Tilley (HTX Technology) and Matt Orcutt (Resolution Labs) for providing technical support of dropletProbe MS. We acknowledge Scott Fauty (Merck & Co., Inc., Kenilworth, NJ, USA) for performing animal experiments.

References

1. Solon EG (2015) Autoradiography techniques and quantification of drug distribution. Cell Tissue Res 360(1):87–107. https://doi.org/10.1007/s00441-014-2093-4
2. Solon EG, Schweitzer A, Stoeckli M, Prideaux B (2010) Autoradiography, MALDI-MS, and SIMS-MS imaging in pharmaceutical discovery and development. AAPS J 12(1):11–26. https://doi.org/10.1208/s12248-009-9158-4
3. Buchberger AR, DeLaney K, Johnson J, Li L (2018) Mass spectrometry imaging: a review of emerging advancements and future insights. Anal Chem 90(1):240–265. https://doi.org/10.1021/acs.analchem.7b04733
4. Duenas ME, Essner JJ, Lee YJ (2017) 3D MALDI mass spectrometry imaging of a single cell: spatial mapping of lipids in the embryonic development of zebrafish. Sci Rep 7(1):14946. https://doi.org/10.1038/s41598-017-14949-x
5. Norris JL, Caprioli RM (2013) Analysis of tissue specimens by matrix-assisted laser desorption/ionization imaging mass spectrometry in biological and clinical research. Chem Rev 113 (4):2309–2342. https://doi.org/10.1021/cr3004295
6. Aichler M, Walch A (2015) MALDI imaging mass spectrometry: current frontiers and perspectives in pathology research and practice. Lab Investig 95(4):422–431. https://doi.org/10.1038/labinvest.2014.156
7. Eikel D, Vavrek M, Smith S, Bason C, Yeh S, Korfmacher WA, Henion JD (2011) Liquid extraction surface analysis mass spectrometry (LESA-MS) as a novel profiling tool for drug distribution and metabolism analysis: the terfenadine example. Rapid Commun Mass Spectrom 25(23):3587–3596. https://doi.org/10.1002/rcm.5274
8. Wu Q, Huang Z, Wang Y, Zhang Z, Lu H (2020) Absolute quantitative imaging of sphingolipids in brain tissue by exhaustive liquid microjunction surface sampling-liquid chromatography-mass spectrometry. J Chromatogr A 1609:460436. https://doi.org/10.1016/j.chroma.2019.460436
9. Kertesz V, Van Berkel GJ (2010) Liquid microjunction surface sampling coupled with high-pressure liquid chromatography-electrospray ionization-mass spectrometry for analysis of drugs and metabolites in whole-body thin tissue sections. Anal Chem 82(14):5917–5921. https://doi.org/10.1021/ac100954p
10. Kertesz V, Cahill JF, Srijanto BR, Collier CP, Vavrek M, Chen B (2020) Absolute quantitation of propranolol from 200-mum regions of mouse brain and liver thin tissues using laser ablation-dropletProbe-mass spectrometry. Rapid Commun Mass Spectrom 35:e9010. https://doi.org/10.1002/rcm.9010
11. Pan N, Rao W, Kothapalli NR, Liu R, Burgett AW, Yang Z (2014) The single-probe: a miniaturized multifunctional device for single cell mass spectrometry analysis. Anal Chem 86 (19):9376–9380. https://doi.org/10.1021/ac5029038
12. Yin R, Burnum-Johnson KE, Sun X, Dey SK, Laskin J (2019) High spatial resolution imaging of biological tissues using nanospray desorption electrospray ionization mass spectrometry. Nat Protoc 14(12):3445–3470. https://doi.org/10.1038/s41596-019-0237-4
13. Chen B, Vavrek M, Cancilla MT (2020) Optimization of dropletProbe-mass spectrometry for whole-body tissue distribution analysis of drug-like molecules. J Am Soc Mass Spectrom 31(12):2558–2562. https://doi.org/10.1021/jasms.0c00168
14. Chen B, Vavrek M, Gundersdorf R, Zhong W, Cancilla MT (2020) Combining MALDI mass spectrometry imaging and droplet-base surface sampling analysis for tissue distribution, metabolite profiling, and relative quantification of cyclic peptide melanotan II. Anal Chim Acta 1125:279–287. https://doi.org/10.1016/j.aca.2020.05.050
15. Kertesz V, Weiskittel TM, Van Berkel GJ (2015) An enhanced droplet-based liquid microjunction surface sampling system coupled with HPLC-ESI-MS/MS for spatially resolved analysis. Anal Bioanal Chem 407 (8):2117–2125. https://doi.org/10.1007/s00216-014-8287-5
16. Kertesz V, Van Berkel GJ (2014) Sampling reliability, spatial resolution, spatial precision, and extraction efficiency in droplet-based liquid microjunction surface sampling. Rapid Commun Mass Spectrom 28(13):1553–1560. https://doi.org/10.1002/rcm.6931

Chapter 13

Quantitative Mass Spectrometry Imaging with Liquid Microjunction Surface Sampling

Qian Wu and Hongmei Lu

Abstract

Mass spectrometry imaging (MSI) could provide chemical spatial distribution within a diverse range of samples, but absolute quantitation with those techniques is still challenging. Recent years, ambient liquid extraction–based MSI techniques, such as liquid microjunction surface sampling (LMJSS), have been largely developed and were found to be favorable to quantitation by directly doping standards in the extraction solvent. Here, we describe the detailed experimental protocols and the data processing methods for quantitative MSI with LMJSS. The new methods could have absolute quantitative MSI of both endogenous lipids and small metabolites from tissue samples.

Key words Quantitative mass spectrometry imaging, Liquid microjunction surface sampling, Lipidomics, Metabolomics, Calibration methods

1 Introduction

Mass spectrometry imaging (MSI) [1] is a sensitive, in situ, and multiplexed approach for the characterization and localization of molecules within a diverse range of samples. In MSI, the spatially resolved ionization/sampling techniques could be classified into vacuum-based ionization methods [2], such as secondary ion MSI [3, 4] and matrix-assisted laser desorption/ionization (MALDI) MSI [1], and ambient ionization methods [5], such as desorption electrospray ionization (DESI) MSI [6], direct liquid extraction-based MSI [7] and infrared matrix assisted laser desorption electrospray ionization (IR-MALDESI) MSI [8]. They have been successfully applied to the detection and localization of small metabolites [9], drugs [10], lipids [11], peptides, and proteins [1] within a diverse range of samples.

Until now, MSI techniques could typically provide spatial distribution and relative abundance of analytes in a sample, but it is still challenging for them to provide accurate and robust absolute

Young-Jin Lee (ed.), *Mass Spectrometry Imaging of Small Molecules: Methods and Protocols*, Methods in Molecular Biology, vol. 2437, https://doi.org/10.1007/978-1-0716-2030-4_13,
© The Author(s), under exclusive license to Springer Science+Business Media, LLC, part of Springer Nature 2022

quantitation results of the analytes in samples [12]. The challenges are mainly attributed to the dependence of the analyte ionization efficiencies on the tissue's physicochemical and morphological properties [13–15], the unknown and locally varied sampling/ extraction efficiencies [13, 14] or matrix cocrystallization states of analytes [16], and the poor reproducibility and robustness of those surface ionization/sampling methods [17, 18]. However, accurate concentrations of analytes other than relative signal intensities in the imaging results are the fundamental for an accurate spatial distribution, which can give additional insight on the degree of turnover or metabolism of target molecules [19], the quantitative pharmacological and toxicological action of a drug [20], or the pharmacokinetic-pharmacodynamic mechanisms [21]. Thus, in the past decade, different quantitative MSI approaches have been proposed to obtain the quantitation results for MSI, which is named quantitative MSI (Q-MSI) [22–26].

Recent years, ambient liquid extraction techniques [7], such as liquid microjunction surface sampling (LMJSS, also called flow-probe) [27], liquid extraction surface analysis (LESA) [28, 29], nanospray desorption electrospray ionization (nanoDESI) [11, 30], and single-probe [31] coupled with MS have been found favoring quantitation[30–31, 33], because internal standards could be easily doped into the liquid extraction solvent for signal calibration. Additionally, the liquid extracts from surface sampling have potential to be transferred for post-sampling liquid chromatography (LC) separation to distinguish isobaric compounds before MS analysis and further improve the repeatability and quantitation ability of the method. Recent years, nano-DESI and single-probe which have the highest spatial resolution (8.5–150 μm) [7] have been used for the quantitation of phosphatidylcholines (PCs) [30] and prostaglandins [11] by doping the extraction solvent with internal standards. However, the two techniques were seldom coupled with an LC system. LESA has been coupled with LC – MS for quantitative spatial resolved MS analysis, but the low spatial resolution (1–3 mm) hinders it to be widely used in MSI [32]. In LMJSS, coaxial capillaries were used to realize continuous flow liquid junction on the surface [33], and the method provided better spatial resolution than LESA (300 μm as reported) [33–35]. Thus, in our previous studies [27, 36], continuous flow LMJSS was coupled with LC-MS to realize quantitative MSI of both lipids and amino acids. Besides coupling LMJSS with LC-MS, we also developed new calibration methods for LMJSS to have in situ MS quantitation on packaging papers without LC separation [37]. In this chapter, we will describe a detailed procedure of quantitative MSI with LMJSS according to our previous studies.

2 Materials

Prepare all solutions using ultrapure water (prepared by purifying deionized water, to attain a resistivity of 18 MΩ-cm at 25 °C) and analytical grade reagents. Prepare and store all reagents at room temperature (unless indicated otherwise). Diligently follow all waste disposal regulations when disposing waste materials. We do not recommend to add sodium azide to reagents.

2.1 LMJSS Extraction Solvents

1. Amino acid standard stock solutions: dissolve the following isotopic amino acid standards into methanol (MeOH) solvent as a mixed standard solution at 1 mg/mL for each amino acid, including 1-^{13}C-L-tryptophan, 1-^{13}C-glycine, 1-^{13}C-L-phenylalanine-1-^{13}C, 1-^{13}C-L-asparagine, 1-^{13}C-L-histidine, 1-^{13}C-L-tyrosine, 1-^{13}C-L-glutamine, 1-^{13}C-L-proline, 1-^{13}C-L-leucine, 1-^{13}C-L-alanine, 1-^{13}C-L-arginine, 1-^{13}C-L-aspartic acid, 1-^{13}C-L-serine, 1-^{13}C-L-lysine, 1-^{13}C-L-threonine, 1-^{13}C-L-glutamic acid, 1-^{13}C-L-isoleucine, 1-^{13}C-L-valine, 1-^{13}C-γ-aminobutyric acid, 1-^{13}C-L-norvaline (*see* **Note 1**).
2. Lipid standard solution: dissolve glucosylceramide (Glc/Cer) (d18:2/16:0) and phosphatidylcholine (PC) (21:0/22:6) in MeOH–chloroform (v/v 1:2) at 1 mg/mL for each standard separately (*see* **Note 1**).
3. LMJSS extraction solvent for lipids: MeOH–isopropanol (IP)–1,1,1,3,3,3-hexafluoroisopropanol (HFIP) (v/v 8:1:1) or MeOH–ACN (v/v 1:1) added with PC (21:0/22:6) at 1 μg/mL, sphingomyelin (SM) (d18:0/20:2) at 1 μg/mL, and GlcCer (d18:2/16:0) at 10 μg/mL (*see* **Note 2**).
4. LMJSS extraction solvent for amino acids: HFIP–MeOH–water (v/v 1:4:5) with 0.5% acetic acid and mixed amino acid standards mentioned in **step 1** diluted to 10 μg/mL for each amino acid (*see* **Note 2**).

2.2 Flow-Probe

1. Capillaries: The outer capillary (fused silica capillary, 250 μm I.D./350 μm O.D.) with 2–5 cm in length and the inner capillary (fused silica capillary, 100 μm I.D./160 μm O.D.) with 10–15 cm in length (*see* **Note 3**).
2. PEEK tee adapting to 1/16 in. O.D. tubes.

2.3 LC-MS System

1. UPLC-ESI-MS instrument: A UPLC system coupled with an ESI mass spectrometer (Shimadzu, Tokyo, Japan).
2. UPLC-ESI-MS system controlling software.
3. UPLC column: an UPLC C18 (2.1 × 50 mm, 1.7 μm) column.
4. The mobile phases: 0.1% formic acid in water (phase A) and 0.1% formic acid in acetonitrile (phase B).

3 Methods

Carry out all procedures at room temperature, unless otherwise specified.

3.1 Assembly of LMJSS

The LMJSS system is shown in Fig. 1.

1. Assemble a flow-probe using two coaxial capillaries. Connect the outer capillary to one end of a PEEK tee with sleeve (1/16 in. O.D. and 380 μm I.D.). Have the inner capillary pass through the tee, and hold the inner capillary in place on the other side of the tee with sleeve (1/16 in. O.D. and 180 μm O.D.). This arrangement provides a path for a continuous flow (*see* **Note 4**).
2. Connect a syringe pump with the third port of the PEEK tee by a capillary (*see* **Note 5**).
3. Connect the inner capillary of the flow-probe to port 1 (Fig. 1b) of a six port-two position valve (*see* **Note 6**).
4. As shown in Fig. 1b, connect the port 2 and port 5 of the valve with an injection loop (11 μL) (*see* **Note 7**).
5. Connect ports 3 and 4 of the valve (Fig. 1c) respectively to a UPLC pump and column or directly to MS inlet.

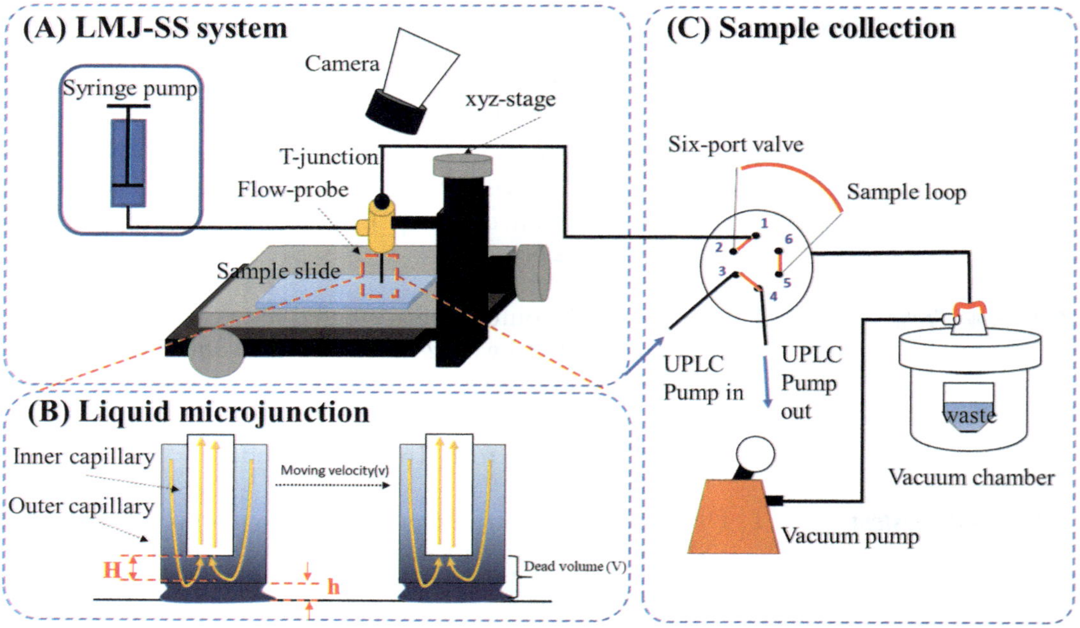

Fig. 1 Schematic illustration of LMJSS system (**a**), structure of flow-probe (**b**) and sample collection system (**c**). *h* indicates probe–surface distance. (Reproduced from Journal of Chromatography A, 1609 (2020) 460436 with permission from Elsevier (Ref. 27))

6. Connect port 6 of the valve (Fig. 1b) with a vacuum chamber by a capillary, and connect the vacuum chamber with a digital vacuum controller, and a diaphragm vacuum pump (*see* **Note 8**).
7. Use a microscopic camera to monitor the stability of LMJ.
8. Attach the flow-probe on a *z* linear stage, and fix the sample slide on *xy* linear stages. Use a moving controller to control all these stages (*see* **Note 9**).

3.2 Mass Spectrometry Imaging with LMJSS

1. Deliver the extraction solvent doped with internal standard via a syringe pump through the flow-probe, and aspirate the solvent into the inner capillary of the probe with a vacuum supplied by diaphragm vacuum pump (*see* **Note 10**).
2. Position the flow-probe vertically upon the tissue surface with a probe–surface distance (*h* in Fig. 1c) of 50 μm, forming liquid microjunction (LMJ) between the probe and surface (*see* **Note 11**).
3. During LMJSS, switch the valve to loading position to have the extraction solvent from the flow-probe flowing through the injection loop and then to the waste.
4. After LMJSS in one pixel, switch the valve to injection position to have solvent flowing directly to the waste, and to have the extracts in the loop injected into the UPLC-MS system or directly MS by the UPLC pump.
5. For imaging the tissue samples, use flow-probe to scan the tissue spot by spot with certain parking time in each spot (2 min for complete extraction and <2 min for incomplete extraction, *see* **Note 12**) and certain step length between spots (*see* **Note 13**) by moving the sample stage along the *xy* axis. Fix the probe–surface distance to 50 μm throughout the spot-by-spot imaging.
6. After extraction of each spot, switch the six port valve to inject the extract collected in the loop into the UPLC system or directly to MS system, and to have the probe stay in that spot until the UPLC–MS analysis (11 min) was completed or MS analysis (5 s) was completed. Then, switch the valve back to make the probe ready to move to the next spot.
7. When the extraction efficiency is not 100% in each spot, use the flow-probe to scan the tissue for the second time spot by spot with the same conditions indicated above (*see* **Note 14**).

3.3 Analytical Conditions for MS or UPLC-MS

1. UPLC for lipids (*see* **Note 15**): Perform UPLC separation with a 3-min gradient from 5% to 100% B, followed by a 9 min isocratic elution with 100% B from 3.0 to 12.0 min at a flow rate of 0.4 mL/min.

2. UPLC for amino acids (*see* **Note 15**): Perform UPLC separation a 9-min gradient from 5% to 10% B, followed by a 6 min gradient from 10% to 25% B at a flow rate of 0.4 mL/min.
3. Operate the mass spectrometer in the full scan mode over an *m/z* range of 50–1000 in positive ion mode. Set capillary voltage at 4500 V. Set nebulizing gas (N_2) at a 1.5 L/min flow rate. Set collision energy at 50%–100% (*see* **Note 16**).
4. Other parameters for MS detection are shown as follows: drying gas (N_2) of ESI at a 100 KPa pressure; detector voltage at 1.59 kV; RP area vacuum at 88.9 Pa; IT area vacuum at 1.8×10^{-2} Pa; TOF area vacuum at 1.3×10^{-4} Pa; The heat block and CDL temperatures at 200 °C; Accumulation time of MS1 at 60 ms; The ion accumulation time at 30 ms for MS2; Precursor ion isolation width as 1 Da (*see* **Note 16**).

3.4 Quantitation Method for LMJSS with Complete Extraction

1. Import the imaging dataset exported from the instrument software (format *.mzdata) into the Mzmine2.0 software (Okinawa, Japan, http://mzmine.github.io/).
2. Perform peak detection, isotopic peak grouping, peak integration, and feature alignment with Mzmine2.0 software. The detailed parameter settings for peak finding, isotopic peak grouping, peak integration, and features alignment with the software Mzmine2.0 are as follows:
   (a) For peak finding, minimum height, 1000; *m/z* tolerance, 0.02000 *m/z*; min-max width time (min), 0.10–1.50; smoothing time (min), 0.500 min; smoothing *m/z*, 0.050.
   (b) For isotopic peak grouping, *m/z* tolerance and retention time tolerance were 0.005 *m/z* and 10.0% relative, respectively.
   (c) For feature alignment, *m/z* tolerance and retention time tolerance were *m/z* 0.05 and 0.05 min, respectively. After that, each peak found by software was manually checked to avoid low signal-to-noise ratio peaks (S/N < 10), error peak finding and blank peaks.
3. Calculate concentrations of targeted analytes in extracts (C_a) by internal standards addition calibration spot-by-spot. In details, calibrate all the amino acids, PCs, cerebrosides (CBs) and SMs by the signal of the isotopic standards of amino acids, PC (21:0/22:6), GlcCer(d18:2/16:0), and SM(d18:0/20:2) added in the extracts respectively as (*see* **Note 17**):

$$C_a = \frac{S_a}{S_s} \times C_s \quad (1)$$

where C_a is the analyte concentration in LMJSS extracts, S_a is the signal (peak area) of targeted analytes in extracts, S_s is the signal of internal standard added in the extracts, and C_s is the concentration of internal standard added in the extracts. C_s/S_s is the calibration coefficient.

4. Calculate the concentrations of all the targeted analytes in the brain tissue using following Eq. (2):

$$C_t = \frac{C_e \times V_e}{A_e \times T_t} \tag{2}$$

where C_t is the concentration of targeted analytes in the specific location of tissue, C_e is the detected analyte concentration in LMJSS extracts, V_e is the volume of the extracts (11 μL), A_e is the sampling area (the spot area observed in microscope on the tissue after LMJ extraction, 400 μm diameter), and T_t is the thickness of the tissue slices (20 μm).

3.5 Quantitation Method for LMJSS with Incomplete Extraction

1. Obtain the peak areas of analytes by LMJSS-MS or LMJSS-LC-MS via software LCMS solution-Browser (Shimadzu, Tokyo, Japan), and export the data as excel files.
2. For quantitation, scan the tissue sample for two times at the same area. Calculate the concentration of analytes (mg/L) in extract from LMJSS in each spot by internal standard calibration. In details, calibrate all the amino acids, PCs, CBs, and SMs by the signal of the isotopic standards of amino acids, PC (21:0/22:6), GlcCer(d18:2/16:0), and SM (d18:0/20:2) added in the extracts respectively with Eq. (1) (*see* **Note 17**).
3. Calculate the extraction amount of analyte (mg) in extract in each spot of brain tissue (M) using following equation:

$$M = C_a \times V_e \tag{3}$$

where C_a is the analyte concentration in LMJSS extracts and V_e is the volume of the extracts (10 μL).

4. Calculate the concentrations of analytes originally in the tissue (q_0, mg/m^2) as follows:

$$q_0 = \frac{M_1^2}{(M_1 - M_2) \times S \times T_t} \tag{4}$$

where M_1 is the extraction amount of analyte (mg) in extract for the first extraction of the spot, M_2 the extraction amount of analyte (mg) for the second extraction of the spot, and S is the spot area (m^2), and T_t is the thickness of the tissue slices (20 μm).

3.6 Reconstruction of Heatmaps

1. Make an array for the calculated concentrations of analytes in all the spots as a matrix according to their locations on the samples.
2. Input the data matrix into OriginPro 8.5 (OriginLab Corporation, Northampton, MA, USA).

3. Draw the heatmaps showing quantitative images of analytes on the samples by using the function of "grey scale map".
4. Adjust the color gradient and the color levels of the map to have a good contrast for the heatmap. The heatmaps showing quantitative spatial distribution of lipids and amino acids in brain tissue and comparison of the results of MSI with that of solvent extraction–LC-MS are shown in Figs. 2 and 3.

4 Notes

1. Store the mixed standard solution in a −20 °C or −80 °C freezer for following experiments.
2. Prepare the extraction solvent, especially HFIP containing solvents, in fume hood to avoid inhalation toxicity of those solvents.
3. If the end of the capillary is not flat enough after the first cutting, use a ceramic chip to gently polish the cross section.
4. The distance between the inner and outer capillary end would have effect on the concentration of the analytes extracted in the extracts and also the stability of the liquid microjunction. In previous studies, 50–100 μm is the optimal distance (H) between the end of inner and outer capillary. Also, burn off the polyimide coating of the inner capillary before use, in case that the coating would cause adsorption of analytes in the probe.
5. Use gastight glass syringe with Teflon plunger to avoid the dissolution of interferents from the syringe into the solvent. There is no requirement on the I.D., O.D. and length of the connecting capillary between syringe pump and PEEK tee, but too large I.D. and length would lead to high dead volume for solvent delivery.
6. Connect inner capillary to the valve tightly to ensure gas tightness and a high and stable vacuum provided by the pump.
7. A capillary (250 μm I.D./350 μm O.D) with 22.4-cm length is used in this method but any sample loop with the desired volume should be fine.
8. The vacuum chamber could be made of a vacuum desiccator with the silicone rubber septum sealing the chamber. Then, insert the capillary from port 6 of the valve into the septum to connect the sampling probe with the vacuum.
9. The spatial resolution of the linear stages would also affect imaging resolution and the precision of the quantitation. In our previous works, The spatial resolution of the linear stages was <2 μm, and the stages could be moved automatically in preset route by programing using the moving controller.

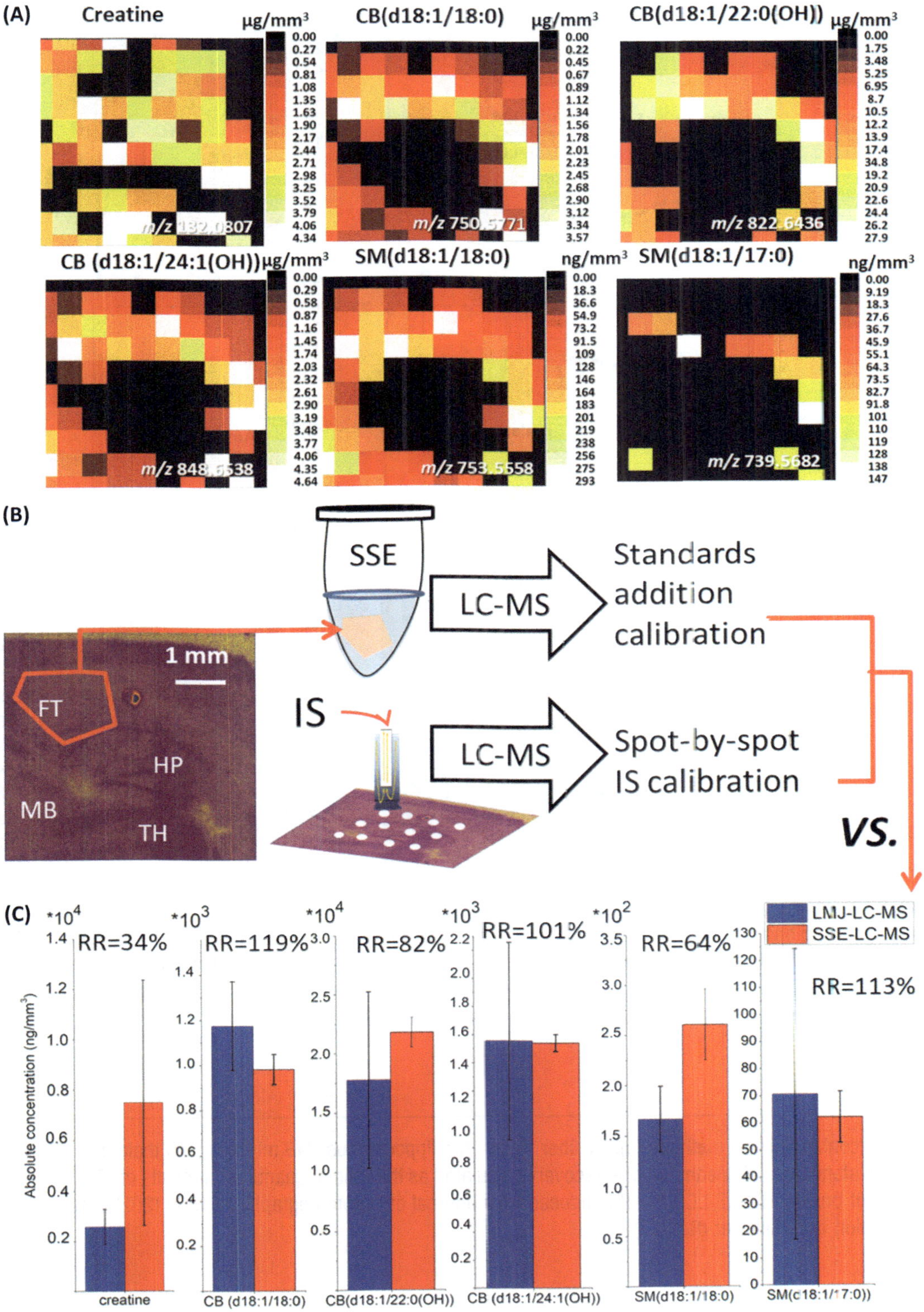

Fig. 2 Quantitative ion images of creatine and sphingolipids with LMJSS-LC-MSI (**a**), schematic illustration of quantitation approaches of LMJSS-LC-MSI and SSE-LC-MS (**b**) and quantitation results comparison of these

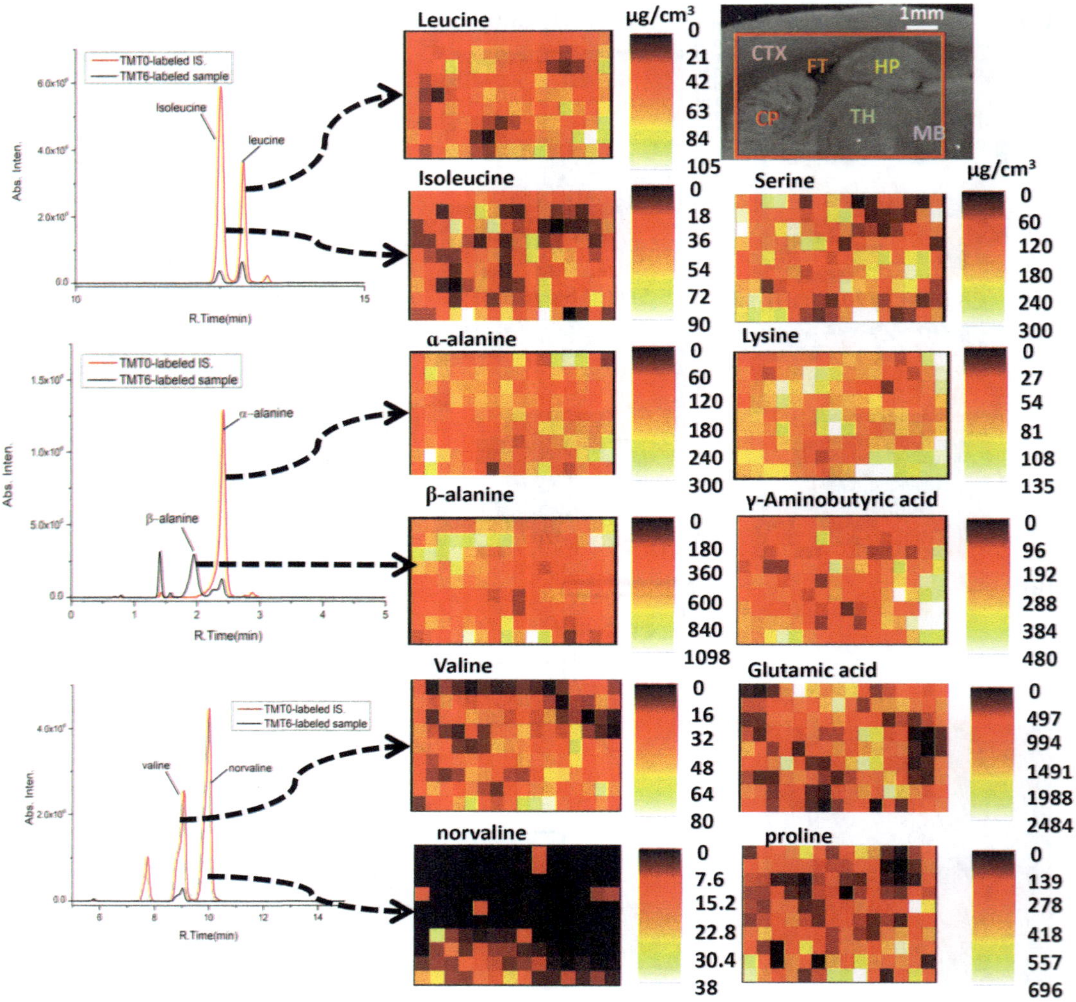

Fig. 3 The chromatograms of three pairs of amino acid isomers distinguished with LC separation, and the heat maps showing absolute quantitative spatial distributions of 11 amino acids in brain tissue with LMJSS-TMT-LC-MS method. CTX indicates cerebral cortex, FT indicates fiber tracts, CP indicates caudoputamen, TH indicates thalamus, MB indicates midbrain; IS indicates internal standard. (Reproduced from Journal of Chromatography A 1621 (2020) 461086 with permission from Elsevier (Ref. 36))

Fig. 2 (continued) two methods (**c**). *FT* fiber tracts, *HP* hippocampus, *MB* midbrain, *TH* thalamus, *IS* means internal standards, *RR* means relative recovery calculated as the ratio of quantitation results of LMJSS-LC-MS to that of traditional SSE-LC-MS. (Reproduced from Journal of Chromatography A, 1609 (2020) 460436 with permission from Elsevier (Ref. 27))

10. For different extraction solvent compositions and given solvent delivering flowrates, the vacuum of the pump should be adjusted to maintain the stability of the liquid microjunction. The solvent flowrate has effect on the concentration of analytes extracted in the solvent. From our previous studies, the analyte concentration in the extracts would be slightly increased with reducing the flow rate. However, from our experiments, if the flowrate is too low, it is difficult to have stable liquid microjunction. It may be due to the reduced vacuum precision of the pump when the supplied vacuum is low.
11. The tip of the flow-probe should be flat enough, and the cross section of the tip should be parallel to the sample surface. Otherwise, the stability of the liquid microjunction would be affected.
12. For incomplete extraction of tissue samples by LMJSS, the extraction time should be >10 s if the kinetic calibration method is used for quantitation, because the experimental kinetic extraction curves in tissues before 10 s could not be fitted to the kinetic model.
13. Set the step length according to the diameter of flow-probe tip, and usually the step length is larger than the O.D. of the tip to minimize the effects of oversampling on image accuracy. In published studies, 500 μm was used for flow-probe with 400-μm in O.D.
14. Since the moving route of the probe is programmed by the computer, for the second scanning of the tissue, the probe could be moved back in the same route without manually finding the starting spot of the first scanning.
15. The separation conditions for lipids and amino acids could be adjusted according to the properties of specific analytes and specific samples, and the conditions indicated here are optimized to have high coverage on all the lipids and amino acids in brain tissues.
16. The MS parameters could be adjusted according to the different instruments and the different properties of analytes, and the parameters indicated here are all for the specific instrument ESI-Ion Trap (IT)-TOF mass spectrometer (Shimadzu, Tokyo, Japan) mentioned in Subheading 2.4.
17. If isotopic standards of some analytes are difficult to be obtained, external standard calibration could also be used. Briefly, prepare standards in corresponding extraction solvent at different concentrations (0.05, 0.1, 0.5, 2, 5 μg/mL). Load the standards in the valve in Fig. 1, and inject them into LC-MS or MS system with conditions described in Subheading 3.2. With the detected signals of standards in different concentrations, the external calibration curves ($y = ax + b$) would be obtained.

For calibration of the analyte signal in extracts from real sample, calculate the concentrations of analytes in the extracts as:

$$C_e = \frac{S_e - b}{a}$$

where C_e is the concentrations of the analytes extracted in solvent and S_e is the signal intensities of the analytes detected in the solvent.

Acknowledgments

The work was supported by the National Natural Science Foundation of China (Nos. 21804142, 22074159, 21873116, and 21675174).

References

1. Caprioli RM, Farmer TB, Gile J (1997) Molecular imaging of biological samples: localization of peptides and proteins using MALDI-TOF MS. Anal Chem 69(23):4751–4760. https://doi.org/10.1021/ac970888i
2. Schwartz SA, Caprioli RM (2010) Imaging mass spectrometry: viewing the future. In: Rubakhin SS, Sweedler JV (eds) Mass spectrometry imaging: principles and protocols. Methods in molecular biology. Humana Press, Totowa, New Jersey, pp 3–19
3. Benninghoven A, Sichtermann WK (1978) Detection, identification and structural investigation of biologically important compounds by secondary ion mass spectrometry. Anal Chem 50(8):1180–1184. https://doi.org/10.1021/ac50030a043
4. Burns MS (1982) Applications of secondary ion mass spectrometry (SIMS) in biological research: a review. J Microsc 127 (Pt 3):237–258. https://doi.org/10.1111/j.1365-2818.1982.tb00419.x
5. Wu C, Dill AL, Eberlin LS, Cooks RG, Ifa DR (2013) Mass spectrometry imaging under ambient conditions. Mass Spectrom Rev 32 (3):218–243. https://doi.org/10.1002/mas.21360
6. Takats Z, Wiseman JM, Gologan B, Cooks RG (2004) Mass spectrometry sampling under ambient conditions with desorption electrospray ionization. Science 306 (5695):471–473. https://doi.org/10.1126/science.1104404
7. Laskin J, Lanekoff I (2016) Ambient mass spectrometry imaging using direct liquid extraction techniques. Anal Chem 88 (1):52–73. https://doi.org/10.1021/acs.analchem.5b04188
8. Robichaud G, Barry JA, Garrard KP, Muddiman DC (2013) Infrared matrix-assisted laser desorption electrospray ionization (IR-MALDESI) imaging source coupled to a FT-ICR mass spectrometer. J Am Soc Mass Spectrom 24(1):92–100. https://doi.org/10.1007/s13361-012-0505-9
9. Goto S, Morikawa T, Kubo A, Takubo K, Fukuda K, Kajimura M et al (2018) Quantitative imaging mass spectroscopy reveals roles of heme oxygenase-2 for protecting against transhemispheric diaschisis in the brain ischemia. J Clin Biochem Nutr 63(1):70–79. https://doi.org/10.3164/jcbn.17-136
10. Bokhart MT, Rosen E, Thompson C, Sykes C, Kashuba ADM, Muddiman DC (2015) Quantitative mass spectrometry imaging of emtricitabine in cervical tissue model using infrared matrix-assisted laser desorption electrospray ionization. Anal Bioanal Chem 407 (8):2073–2084. https://doi.org/10.1007/s00216-014-8220-y
11. Lanekoff I, Thomas M, Laskin J (2014) Shotgun approach for quantitative imaging of phospholipids using Nanospray desorption electrospray ionization mass spectrometry. Anal Chem 86(3):1872–1880. https://doi.org/10.1021/ac403931r
12. Buchberger AR, DeLaney K, Johnson J, Li L (2018) Mass spectrometry imaging: a review of emerging advancements and future insights.

Anal Chem 90(1):240–265. https://doi.org/10.1021/acs.analchem.7b04733

13. Tomlinson L, Fuchser J, Fuetterer A, Baumert M, Hassall DG, West A et al (2014) Using a single, high mass resolution mass spectrometry platform to investigate ion suppression effects observed during tissue imaging. Rapid Commun Mass Spectrom 28 (9):995–1003. https://doi.org/10.1002/rcm.6869
14. Taylor AJ, Dexter A, Bunch J (2018) Exploring ion suppression in mass spectrometry imaging of a heterogeneous tissue. Anal Chem 90 (9):5637–5645. https://doi.org/10.1021/acs.analchem.7b05005
15. Hansen HT, Janfelt C (2016) Aspects of quantitation in mass spectrometry imaging investigated on Cryo-sections of spiked tissue homogenates. Anal Chem 88 (23):11513–11520. https://doi.org/10.1021/acs.analchem.6b02711
16. Park S, Yoon S, Min H, Moon SM, Choi YJ, Kim IS et al (2020) Compartmentalized arrays of matrix droplets for quantitative mass spectrometry imaging of adsorbed peptides. Anal Chem 92(13):8715–8721. https://doi.org/10.1021/acs.analchem.9b05316
17. Abbassi-Ghadi N, Jones EA, Veselkov KA, Huang J, Kumar S, Strittmatter N et al (2015) Repeatability and reproducibility of desorption electrospray ionization-mass spectrometry (DESI-MS) for the imaging analysis of human cancer tissue: a gateway for clinical applications. Anal Methods 7(1):71–80. https://doi.org/10.1039/c4ay01770f
18. Morikawa-Ichinose T, Fujimura Y, Murayama F, Yamazaki Y, Yamamoto T, Wariishi H et al (2019) Improvement of sensitivity and reproducibility for imaging of endogenous metabolites by matrix-assisted laser desorption/ionization-mass spectrometry. J Am Soc Mass Spectrom 30(8):1512–1520. https://doi.org/10.1007/s13361-019-02221-7
19. Nazari M, Bokhart MT, Loziuk PL, Muddiman DC (2018) Quantitative mass spectrometry imaging of glutathione in healthy and cancerous hen ovarian tissue sections by infrared matrix-assisted laser desorption electrospray ionization (IR-MALDESI). Analyst 143 (3):654–661. https://doi.org/10.1039/c7an01828b
20. Tang W, Chen J, Zhou J, Ge J, Zhang Y, Li P et al (2019) Quantitative MALDI imaging of spatial distributions and dynamic changes of Tetrandrine in multiple organs of rats. Theranostics 9(4):932–944. https://doi.org/10.7150/thno.30408
21. Abu Sammour D, Marsching C, Geisel A, Erich K, Schulz S, Guevara CR et al (2019) Quantitative mass spectrometry imaging reveals mutation status-independent lack of Imatinib in liver metastases of gastrointestinal stromal tumors. Sci Rep 9:10698. https://doi.org/10.1038/s41598-019-47089-5
22. Ellis SR, Bruinen AL, Heeren RMA (2014) A critical evaluation of the current state-of-the-art in quantitative imaging mass spectrometry. Anal Bioanal Chem 406(5):1275–1289. https://doi.org/10.1007/s00216-013-7478-9
23. Hamm G, Heron A, Legouffe R, Pamelard F, Bonnel D, Stauber J (2014) Contribution of quantitative mass spectrometry imaging (msi) in pharmaceutical field: application to drug and peptide analysis in tissue. Drug Metabolism Rev 45:35
24. Pirman D (2015) Quantitative profiling of tissue drug distribution by MS imaging. Bioanalysis 7(20):2649–2656. https://doi.org/10.4155/bio.15.179
25. Lanekoff I, Laskin J (2018) Quantitative mass spectrometry imaging of molecules in biological systems. In: Grushka E, Grinberg N (eds) Advances in chromatography, vol 54. CRC Press, Boca Raton, Florida, pp 43–72
26. Tobias F, Hummon AB (2020) Considerations for MALDI-based quantitative mass spectrometry imaging studies. J Proteome Res 19 (9):3620–3630. https://doi.org/10.1021/acs.jproteome.0c00443
27. Wu Q, Huang Z, Wang Y, Zhang Z, Lu H (2020) Absolute quantitative imaging of sphingolipids in brain tissue by exhaustive liquid microjunction surface sampling-liquid chromatography-mass spectrometry. J Chromatogr A 1609:460436. https://doi.org/10.1016/j.chroma.2019.460436
28. Chen W, Wang L, Van Berkel GJ, Kertesz V, Gan J (2016) Quantitation of repaglinide and metabolites in mouse whole-body thin tissue sections using droplet-based liquid microjunction surface sampling-high-performance liquid chromatography-electrospray ionization tandem mass spectrometry. J Chromatogr A 1439:137–143. https://doi.org/10.1016/j.chroma.2015.10.093
29. Kertesz V, Weiskittel TM, Vavrek M, Freddo C, Van Berkel GJ (2016) Extraction efficiency and implications for absolute quantitation of propranolol in mouse brain, liver and kidney tissue sections using droplet-based liquid microjunction surface sampling high-performance liquid chromatography/electrospray ionization tandem mass spectrometry. Rapid Commun Mass

Spectrom 30(14):1705–1712. https://doi.org/10.1002/rcm.7607

30. Duncan KD, Fang R, Yuan J, Chu RK, Dey SK, Burnum-Johnson KE et al (2018) Quantitative mass spectrometry imaging of prostaglandins as silver ion adducts with Nanospray desorption electrospray ionization. Anal Chem 90 (12):7246–7252. https://doi.org/10.1021/acs.analchem.8b00350
31. Pan N, Rao W, Kothapalli NR, Liu R, Burgett AW, Yang Z (2014) The single-probe: a miniaturized multifunctional device for single cell mass spectrometry analysis. Anal Chem 86 (19):9376–9380. https://doi.org/10.1021/ac5029038
32. Kertesz V, Van Berkel GJ (2014) Sampling reliability, spatial resolution, spatial precision, and extraction efficiency in droplet-based liquid microjunction surface sampling. Rapid Commun Mass Spectrom 28(13):1553–1560. https://doi.org/10.1002/rcm.6931
33. Griffiths RL, Randall EC, Race AM, Bunch J, Cooper HJ (2017) Raster-mode continuous-flow liquid microjunction mass spectrometry imaging of proteins in thin tissue sections. Anal Chem 89(11):5683–5687. https://doi.org/10.1021/acs.analchem.7b00977
34. Feider CL, Elizondo N, Eberlin LS (2016) Ambient ionization and FAIMS mass spectrometry for enhanced imaging of multiply charged molecular ions in biological tissues. Anal Chem 88(23):11533–11541. https://doi.org/10.1021/acs.analchem.6b02798
35. Gill EL, Marks M, Yost RA, Vedam-Mai V, Garrett TJ (2017) Monitoring dopamine ex vivo during electrical stimulation using liquid-microjunction surface sampling. Anal Chem 89(24):13658–13665. https://doi.org/10.1021/acs.analchem.7b04463
36. Wu Q, Li Y, Wang Y, Lu H (2020) Quantitative mass spectrometry imaging of amino acids with isomer differentiation in brain tissue via exhaustive liquid microjunction surface sampling-tandem mass tags labeling-ultra performance liquid chromatography-mass spectrometry. J Chromatogr A:1621. https://doi.org/10.1016/j.chroma.2020.461086
37. Mei Y, Wu Q, Zhou S, Wang Z, Liang Q, Li Y et al (2020) Rapid in situ quantitation of photoinitiators in packaging by two-points kinetic calibration of liquid microjunction surface sampling-mass spectrometry. Talanta 216:121017. https://doi.org/10.1016/j.talanta.2020.121017

Part IV

New Types of Applications

Chapter 14

Enhancing Metabolite Coverage for Matrix-Assisted Laser Desorption/Ionization Mass Spectrometry Imaging Through Multiple On-Tissue Chemical Derivatizations

Kelly C. O'Neill, Maria Emilia Dueñas, Evan Larson, Trevor T. Forsman, and Young-Jin Lee

Abstract

The ability to study and visualize metabolites on a cellular and sub-cellular level is important for gaining insights into biological pathways and metabolism of multicellular organisms. Matrix-assisted laser desorption/ionization (MALDI) mass spectrometry imaging (MSI) is a powerful analytical tool for metabolomics experiments due to its high sensitivity and small sampling size. The spatial resolution in MALDI-MSI is mainly limited by the number of molecules available in a small sampling size. When the sampling size is low enough to achieve cellular or subcellular spatial resolution, signal intensity is sacrificed making poorly ionized metabolites difficult to detect. To overcome this limitation, on-tissue chemical derivatization reactions have been used to enhance the desorption/ionization efficiency of selected classes of compounds by adding a functional group with a permanent positive charge or one that can be easily ionized. By utilizing several chemical derivatizations in parallel, metabolite coverage can be drastically improved. This chapter outlines methodology for sample preparation and data analysis for on-tissue chemical derivatization using various derivatization reagents.

Key words Metabolomics, On-tissue derivatization, Mass spectrometry imaging, Matrix-assister laser desorption/ionization, Metabolite coverage

1 Introduction

Mass spectrometry imaging (MSI) has become a widely used analytical tool for mapping metabolites and has led to a deeper biological understanding of various multicellular organisms. MSI has led to innovations in drug discovery and development [1, 2], disease research [3–5], biomarker discovery [6], plant biology [7], and forensic science [8]. Each of these fields involves highly complex organisms and biological systems; thus, metabolomics experi-

Young-Jin Lee (ed.), *Mass Spectrometry Imaging of Small Molecules: Methods and Protocols*,
Methods in Molecular Biology, vol. 2437, https://doi.org/10.1007/978-1-0716-2030-4_14,
© The Author(s), under exclusive license to Springer Science+Business Media LLC, part of Springer Nature 2022

ments should be able to determine differences on cellular and subcellular levels in order to get the most accurate information. Matrix-assisted laser desorption/ionization (MALDI)-MSI can achieve small sampling size, which allows for imaging at the cellular and subcellular level. Along with high chemical versatility and sensitivity, these traits have allowed MALDI-MSI to become increasingly popular for spatial mapping of metabolites across different biological systems.

Laser spot size is most critical in limiting the spatial resolution in MALDI-MSI. It has been reduced down to 5–10 μm by simply modifying optical systems in commercial instruments [9, 10]. Additionally, transmission mode geometry MALDI (t-MALDI) and t-MALDI-2 technology has recently been implemented to reduce the laser spot size below 1 μm [11, 12]. Our group has established a MALDI-MSI platform with 5–10 μm spatial resolution, and this has been applied to visualize metabolites in maize roots [13, 14], leaves [9, 14, 15], and seeds [16, 17]. Moreover, this platform has shown that metabolite distribution is sometimes heterogeneous even between the same type of cells [9, 18] and the sub-cellular localizations can have important biological implications [19].

Although MALDI-MSI provides high-spatial resolution information that is unprecedented in traditional metabolomics, there are some issues such as a reduction in analyte signal owing to small sampling size, a limited number of analytes due to the selectivity of the matrix, and lack of chromatographic separation. Chemical derivatization strategies can be used to overcome these limitations to enhance the ionization efficiency of metabolites. In a recent report by Dueñas et al., a combination of chemical derivatization reactions was adopted to selectively enhance metabolite signals of specific functional groups in maize. For instance, 4-hydroxy-3-methoxycinnamaldehyde (coniferyl aldehyde; CA) was used for the derivatization of primary amines [20, 21], Girard's reagent T (GT) for carbonyl groups [22–24], and 2-picolylamine (2-PA) for carboxylic acids [25]. The general reactions used for these modifications are shown in Scheme 1. Implementing chemical derivatization reactions such as these on serial tissue sections dramatically expands the metabolite coverage by enhancing the signal of certain classes of compounds at a time. This strategy has been demonstrated by our group and proven to be beneficial for greater metabolite coverage on maize roots and leaves [14].

In this chapter, we describe a sample preparation method for on-tissue chemical derivatization using CA, GT, and/or 2-PA as derivatization reagents. This strategy is applicable to any tissue; however, this chapter uses plant tissue as an example. Figure 1 shows a general workflow using maize roots as the example tissue. Briefly, maize roots are grown hydroponically in wetted paper

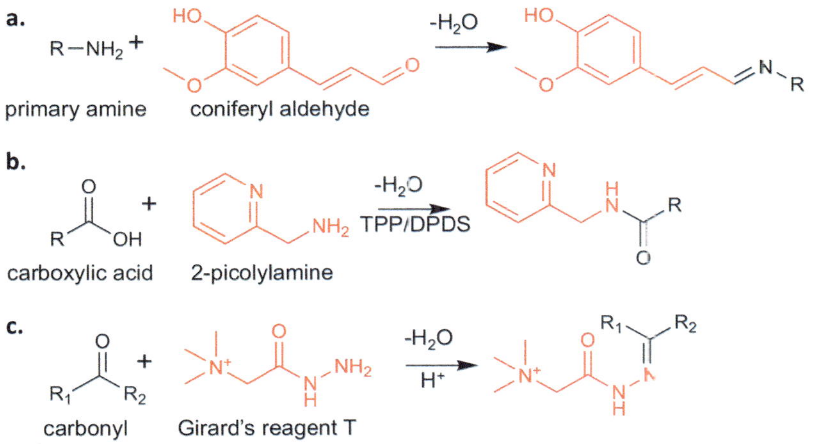

Scheme 1 Three on-tissue chemical derivatization reactions used in this method: (**a**) coniferyl aldehyde (CA) for a primary amine, (**b**) 2-picolylamine (2-PA) for a carboxylic acid, and (**c**) Girard's reagent T (GT) for a carbonyl group. TPP: triphenylphosphine. DPDS: 2,2-dipyridyldisulfide

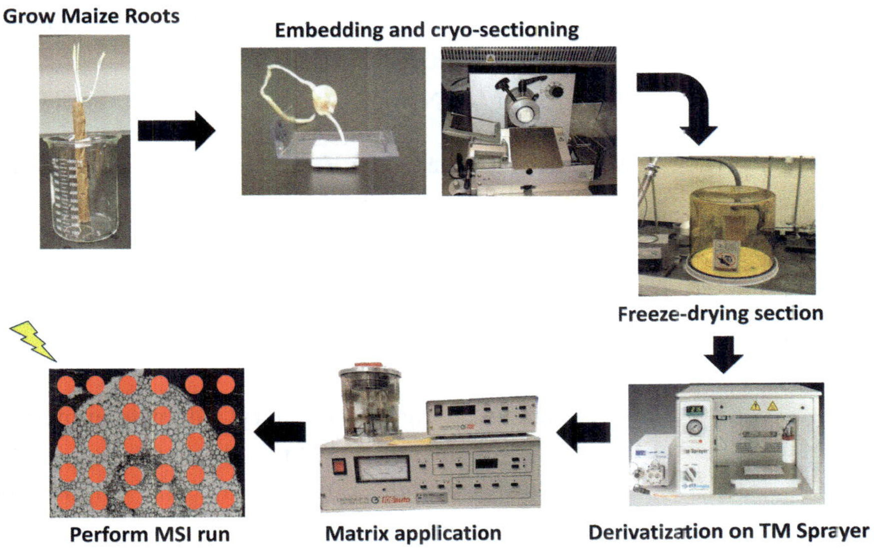

Fig. 1 Workflow for MSI experiment using maize roots as an example tissue sample

towels. Once the desired length is reached, the roots are harvested, embedded in gelatin, and flash-frozen in liquid nitrogen. Then the roots are cryosectioned to 10 μm thickness and vacuum-dried. Next, the derivatization agent and matrix are applied to the tissue sections prior to the MSI experiment. The end result is a dramatic increase in ionization efficiency for many analytes and visualization for compounds that was not previously possible.

2 Materials

2.1 Cryosectioning

1. Cryomicrotome: set at −20 °C.
2. Glass microscope slides and a microscope slide box (*see* **Note 1**).
3. Scotch tape.
4. Plant tissue (e.g., maize root tissue as shown in Fig. 1).
5. 300 bloom gelatin.
6. High-purity water.
7. Cryomold.
8. CryoJane® adhesive tape (Leica Biosystems, Buffalo Grove, IL, USA; *see* **Note 2**).
9. Liquid nitrogen.
10. Disposable pipettes.
11. 70% (v/v) ethanol solution.
12. Optimal Cutting Temperature (OCT) compound.
13. Styrofoam container with dry ice.
14. Roller tool (*see* **Note 3**).

2.2 Tissue Drying

1. Vacuum chamber with a vacuum pump capable of reaching pressures in the mTorr range (*see* **Note 4**).
2. Heat sink: aluminum block precooled to −80 °C.

2.3 Derivatization and Matrix Application Via TM Sprayer

1. TM Sprayer (HTX Technologies, LLC. Chapel Hill, NC, USA) (*see* **Note 5**).
2. Syringe pump.
3. CA reagent: 20 mg/mL coniferyl aldehyde in methanol.
4. 2-PA reagent: 6 mM 2-picolylamine, 30 mM 2,2-dipyridyldisulfide (DPDS), 30 mM triphenylphosphine (TPP) in acetonitrile.
5. Matrix solution: 40 mg/mL 2,5- dihydroxybenzoic acid (DHB) in 70% methanol.
6. 1 mL Luer lock glass syringes with fill needles.
7. Tank of compressed nitrogen.

2.4 Derivatization Application Via ESI Deposition

1. Power supply (up to 5 or 10 kV).
2. Syringe pump.
3. Fused silica capillary: 363 μm O.D, 150 μm I.D., 40 mm long.
4. Capillary sleeve: 0.025″ OD, 0.0155″ ID.
5. PEEK tubing: 1/16″ OD, 0.010″ ID, 3–5′ long.

6. Finger tight PEEK tubing fitting (×2): 1/16″ OD.
7. Luer lock syringe.
8. Luer lock adaptor for 1/16″ OD PEEK tubing.
9. Zero volume stainless steel connector for 1/16″ PEEK tubing.
10. Ring stand (or anything that can hold the ESI capillary sprayer).
11. GT reagent: 10 mg/mL Girard's Reagent T, 2% trifluoroacetic acid (TFA) in methanol.
12. 1 mL glass syringe with needles.

2.5 Matrix Application Via Sputter Coater

1. Sputter coater.
2. Methanol.
3. Tank of compressed argon (minimum 99.9% purity).
4. Vacuum pump.
5. Gold sputter targets.

3 Methods

3.1 Tissue Sectioning

This sub-section aims to explain how to prepare tissue samples prior to derivatization, matrix application, and MSI.

1. Prepare a 10% w/v solution of gelatin in high-purity water and heat at ~40 °C until the gelatin has completely dissolved and the solution is clear.
2. Using a disposable pipette, fill a cryomold with the gelatin solution. Place the tissue of interest in the cryomold, and immediately flash-freeze the sample, by placing the cryomold on the liquid nitrogen.
3. When the gelatin is ~80% opaque (*see* **Note 6**), place the mold in a −20 °C freezer and allow the sample to thermally equilibrate (wait at least 30 min, but 24 h is ideal for best results).
4. Transport the mold to the cryostat in a Styrofoam cooler filled with dry ice to prevent melting. Make sure the cryostat has been prechilled to −20 °C.
5. Remove the sample from the mold and affix it to the aluminum cryostat chuck using OCT compound (*see* **Note 7**). Wait until the OCT compound turns white before cryosectioning the tissue.
6. Place the cryostat chuck onto the microtome sample stage, and orient the sample in the desired position.
7. Clean the blade with a 70% ethanol solution before sectioning and cryosection the tissue at the desired thickness (*see* **Notes 8–10**).

8. Collect the tissue on CryoJane adhesive tape by first removing the backing from the CryoJane tape and then placing the tape over the tissue sample. Use a roller tool to press the tape against the gelatin to ensure adhesion. Cryosection the tissue at 10–15 μm thickness.
   (a) Place the CryoJane tape with section attached face-up on a prechilled glass slide. Attach the CryoJane tape to the glass slide by taping both ends to the slide with Scotch tape (*see* **Note 11**).
   (b) To thaw mount, place the CryoJane tape with section attached face-down on a prechilled glass slide, use a roller tool to ensure the sample is flat, and then hold your finger firmly on the bottom of the glass slide under the sample. Once the sample melts, place the glass slide on a cold metal block. When the sample refreezes (gelatin will turn opaque) slowly peel the CryoJane tape from the glass slide. The sample and gelatin should remain on the glass slide.
9. Repeat **step 8**, until the desired number of tissue sections have been collected.
10. Store all collected tissue sections at a −80 °C freezer until analysis. Once again, transport sections in a Styrofoam cooler with dry ice.

3.2 Tissue Drying

This sub-section describes how to dry tissue samples and bring them up to room temperature while minimizing compound delocalization.

1. Connect vacuum chamber to a vacuum pump.
2. Remove tissue samples from the −80 °C freezer, and place the glass slides with the tissue sections onto the heat sink (*see* **Note 12**).
3. Quickly place the heat sink in the vacuum chamber and turn on vacuum pump to evacuate chamber.
4. Once the samples have dried and the heat sink has warmed to room temperature, vent the chamber and remove the tissue samples.

3.3 On-Tissue Derivatization with TM Sprayer

This sub-section explains how to derivatize tissue sections using a TM Sprayer. This technique is optimal for CA and 2-PA derivatization. Figure 2 includes screenshots of the TM Sprayer software with numbered labels for features mentioned. To further improve ionization through cation adduct formation, cation solutions (e.g., potassium acetate) can be separately applied to the tissues using TM sprayer.

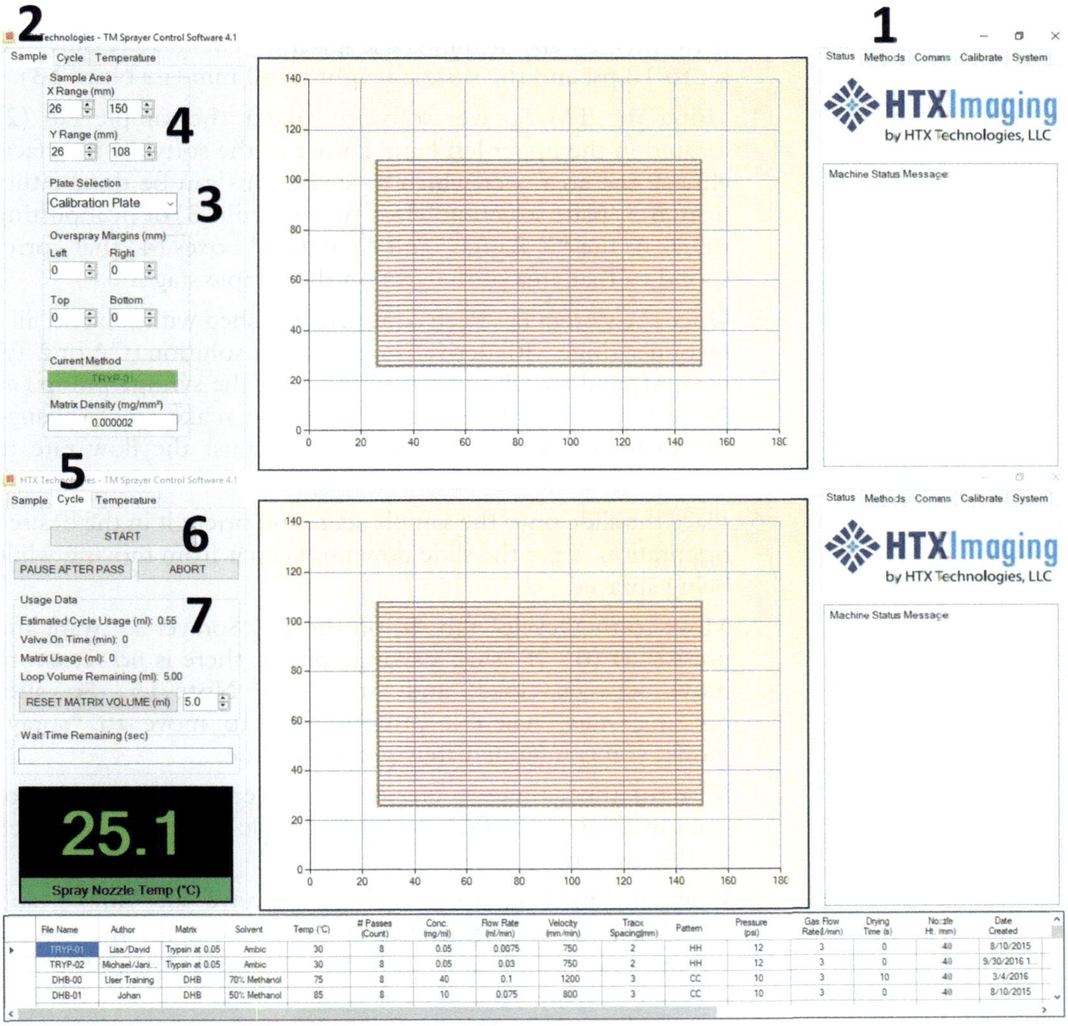

Fig. 2 Screenshots from TM Sprayer software with reference numbers corresponding to the text

1. Fill a syringe with 50:50 methanol–water and place it in the syringe pump.
2. Turn on the syringe pump. Make sure the diameter is correctly set to match the diameter of the syringe you are using. Set the flow rate between 0.05 and 0.2 mL/min, start the flow, and turn on the nitrogen gas. Allow the solution to run for several minutes until the system is flushed with solvent.
3. Meanwhile, create a method using the TM Sprayer software using the methods tab (**1**) located in the upper right hand corner of the software interface. The following parameters are used to coat the tissue section for both derivatization reagents: 30 °C nozzle spray temperature, 0.03 mL/min flow rate,

1200 mm/min linear velocity, and 8 passes in a crisscross pattern (CC) and 3 mm track spacing. The gas pressure was set to 10 psi and the nozzle height to 40 mm (*see* **Note 13**).

4. Using the TM Sprayer software, under the sample tab (**2**) located in the upper left hand corner of the software interface, choose the area over which to spray. This can be done either from the "plate selection" drop down menu (**3**) or by inputting values to the "X Range" and "Y Range" boxes (**4**) that correspond to that area of the grid on the sample stage.
5. Once the system has been completely flushed with solvent, fill a second syringe with derivatization regent solution (CA or 2-PA reagent) and switch out the syringes on the syringe pump (*see* **Note 14**). If the syringe is a different size, make sure to change the diameter on the syringe pump. Adjust the flow rate to 0.03 mL/min to match the method.
6. Place the slide onto the sample stage and orient it in the desired orientation. Tape the slide down to keep it from moving while being sprayed.
7. Move to the "cycle" tab (**5**) on the TM Sprayer software and press start (**6**). If using a syringe pump, there is no reason to move between "load" and "spray" (*see* **Note 15**). So, after pressing start, when it prompts you to move to "spray" mode, you can immediately press continue.
8. Monitor the syringe to ensure the reagent solution does not run out. Make use of the "estimated cycle usage" (**7**) provided by the software.
9. When the method has completed, remove the sample, and switch the derivatization reagent syringe out for the solvent syringe once again, and flush the system before exiting the system (*see* **Note 16**).
10. Turn off the nitrogen gas and wait for the low pressure light to come on before exiting the software. Confirm that a drop of solvent forms on the spray tip to ensure there are no clogs. Exit the software before turning off the TM Sprayer.

3.4 Derivatization Application Via ESI Deposition

This section will explain how to apply a derivatization agent via Electrospray Ionization (ESI) deposition. ESI deposition is another method of on-tissue derivatization [25] where the high voltage (*see* **Note 17**) applied to the spray facilitates the formation of micro-droplets. ESI deposition works very well for charged derivatization agents (e.g., GT) [14]. The first step is to construct the ESI deposition apparatus as shown in Fig. 3.

1. Insert the fused silica capillary (*see* **Note 18**) into the capillary sleeve and then into a PEEK tubing fitting.

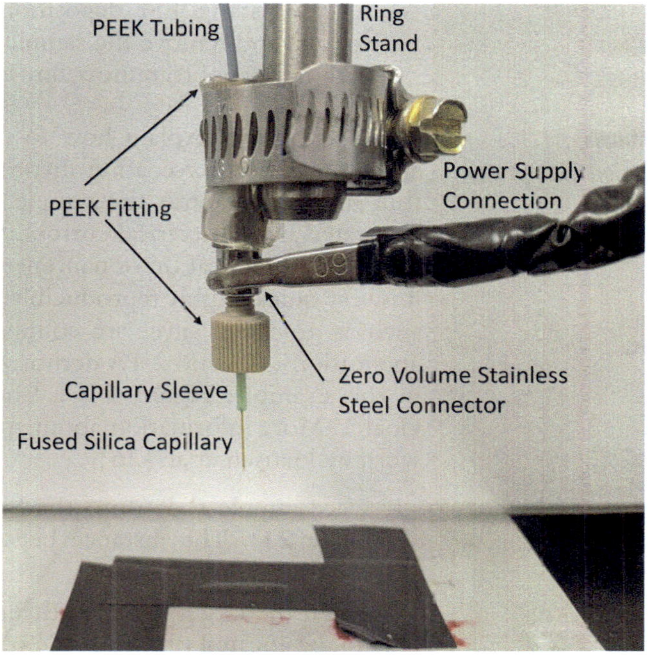

Fig. 3 Image of a homemade ESI deposition setup

2. Connect a PEEK tubing to a syringe luer adaptor on one side and another Peek tubing fitting on the other side.
3. Insert both PEEK tubing fittings into a zero volume stainless steel connector. This metal connection will provide electrical contact to the liquid flow by connecting to the power supply.
4. Connect the ground connection to the base so there is an electric field between the sprayed surface (*see* **Note 19**) and the sprayer. Mount the PEEK tubing setup to a ring stand but keep the metal connection shielded from the ring stand to avoid grounding out the electrical connection. The fused silica capillary should be 3 cm above the surface that will be sprayed.
5. Fill the 1 mL syringe with a derivatization solution (e.g., GT reagent) and connect to the PEEK tubing. Place the syringe into the syringe pump and set the flow rate to 0.3 mL/h. Make sure the diameter is correctly set to match the diameter of the syringe you are using.
6. Allow the solution to completely fill the tubing and begin to form droplets at the fused silica capillary tip. Once droplets begin to form, turn on the 5 kV power supply.
7. Carefully place the sample (*see* **Note 20**) to be sprayed under the capillary and spray around 200 μL of derivatization agent (will need to adjust on a per sample basis).

8. After finishing ESI deposition, stop the flow, turn off the voltage and remove the sample. Make sure to flush the tubing with methanol to remove any leftover reagent.

3.5 Matrix Application Via Sputter Coater

This section will explain how to use a sputter coater for matrix application. Sputter coating involves the physical deposition of a thin metal layer onto the sample surface through argon plasma generated by an electrical current. Sputter coating is a convenient method for application of nanometer-thick metal matrices because it is very quick, highly reproducible, and homogenous. Inert metals such as gold and silver are commonly used matrices for various applications. CA and 2-PA derivatization both use gold as a matrix in this example. Figure 4 is a labelled schematic of the Sputter Coater. More information about sputter coating is available in the work by Hansen et al. [26]

1. Place sample slide onto the Sputter Coater sample table (*see* **Note 21**). The distance between the metal target and the sample is about 4 cm.
2. Clean the target using methanol and place the target into the holder located on the inside of the chamber top plate. This is done by removing the dark space shield, unscrewing the threaded clamp ring, placing the target inside, and replacing the threaded clamp ring.
3. Turn on the sputter coater, then the vacuum pump, and finally open the argon gas tank to 5 psi.
4. Set the sputter time by holding down the pause button while pressing the up or down arrows until the timer display reads the desired time. The optimal time for our setup is 20 s.
5. Set the current to 40 mA by holding down the "set mA" button while pressing the up or down arrows until the display reads 40 mA.

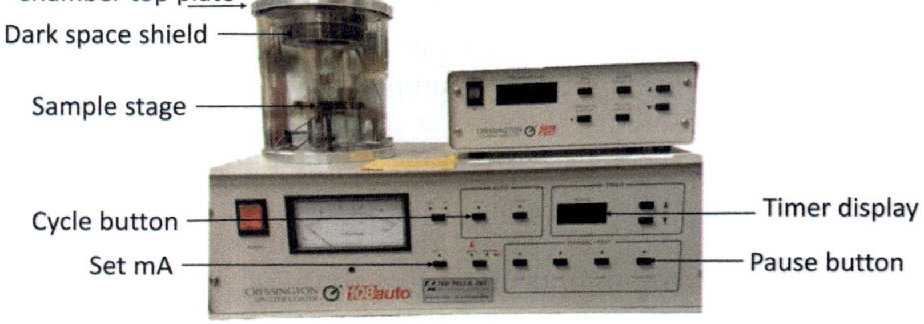

Fig. 4 Photo of sputter coater with labels for the major components

6. Wait until the pressure in the chamber is between 0.05 and 0.1 mbar, and then hit the cycle button, which will automatically move through the steps for sputtering. It will begin by flushing out the chamber twice with argon gas (pressure ~ 8 psi) before bringing it back down to vacuum pressure. Then the plasma will form and the timer will begin to count down (*see* **Note 22**).
7. Once the sputter coating has been finished, turn off the sputter coater, vacuum pump, and close the argon gas tank.
8. Allow the chamber to vent and remove samples (*see* **Note 23**).

3.6 Matrix Application Via TM Sprayer

This section discusses matrix application using a TM Sprayer. It is important to note that there are other methods of matrix application for organic matrices, such as sublimation or other nebulized sprayer alternatives [27]. Charged analytes, such as those derivatized with GT, work the best with DHB matrix according to our experience.

1. Follow the same steps in Subheading 3.3 except for the spray parameters.
2. Choose a method from the existing list or create a new one.
3. Our method for matrix application included the following parameters: 75 °C nozzle spray temperature, 0.1 mL/min flow rate, 1200 mm/min linear velocity, 8 passes in a crisscross pattern (CC), and 3 mm track spacing. The gas pressure was set to 10 psi and the nozzle height to 40 mm.

3.7 Data Acquisition

This section will describe how to perform a mass spectrometry imaging experiment on a Thermo MALDI Linear Ion Trap (LIT)- Orbitrap Instrument (MALDI-LTQ-Orbitrap Discovery; Thermo Finnigan, San Jose, CA, USA). Similar data analysis can be performed for other types of imaging mass spectrometers following the vendor-specific user manual.

1. Scan an image of the tissue to load into the TunePlus instrument software.
2. Optimize instrument parameters such as laser energy and number of laser shots in the TunePlus software. Save a tune file containing this information.
3. Open the MALDI window of TunePlus and select the desired area to image using the scanned optical image. Select the raster step size and save this information as a position file.
4. Set up a new instrument method using Xcaliber software. The method should include the desired mass range, resolution, polarity, and the saved tune file from **step 2**.

5. Set up a new sequence using Xcaliber software. Include a name for the data file, a file destination folder, the saved instrument method, and saved position file. Repeat this with all the tissue sections you wish to run.
6. Select all of the samples to be analyzed and select "run sequence."

3.8 Analysis of Derivatized Data Sets

This section will explain how to analyze derivatized data sets, identify derivatized peaks, and address some challenges. The exact masses added due to derivatization (Δm_{der}) are 160.052 u, 114.103 u, and 90.058 u for CA, GT, and 2-PA derivatization, respectively. Subtracting these masses from the mass values of the peaks only present in derivatized samples may help identify the derivatized metabolites. GT contains a permanently positive charge, so derivatized compounds are detected as Δm_{der} adducts. Products from CA and 2-PA derivatization commonly form potassium adducts in plant tissues and are therefore detected as Δm_{der} + K adducts.

1. Export the mass list to Excel from Xcaliber (Thermo).
2. Unique features not present in the underivatized data sets can be identified in the derivatized data sets using several filtering steps. Filtering can be done manually in Excel as described in **steps 3–5** or using a python script on the text file.
3. Remove *m/z* values with a signal intensity of less than 500 (or signal-to-noise-ratio of 20) from the peak list (*see* **Note 24**).
4. Remove all ^{13}C isotope peaks, alkaline ion adducts, and known contamination peaks.
5. Generate MS images for the remaining *m/z* values using MSiReader [28] to ensure they are localized to the tissue; remove any peaks localized off-tissue.
6. Identify unique features using database search with accurate masses of experimental data and a mass tolerance of 5 ppm. Make sure to use the underivatized mass for database search. The databases utilized for maize application were METLIN (https://metlin.scripps.edu) and Maize Genetics and Genomics Database (https://corncyc-b73-v4.maizegdb.org).
7. Filter mass values based on the functional groups present. For example, a match for a feature from a CA dataset should contain a primary amine group.
8. If applicable, perform MS/MS experiments on tissue sections to confirm derivatization. For example, a fragment at *m/z* 161 indicates derivatization with CA.

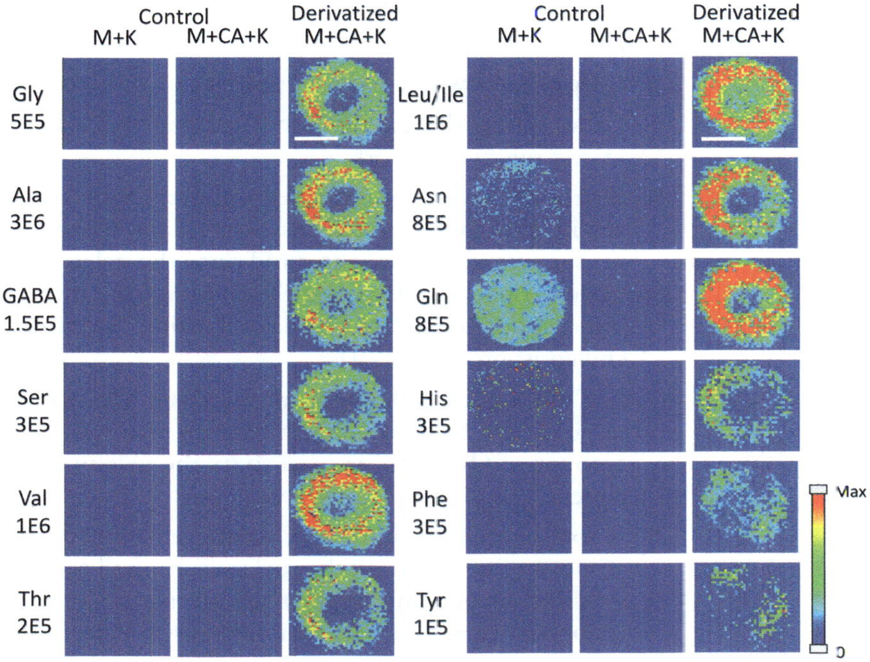

Fig. 5 MS images for the 12 amino acids detected in Mo17 maize roots. The first 2 columns are control roots without CA derivatization and the last column is with CA derivatization. "M + K" indicates the nonderivatized amino acids detected as potassium ion adduct. "M + CA + K" indicates the derivatized amino acids detected as potassium ion adducts. The numbers below each amino acid label correspond to the maximum intensity scale used to produce the false color image. The scale bar is 500 μm

9. Use imaging software to create the images for unique features and label with identifications as seen in Fig. 5, which shows the MS images of several amino acids in maize root tissue with and without derivatization with CA. Without derivatization, only three amino acids were detected, while after derivatization, twelve were detected with good intensity.

4 Notes

1. For MALDI-Linear Ion Trap (LIT)-Orbitrap instruments it is not necessary to use only conductive materials because of the relatively low voltage needed to extract ions. Therefore, normal scotch tape and glass microscope slides are acceptable. If using another system, such as a MALDI-TOF, conductive materials such as indium tin oxide (ITO) glass slides will be needed.
2. CryoJane® adhesive tape is removable from tissue sections after flash UV curing. In this method, we use CryoJane tape without UV curing, but CryoJane Tape-Transfer System can be used to produce high-quality frozen sections of difficult tissues.

3. The roller tool can be anything small and cylindrical. A small paintbrush handle, pen, or pencil would all work.
4. A lyophilizer with a vacuum system can be used instead.
5. Other similar devices that can homogeneously deposit organic matrix might also work.
6. When embedding samples, leaving the gelatin completely frozen in liquid nitrogen can cause the gelatin to crack. Therefore, we suggest to remove the mold once the gelatin is about 80% frozen (opaque). Alternatively, the sample can be embedded under milder conditions by flash-freezing the sample in dry ice chilled isopropanol [29].
7. As OCT can cause background signals in the mass spectrometer, it cannot be used as an embedding material but a minimum amount of OCT can be used to fix the gelatin to the cryostat chuck.
8. Before collecting your tissue, use the cryostat to section several sections of gelatin until the surface is flat and the tissue sample is visible. Depending on the tissue, it may be helpful to look at an optical image of the sectioned tissue before beginning to collect sections. If there is a specific morphology or cell type you are interested in, make sure it is present in the portion of the tissue sectioned. To ensure the region of interest is collected, we typically observe one or two sections under a bright field microscope before beginning collection. Since the tissue goes from −20 °C to room temperature when visualizing under the microscope, the tissue will break and the cells will burst. Discard this section as it cannot be used for MSI.
9. If you choose to look at tissue sections under a bright field microscope before collecting sections, it helps to put an additional chilled glass slide under the one that has the tissue attached. This slows the amount of time it takes for the gelatin to melt completely. In order to visualize an intact sample, complete this step within 10 s after removing from the cryostat.
10. When sectioning, it can be helpful to use a forceps to collect the section. Holding the bottom edge of the tape with a forceps while using the microtome is more efficient than using your hands.
11. In this method, we attach the tape with the tissue on a MALDI glass slide, which may not work for MALDI-MS with high extraction field. High extraction field instruments require a coherent field on the sample, so samples directly mounted to ITO slides by thaw mounting or thermal adhesion are necessary. Samples on nonconductive tape attached to the slides would not work.

12. The heat sink could be a variety of different materials; however, we use an aluminum block that is 18 cm × 6 cm × 0.65 cm. With this heat sink, the vacuum drying and thermalization process for plant tissues takes 75–90 min.
13. While we suggest conditions that worked best for our system, some additional optimization may be necessary depending on the apparatus condition and sample.
14. If using the syringe pump with the TM Sprayer (as opposed to a HPLC pump), switch between syringes as quickly as possible to limit the amount of time there is nothing running through the lines.
15. Since the sample is in the syringe and constantly being pumped, the sample loop does not need to be filled. Hence, it does not matter whether the sample loop is in the "Load" or "Spray" position. This will not be true if a HPLC pump is used instead of a syringe pump.
16. It is necessary to flush the TM sprayer with several milliliters of solvent. If any derivatization agent remains in the tubing or nozzle, the nozzle may become clogged.
17. As a note of caution, the ESI setup requires a high voltage power supply and can be dangerous to operate. Be careful not to touch the spray or any metal connected to the voltage with bare skin.
18. The end tip of the capillary should be cut clean to achieve a good spray. Alternatively, commercially available precut capillary can be used.
19. Due to the necessity of a stable electric field, it may be helpful to mount the whole setup onto a square aluminum, or other conductive metal, slab. This allows you to easily connect the ground as well as a more consistent field.
20. The ESI setup requires a strong electric field between the spray surface, so the sample slide must also be conductive, stainless steel slides or ITO glass slides are good options.
21. When sputtering samples, make sure slide is placed as close to the center of the sample stage as possible.
22. In some instances, the plasma does not form immediately when using the sputter coater. If this happens, wipe down the target and metal holder once again with methanol as metal residue will sometimes form and hinder the plasma formation.
23. It is useful to wipe down the inside of the sputter coater chamber immediately after sputtering with methanol to keep everything clean. If metal residue accumulates over time in the chamber, it is much harder to remove.
24. This threshold can be changed depending on the specific application.

Acknowledgements

This work is supported by National Science Foundation (NSF) (Award# 1905335).

References

1. Nilsson A, Fehniger TE, Gustavsson L, Andersson M, Kenne K, Marko-Varga G, Andrén PE (2010) Fine mapping the spatial distribution and concentration of unlabeled drugs within tissue micro-compartments using imaging mass spectrometry. PLoS One 5(7):e11411
2. Castellino S, Groseclose MR, Wagner D (2011) MALDI imaging mass spectrometry: bridging biology and chemistry in drug development. Bioanalysis 3(21):2427–2441. https://doi.org/10.4155/bio.11.232
3. Shimma S, Sugiura Y, Hayasaka T, Hoshikawa Y, Noda T, Setou M (2007) MALDI-based imaging mass spectrometry revealed abnormal distribution of phospholipids in colon cancer liver metastasis. J Chromatogr B 855(1):98–103. https://doi.org/10.1016/j.jchromb.2007.02.037
4. Gónzalez de San Román E, Manuel I, Giralt MT, Ferrer I, Rodríguez-Puertas R (2017) Imaging mass spectrometry (IMS) of cortical lipids from preclinical to severe stages of Alzheimer's disease. Biochim Biophys Acta 1859 (9, Part B):1604–1614. https://doi.org/10.1016/j.bbamem.2017.05.009
5. Roux A, Muller L, Jackson SN, Post J, Baldwin K, Hoffer B, Balaban CD, Barbacci D, Schultz JA, Gouty S, Cox BM, Woods AS (2016) Mass spectrometry imaging of rat brain lipid profile changes over time following traumatic brain injury. J Neurosci Methods 272:19–32. https://doi.org/10.1016/j.jneumeth.2016.02.004
6. Black AP, Angel PM, Drake RR, Mehta AS (2019) Antibody panel based N-glycan imaging for N-glycoprotein biomarker discovery. Curr Protoc Protein Sci 98(1):e99. https://doi.org/10.1002/cpps.99
7. Lee YJ, Perdian DC, Song Z, Yeung ES, Nikolau BJ (2012) Use of mass spectrometry for imaging metabolites in plants. Plant J 70 (1):81–95. https://doi.org/10.1111/j.1365-313X.2012.04899.x
8. Francese S, Bradshaw R, Ferguson LS, Wolstenholme R, Clench MR, Bleay S (2013) Beyond the ridge pattern: multi-informative analysis of latent fingermarks by MALDI mass spectrometry. Analyst 138(15):4215–4228. https://doi.org/10.1039/c3an36896c
9. Korte AR, Yandeau-Nelson MD, Nikolau BJ, Lee YJ (2015) Subcellular-level resolution MALDI-MS imaging of maize leaf metabolites by MALDI-linear ion trap-Orbitrap mass spectrometer. Anal Bioanal Chem 407 (8):2301–2309. https://doi.org/10.1007/s00216-015-8460-5
10. Zavalin A, Yang J, Caprioli R (2013) Laser beam filtration for high spatial resolution MALDI imaging mass spectrometry. J Am Soc Mass Spectrom 24(7):1153–1156. https://doi.org/10.1007/s13361-013-0638-5
11. Zavalin A, Todd EM, Rawhouser PD, Yang J, Norris JL, Caprioli RM (2012) Direct imaging of single cells and tissue at sub-cellular spatial resolution using transmission geometry MALDI MS. J Mass Spectrom 47(11):i. https://doi.org/10.1002/jms.3132
12. Niehaus M, Soltwisch J, Belov ME, Dreisewerd K (2019) Transmission-mode MALDI-2 mass spectrometry imaging of cells and tissues at subcellular resolution. Nat Methods 16 (9):925–931. https://doi.org/10.1038/s41592-019-0536-2
13. Feenstra AD, Dueñas ME, Lee YJ (2017) Five micron high resolution MALDI mass spectrometry imaging with simple, interchangeable, multi-resolution optical system. J Am Soc Mass Spectrom 28(3):434–442. https://doi.org/10.1007/s13361-016-1577-8
14. Dueñas ME, Larson EA, Lee YJ (2019) Toward mass spectrometry imaging in the metabolomics scale: increasing metabolic coverage through multiple on-tissue chemical modifications. Front Plant Sci 10:860
15. Dueñas ME, Klein AT, Alexander LE, Yandeau-Nelson MD, Nikolau BJ, Lee YJ (2017) High spatial resolution mass spectrometry imaging reveals the genetically programmed, developmental modification of the distribution of thylakoid membrane lipids among individual cells of maize leaf. Plant J 89(4):825–838. https://doi.org/10.1111/tpj.13422

16. Feenstra AD, Alexander LE, Song Z, Korte AR, Yandeau-Nelson MD, Nikolau BJ, Lee YJ (2017) Spatial mapping and profiling of metabolite distributions during germination. Plant Physiol 174(4):2532. https://doi.org/10.1104/pp.17.00652
17. Dueñas ME, Carlucci L, Lee YJ (2016) Matrix recrystallization for MALDI-MS imaging of maize lipids at high-spatial resolution. J Am Soc Mass Spectrom 27(9):1575–1578. https://doi.org/10.1007/s13361-016-1422-0
18. Dueñas ME, Feenstra AD, Korte AR, Hinners P, Lee YJ (2018) Cellular and subcellular level localization of maize lipids and metabolites using high-spatial resolution MALDI mass spectrometry imaging. In: Lagrimini LM (ed) Maize: methods and protocols. Springer New York, New York, NY, pp 217–231. https://doi.org/10.1007/978-1-4939-7315-6_13
19. Hansen RL, Guo H, Yin Y, Lee YJ (2019) FERONIA mutation induces high levels of chloroplast-localized Arabidopsides which are involved in root growth. Plant J 97 (2):341–351. https://doi.org/10.1111/tpj.14123
20. Manier ML, Spraggins JM, Reyzer ML, Norris JL, Caprioli RMCJMSR (2014) A derivatization and validation strategy for determining the spatial localization of endogenous amine metabolites in tissues using MALDI imaging mass spectrometry. J Mass Spectrom 49 (8):665–673. https://doi.org/10.1002/jms.3411
21. Esteve C, Tolner EA, Shyti R, van den Maagdenberg AMJM, McDonnell LA (2016) Mass spectrometry imaging of amino neurotransmitters: a comparison of derivatization methods and application in mouse brain tissue. Metabolomics 12(2):30. https://doi.org/10.1007/s11306-015-0926-0
22. Shimma S, Kumada H-O, Taniguchi H, Konno A, Yao I, Furuta K, Matsuda T, Ito S (2016) Microscopic visualization of testosterone in mouse testis by use of imaging mass spectrometry. Anal Bioanal Chem 408 (27):7607–7615. https://doi.org/10.1007/s00216-016-9594-9
23. Cobice DF, Livingstone DEW, Mackay CL, Goodwin RJA, Smith LB, Walker BR, Andrew R (2016) Spatial localization and quantitation of androgens in mouse testis by mass spectrometry imaging. Anal Chem 88 (21):10362–10367. https://doi.org/10.1021/acs.analchem.6b02242
24. Barré FPY, Flinders B, Garcia JP, Jansen I, Huizing LRS, Porta T, Creemers LB, Heeren RMA, Cillero-Pastor B (2016) Derivatization strategies for the detection of triamcinolone acetonide in cartilage by using matrix-assisted laser desorption/ionization mass spectrometry imaging. Anal Chem 88(24):12051–12059. https://doi.org/10.1021/acs.analchem.6b02491
25. Wu Q, Comi TJ, Li B, Rubakhin SS, Sweedler JV (2016) On-tissue derivatization via electrospray deposition for matrix-assisted laser desorption/ionization mass spectrometry imaging of endogenous fatty acids in rat brain tissues. Anal Chem 88(11):5988–5995. https://doi.org/10.1021/acs.analchem.6b01021
26. Hansen RL, Dueñas ME, Lee YJ (2019) Sputter-coated metal screening for small molecule analysis and high-spatial resolution imaging in laser desorption ionization mass spectrometry. J Am Soc Mass Spectrom 30 (2):299–308. https://doi.org/10.1007/s13361-018-2081-0
27. Korte AR, Yagnik GB, Feenstra AD, Lee YJ (2015) Multiplex MALDI-MS imaging of plant metabolites using a hybrid MS system. In: Mass spectrometry imaging of small molecules. Springer New York, New York, NY, pp 49–62. https://doi.org/10.1007/978-1-4939-1357-2_6
28. Robichaud G, Garrard KP, Barry JA, Muddiman DC (2013) MSiReader: an open-source Interface to view and analyze high resolving power MS imaging files on Matlab platform. J Am Soc Mass Spectrom 24(5):718–721. https://doi.org/10.1007/s13361-013-0607-z
29. Dannhorn A, Kazanc E, Ling S, Nikula C, Karali E, Paola Sera M, Vorng J, Inglese P, Maglennon G, Hamm G, Swales J, Strittmatter N, Barry ST, Sansom OJ, Poulogiannis G, Brunch J, Goodwin RJA, Takats Z (2020) Universal sample preparation unlocking multimodal molecular tissue imaging. Anal Chem 92(16):11080–11088. https://doi.org/10.1021/acs.analchem.0c00826

Chapter 15

Matrix-Assisted Laser Desorption/Ionization Mass Spectrometry Imaging of Glycogen In Situ

Tara R. Hawkinson and Ramon C. Sun

Abstract

Histopathological detection and quantitation of glycogen in situ are important for the assessment of glycogen storage diseases and different types of cancer. The current standard method for defining the regionality of glycogen rely almost exclusively on Periodic Acid-Schiff (PAS) staining, a workflow that lacks specificity and sensitivity. Herein, we describe a new and much improved workflow to detect microenvironmental glycogen in situ using enzyme-assisted matrix-assisted laser desorption/ionization-mass spectrometry imaging (MALDI-MSI). This method provides superior sensitivity and can elucidate the molecular features of glycogen structure, with 50 μm spatial resolution for a next-generation histopathological assessment of glycogen.

Key words Glycogen, Isoamylase, MALDI-MSI, FFPE tissue, Alzheimer's disease, Glycogen chain length

1 Introduction

Glycogen is a branched oligosaccharide that consists of linear glucose units covalently linked by linear α-1,4-linkages and additional α-1,6-linkages for branching to increase packing in a globular structure [1]. Glycogen structure is maintained through the optimal number of branch points and the number of glucose units per branch [2]. The length of each of these branches is known as the degree of polymerization (DP) [3]. Typical DP within a glycogen molecule range from approximately 3 to 21 glucose units in length in a healthy state, with an average chain length of 13 units [2, 4, 5]. Glycogen is primarily stored in the liver or muscle as a glucose reserve [6]. In addition, glycogen has been reported in the kidney [7], lung [8], bladder [9], heart [10], and brain [11] but their physiological roles are yet to be revealed. The importance of glycogen in organismal physiology is especially highlighted in glycogen storage diseases (GSD); inherited disorders that contain defects

Young-Jin Lee (ed.), *Mass Spectrometry Imaging of Small Molecules: Methods and Protocols*,
Methods in Molecular Biology, vol. 2437, https://doi.org/10.1007/978-1-0716-2030-4_15,
© The Author(s), under exclusive license to Springer Science+Business Media, LLC, part of Springer Nature 2022

with enzymes involved in glycogen synthesis and degradation [12]. In such diseases, stored glycogen can become aberrant with overt branches or phosphorylation [12–14]. Hyper-branched and hyper-phosphorylated glycogen form aggregates that present as histological plaques which are no longer accessible to degradation enzymes [15]. Formation of glycogen plaques, also known as polyglucosan bodies (PGB) lead to cardiomyopathy [16], muscle weakness [17, 18], and in extreme cases, rapid neurodegeneration and early death [19–21].

Current biochemical assays to detect glycogen in pulverized tissues or cells rely primarily on indirect measurements such as $NADP^+$/NADPH (or ATP/ADP) exchange to measure glucose-6-phosphate (G6P) [22, 23]. These assays provide modest sensitivity but lack the ability to visualize DP localization in situ. Recently, we developed a robust method of quantitating glycogen by gas chromatography mass-spectrometry (GCMS) with fmol detection limit and provided a mid to high throughput [24]. However, this methodology also relies on pooled tissue samples and the hydrolysis of glycogen to glucose, which compromise tissue heterogeneity and glycogen structure. While glycogen DP can be determined by high-performance anion-exchange chromatography with pulsed amperometric detection (HPAEC-PAD) [25]; ultimately, spatial distribution of total glycogen still relies on coupling samples to immunohistochemical staining methods such as Periodic Acid-Schiff (PAS). Further, PAS staining is nonspecific to N-linked glycans and all oxidizable functional groups including aldehydes and ketones [26]. Currently, the field lacks a robust, high throughput assay to profile glycogen structure in tissue. With these limitations in mind, we designed a novel enzyme-assisted mass-spectrometry imaging method to visualize glycogen accumulation in situ that also offers biochemical insights of glycogen DP (Fig. 1).

In recent years, technological advances in matrix-assisted laser desorption/ionization-mass spectrometry imaging (MALDI-MSI) have modernized the field of digital pathology with the inclusion of metabolic processes, and we are now capable of visualizing metabolism in situ with near single-cell resolution [27]. MALDI-MSI can be used to image proteins, lipids, and carbohydrates from plants to biofluids to whole tissue slices [27–32]. Enzyme-assisted MALDI-MSI is one of the latest additions to this rapidly advancing field [33]. Drake et al. revolutionized enzyme-assisted MALDI-MSI analysis through the development of on-tissue digestion and release of N-linked glycans by utilizing the enzyme peptide N-glycosidase F (PNGase F) [34–36]. We adapted this workflow for the spatial imaging of glycogen structure and employed the highly specific glycogen 6-alpha-D glucanohydrolase (isoamylase) that cleaves glycogen at α-1,6 branching points to produce a gradient of glucose polymers (Fig. 2) [37]. These glycogen-derived glucose polymers and their spatial distribution can be detected by MALDI-MSI

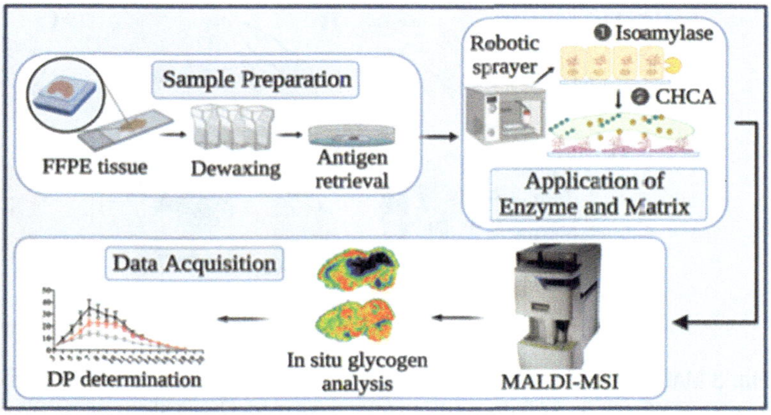

Fig. 1 Workflow for on-tissue isoamylase digestion and imaging of glycogen DP. FFPE tissues are cut onto glass slides and dewaxed, followed by antigen retrieval. Then, the slides are sprayed with isoamylase using a robotic sprayer and incubated at 37 °C for 2 h to allow for the digestion of glycogen chains. Next, a CHCA matrix is sprayed to facilitate the ionization of the sample so that it can be detected by the mass spectrometer. Slides are scanned and data is acquired by the MALDI-MSI and then exported to the HDI software by Waters Corporation for analysis. FFPE = formalin-fixed, paraffin-embedded, CHCA = alpha-cyano-4-hydroxycinnamic acid, DP = degree of polymerization, MALDI-MSI = matrix-assisted laser desorption/ionization mass spectrometer imager, HDI = high-definition imaging

A

DP	*m/z*
3	527
4	689
5	851
6	1013
7	1175
8	1337
9	1499
10	1661
11	1823
12	1985
13	2147
14	2309
15	2471
16	2633
17	2795
18	2957

B

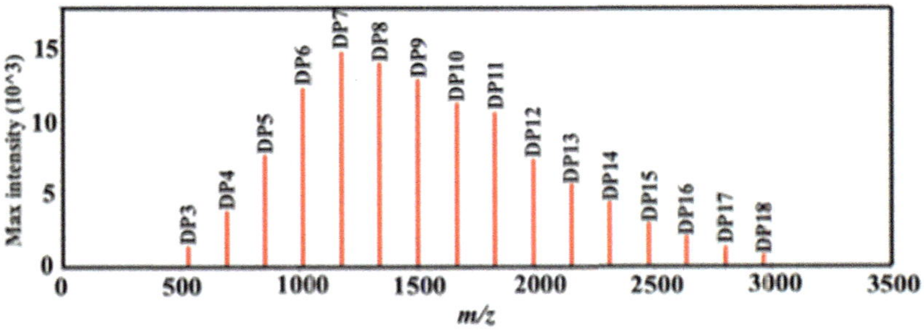

Fig. 2 Glycogen-derived DP presented as mass list and ion chromatogram. (**a**) DP length and its corresponding singularly charged mass (*m/z*, rounded to the nearest 1) derived from glycogen after isoamylase digestion. (**b**) Representative ion chromatogram of DP polymers observed by MALDI-MSI in positive mode, only DP3–18 were shown using the HDI software, Waters Corporation. DP = degree of polymerization, MALDI-MSI = matrix-assisted laser desorption/ionization mass spectrometer imager

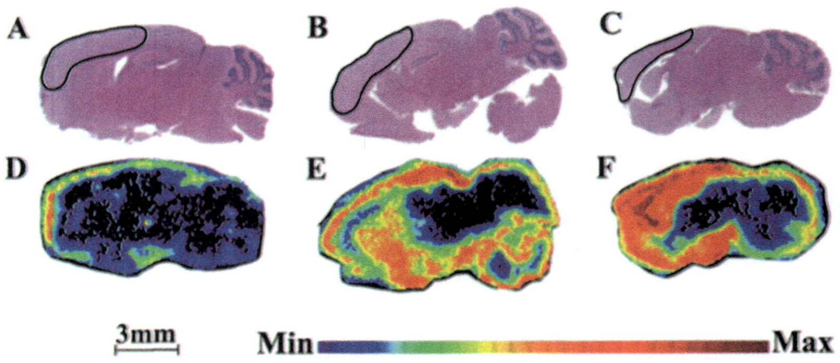

Fig. 3 MALDI-MSI of glycogen in WT and Alzheimer's disease mouse models. (**a–c**) H&E-stained histological images of WT (**a**), 5×FAD (**b**), and Tau (**c**) mouse model brain slices with ROIs circled in the frontal cortex. D-F. Spatial distribution of microenvironmental glycogen in brain slices shown in **a–c**. Isoamylase-assisted, cleaved glucose chains measured on Waters XS MALDI-MSI and exported for analysis using HDI software from Waters Corporation. Representative images contain chain length profiles from DP6-DP9 normalized to TIC. N = 3. H&E = hematoxylin and eosin, WT = C57BL/6, Tau = rTg(TauP301L), ROI = region of interest, MALDI-MSI = matrix-assisted laser desorption/ionization mass spectrometer imager, HDI = high-definition imaging, DP = degree of polymerization, TIC = total ion chromatogram

at resolution of 50 μm (Fig. 3). Using this workflow, we can resolve the glycogen DP differences between healthy and diseased tissue (Fig. 4). Here, we used a Waters Synapt XS quadrupole time-of-flight (QTOF) instrument, but this methodology can be adapted to other MALDI-enabled mass spectrometer imagers.

2 Materials

2.1 General Materials and Equipment

1. Vortex mixer.
2. Microcentrifuge.
3. Positively charged microscope glass slides.
4. Histology cassettes.
5. Paraffin wax.
6. Slide-A-Lyzer dialysis column.
7. Incubator.
8. Histological slide staining chambers.
9. Vertical slide staining rack.
10. Desiccator.
11. Water bath steamer.
12. Five-slide mailers for microscope glass slides.
13. TM Sprayer (HTX Technologies LLC, Chapel Hill, NC).
14. Luer lock sterile 3 mL syringe.

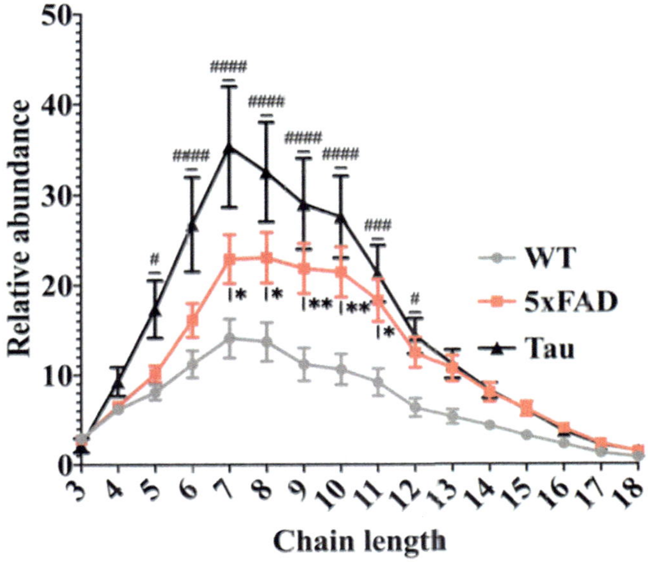

Fig. 4 Glycogen chain length distribution in WT and Alzheimer's disease mouse models. Glycogen chain length distribution in the frontal cortex of WT compared to 5×FAD and Tau mouse model brains. Glycogen-derived glucose polymer abundance normalized to TIC and number of pixels. Data was exported for analysis using HDI software from Waters Corporation. Graphs were analyzed using Prism. Error bars represent SEM, and p-value significance were determined by a two-way ANOVA with Tukey's post hoc comparison against WT for each chain length. N = 3. WT = C57BL/6, Tau = rTg(TauP301L), TIC = total ion chromatograph, HDI = high-definition imaging, SEM = standard error of the mean, ANOVA = analysis of variance

15. 18-gauge disposable needles
16. Polyetheretherketone (PEEK) tubing: strongly encouraged to use a Trypsin line (HTX technologies).
17. Temperature- and humidity-controlled metal chamber: strongly encouraged to use an M.5 HumidChamber (HTX technologies).
18. Microspatula.
19. Sonicator.
20. Gastight glass plunger 5 mL syringe.
21. 0.2 μm syringe filter
22. High-pressure HPLC compact pump.
23. External syringe pump.
24. Synapt XS MALDI mass spectrometer (Waters Corp, Milford, MA).
25. High-Definition Imaging (HDI) Software (Waters Corp, Milford, MA).

2.2 Reagents

1. Phosphate buffered saline (PBS).
2. Xylenes.
3. 200 proof ethanol.
4. High performance liquid chromatography (HPLC) grade water.
5. Isoamylase enzyme.
6. Alpha-cyano-4-hydroxycinnamic acid (CHCA) powder.

2.3 Solutions

All solutions are stored at room temperature unless otherwise noted.

1. 10% buffered formalin in 1× PBS (v/v).
2. 95%, 90%, 70% (w/v) ethanol (EtOH) solutions: dilute 200 proof ethanol with HPLC water.
3. 0.1× PBS: Dilute with HPLC water.
4. Citraconic buffer: 25 μL citraconic anhydride and 2 μL 12 M HCl in 50 mL HPLC water. Ensure the pH of the final solution is around 3.0, adjust if necessary using HCl or NaOH. Make fresh each use.
5. 50% volume acetonitrile: Dilute with HPLC water.
6. 50% volume methanol: Dilute with HPLC water. Degas solution using sonicator.
7. Alpha-cyano-4-hydroxycinnamic acid (CHCA) stock solution: 100 μL trifluoroacetic acid (TFA) into 100 mL, 50% acetonitrile.
8. Glycogen standard: 10 mg/mL dilute with HPLC water. Store at 4 °C.

2.4 Optional Materials

1. Metal tongs.
2. Antistatic gun.

2.5 Mouse Information

Male, 1 year old wild type (WT) and transgenic mice bred into C57BL/6 background purchased from Jackson Laboratories are used in this method. All mouse studies are approved by the institutional animal care and use committee (IACUC).

1. WT mice: heterozygous littermate control.
2. 5×FAD mice: express 5 mutations linked to familial Alzheimer's disease (AD) and exhibit amyloid pathology.
3. rTg(TauP301L) mice: express a single mutation linked to familial Alzheimer's disease and exhibit tauopathy.

3 Methods

3.1 Preparation of Formalin-Fixed, Paraffin-Embedded (FFPE) Tissue Sections

1. After harvesting tissues, wash tissue or tissue sections (such as a mouse brain) once in 1× PBS and twice with water.
2. Completely submerge fresh tissue in 10% buffered formalin for at least 24, but up to 48 h at room temperature.
3. Pour out 10% buffered formalin and replace with 70% EtOH.
4. Place tissue flat in histology cassette and dehydrate the tissue by soaking in different concentrations of EtOH, followed by xylenes, and paraffin wax, each for 40 min: 70% EtOH, 90% EtOH, 95% EtOH, 100% EtOH (×3), xylene, paraffin wax (×3) (*see* **Note 1**).
5. Place histology cassette with paraffin wax on a cold plate to solidify the tissue block.
6. Cut FFPE tissue blocks using microtome at a thickness of 4 μm.
7. Collect tissue that has been ribboned in wax, and place sliced tissue sections directly onto the positively charged side of glass slide. Store slides at 4 °C until they are ready to use; up to 1 month.

3.2 Desalt Isoamylase

Isoamylase enzyme is likely stored in a buffer with a high salt concentration. The crystallization of this salt on a tissue sample interferes in MALDI-MSI analysis. It is highly recommended to exchange the storage buffer with a low-salt buffer before spraying the enzyme onto the tissues.

1. Equilibrate a dialysis membrane in 0.01× PBS for 15 min. Make sure 0.01× PBS is added to both sides of the membrane.
2. Discard the 0.01× PBS on both sides of membrane.
3. Add fresh 0.01× PBS to the dialysis tube.
4. Add a maximum of 200 μL isoamylase stock to the dialysis chamber.
5. Pour out and add fresh 0.01× PBS to the dialysis tube at 2 and 18 h.
6. Collect dialyzed isoamylase from dialysis chamber at 24 h.
7. Aliquot isoamylase at 3 U, which can be determined using the formula $M_1V_1 = M_2V_2$ according to the initial enzyme Unit concentration and the final volume. Store at −80 °C until use.

3.3 Dewax FFPE Slides

Tissues are preserved in paraffin to increase the sample's shelf life. A FFPE tissue block can be stored long-term for up to a decade or longer [38]. However, paraffin removal is a crucial part of the MALDI-MSI sample preparations.

1. Heat slides with the tissue facing up on the base of a 60 °C incubation chamber with adequate humidity for 1 h (*see* **Note 2**).
2. Remove slides from the incubator and allow them to cool to room temperature for 1 min.
3. Dewax slides (*see* **Note 3**) by completely submerging slides in xylene coplin jar for 3 min (×2), in 100% ethanol for 1 min (×2), in 95% ethanol for 1 min, in 70% ethanol for 1 min, and finally in water for 3 min (×2).
4. Wipe the back and edges of the slides with a Kimwipes and place in a vacuum desiccator to dry for 8 min, or until tissue is opaque (*see* **Note 4**).

3.4 Antigen Retrieval

1. Preheat water bath steamer for approximately 10 min, or until water is boiling.
2. Place no more than two slides in a five-slide mailer, in the outermost slots with the tissue facing inward.
3. Pour citraconic buffer into the mailers until they overflow.
4. Snap one corner of the mailer closed and leave the other corner open for ventilation.
5. Incubate slides in the middle of the preheated water bath steamer for 30 min.
6. Take mailers out of water bath steamer carefully (*see* **Note 5**).
7. Slowly cool the slides by placing the mailer in a clean container with lukewarm tap water. Ensure the tap water only fills up to half of the mailer height when the mailer is set in the container of water (*see* **Note 6**).
8. Allow slides to cool for 5 min.
9. Pour out half the citraconic buffer into a waste container.
10. Fill up the mailer by replacing the buffer with HPLC water.
11. Allow slides to cool for 5 min.
12. Repeat steps 9–11 (×2) for a total of 3 replacements, allowing 5 min for the slides to cool in between water replacements.
13. Finish by rinsing the slides with 100% HPLC water.
14. Wide off excess condensation on the backs and edges of slides with a Kimwipes and add a 1 μL of positive control (glycogen standard) in a designated, dry spot on each slide (*see* **Note 7**).
15. Dry slides in a vacuum desiccator for 15 min at room temperature.

3.5 Isoamylase Application

1. Remove frozen, desalted 3 U isoamylase stock from −80 °C.
2. Allow aliquot to thaw at room temperature.

3. Resuspend isoamylase aliquot to a final volume of 1 mL with HPLC water.
4. Vortex for 10 s and spin down.
5. Using a sterile 18-gauge needle, collect isoamylase in a 3 mL syringe.
6. Remove all bubbles and set syringe aside (*see* **Note 8**).
7. Turn on TM sprayer, turn on nitrogen gas, and stabilize pressure to 10 psi.
8. Open TM software and select the following sprayer settings—Nozzle temperature: 45 °C, heated tray temperature: ambient.
9. Turn off HPLC pump and disconnect sprayer line from TM sprayer.
10. Clean PEEK enzyme line with HPLC water using a 3 mL syringe, and flush water into waste container. Repeat 2×.
11. Clean TM sprayer using an external syringe pump. Connect 3 mL syringe with HPLC water to PEEK enzyme line and the other end of peek enzyme line to TM sprayer.
12. Lock in the HPLC water syringe to the external syringe pump and turn it on.
13. Push "start" on the external syringe pump and let HPLC water flow through TM sprayer for 5 min at a rate of 95 μL/min as temperature is stabilizing.
14. Stop the external syringe pump and disconnect the PEEK enzyme line and 3 mL syringe with of HPLC water.
15. Expel remaining HPLC water into waste container.
16. Push air through peek enzyme line until there is no remaining HPLC water.
17. Connect the 3 mL syringe with isoamylase to the peek enzyme line and the other end of the peek line to the TM sprayer. Lock the syringe in place using the external syringe pump.
18. Set up samples on sprayer stage and tape the slides down to the stage.
19. Change X and Y parameters to define the spraying area.
20. Confirm nozzle temperature has reached 45 °C, heated tray is ambient, and flow rate through external syringe pump is 25 μL/min.
21. Push "start" on the external syringe pump and begin spraying isoamylase at 25 μL/min.
22. Confirm there is an even application of isoamylase using a test slide prior to starting the spray on the tissue samples.
23. Using controlling computer, click "start" and spray 15 passes of isoamylase onto slides with tissues and standards.

24. Once the spray is complete, allow the slides to incubate in temperature- and humidity- controlled incubation chamber inside a 37 °C incubator for 2 h (*see* **Note 9**).
25. Desiccate slides for 15 min, or overnight at room temperature.

3.6 Matrix Application

1. Take CHCA out of −20 °C storage and allow the power to warm up to room temperature (*see* **Note 10**).
2. Prepare CHCA at 7 mg/mL using CHCA powder and CHCA stock solution.
3. Weigh CHCA powder directly into a 15 mL conical using a micro spatula designated for MALDI sample preparation (*see* **Note 11**).
4. Pipette CHCA stock solution into 15 mL conical.
5. Vortex briefly (*see* **Note 12**).
6. Sonicate for 10 min at 89 s on and 30 s off.
7. Pull CHCA solution up into glass syringe using a clean 18-gauge needle.
8. Remove needle and replace with 0.2 μm filter.
9. Push solution into a 50 mL conical and set aside.
10. Turn on TM sprayer, turn on nitrogen gas, and stabilize pressure to 10 psi.
11. Open TM controlling computer software and confirm sprayer settings—Nozzle temperature: 79 °C, heated tray temperature: 50 °C, flow rate: 100 μL/min.
12. Prime Azura Knauer pump by unscrewing the black cap on the front of the pump, then pull approximately 5 mL of degassed 50% methanol through the pump before securing the black cap back in place (*see* **Note 13**).
13. While the sprayer is in the "LOAD" position, flush the central line with 5 mL of 50% acetonitrile.
14. Using a clean glass syringe and a clean needle, pull up the pre-filtered CHCA solution.
15. While sprayer is still in "LOAD" position, inject CHCA matrix slowly into the central line (*see* **Note 8**).
16. Set up samples on sprayer stage and tape the slides down to the stage.
17. Change X and Y parameters to define the spraying area using controlling computer.
18. When nozzle temperature has reached 79 °C, heated tray is at 50 °C, and flow rate is 100 μL/min, switch sprayer from "LOAD" to "SPRAY."
19. Confirm there is an even application of matrix on a test slide prior to starting the spray on the tissue samples.

20. Press "start" on TM controlling computer to begin the matrix application and spray 10 passes of CHCA at 100 μL/min onto tissue and standards.
21. Monitor spray to ensure matrix application is even across all samples.
22. Switch sprayer back to "LOAD" when spray is complete.
23. Remove slides and lay flat to dry at room temperature.
24. Let slides sit for at least 5 min before loading them into the mass spectrometer.
25. Slides can be scanned on MALDI immediately or stored in a vacuum desiccator at room temperature for up to 1 week.

3.7 MALDI-MSI Settings

Waters Synapt XS MALDI-MSI is equipped with a 6-month old, Nd:YAG UV laser. Laser energy set to 250μjoules, and the firing rate set to 1000 Hz/s. The raster size can be 50 μm–200 μm, and time per pixel as well as the scan rate are both set to 0.5 s. The mass spectrometer trap collision energy is set to 4 V and the transfer cell collision energy set to 2 V. Mass spectrometer polarity set to positive mode, and analyzer set to sensitive mode. Data acquisition and spectrums were uploaded to High-Definition Imaging (HDI) software for mass range analysis of 500–3500 *m/z*.

4 Notes

1. To create a molten consistency, the paraffin wax should be heated between 58–60 °C. This step is best carried out in a histology embedding station that has heating plates to maintain the temperature of the wax over time.
2. Use a clean tray filled with HPLC water at the base of the incubator to maintain adequate moisture. As an alternative to the oven, a traditional heat-block with a domed-lid can be used.
3. For all dewaxing steps, fill individual coplin jars with designated solution and submerge the slides sequentially in each solution. Dewaxing solutions should be changed every 8 to 10 slides prepared.
4. The same desiccator is used for all desiccation steps, including long-term storage of slides.
5. For personal safety, optional metal tongs are used to remove the mailers after antigen presentation. If tongs are not available, use heat-resistant gloves to remove the mailers.
6. The water level will vary depending on the number of slides prepared.

7. Trace the area around the standards with a marker on the reverse side of a slide to distinguish where the standard was deposited, standards will be transparent once dried.
8. It is critical to ensure there are no bubbles in the enzyme or matrix syringe prior to injecting solution into the TM sprayer. Bubbles in the sprayer can cause sputtering of enzyme or matrix onto the tissue samples, leaving an uneven application and inconsistent results.
9. Temperature- and humidity-controlled enzyme incubation chamber should be placed in a 37 °C incubator with a moist paper towel in the base of each chamber. Here, it is crucial to ensure slides are flat during enzyme incubation, otherwise glucose polymer release will be uneven across the tissue.
10. Allow the CHCA powder to warm up to room temperature before opening the container or condensation will introduce errors in molar calculation and destabilize CHCA over time. When CHCA is first opened, the powder has high static charge within the glass jars. Optional anti-static device used prior to opening will mitigate the CHCA powder from dispersion. Repeated thawing affects the integrity of the CHCA, and we recommend splitting the CHCA stock to single-use aliquots.
11. Micro spatulas are to be cleaned only with acetonitrile or methanol followed by HPLC water, and not detergent. Trace amounts of detergent are not compatible with MALDI-MSI.
12. Vortexing this solution will not be sufficient to dissolve the CHCA powder. This step is to ensure CHCA stuck to the sides of the 15 mL conical are within solution and will be fully dissolved during sonication.
13. This protocol can be adjusted if sputtering occurs during the matrix spray. Draw another 5 mL of degassed, 50% methanol through the Knauer pump and test on a blank slide. Repeat if necessary.

Acknowledgments

R.C.S. is supported through NIH R01 grant AG066653-01; St. Baldrick's career development award; Rally foundation independent investigator grant; V-scholar foundation award; and University of Kentucky College of Medicine and Markey Cancer Center start-up funds. This research was also supported by funding from the University of Kentucky Markey Cancer Center and the NIH-funded Biospecimen Procurement & Translational Pathology Shared Resource Facility of the University of Kentucky Markey Cancer Center P30CA177558.

References

1. Brewer MK, Gentry MS (2019) Brain glycogen structure and its associated proteins: past, present and future. Adv Neurobiol 23:17–81. https://doi.org/10.1007/978-3-030-27480-1_2
2. Sullivan MA, Nitschke S, Skwara EP, Wang P, Zhao X, Pan XS, Chown EE, Wang T, Perri AM, Lee JPY, Vilaplana F, Minassian BA, Nitschke F (2019) Skeletal muscle glycogen chain length correlates with insolubility in mouse models of polyglucosan-associated neurodegenerative diseases. Cell Rep 27(5): 1334–1344.e6. https://doi.org/10.1016/j.celrep.2019.04.017
3. Shrivastava A (2018) Introduction to plastics engineering. William Andrew, Cambridge, MA
4. Melendez R, Melendez-Hevia E, Cascante M (1997) How did glycogen structure evolve to satisfy the requirement for rapid mobilization of glucose? A problem of physical constraints in structure building. J Mol Evol 45:446–455
5. Nitschke F, Sullivan MA, Wang P, Zhao X, Chown EE, Perri AM, Israelian L, Juana-Lopez L, Bovolenta P, Rodriguez de Cordoba S, Steup M, Minassian BA (2017) Abnormal glycogen chain length pattern, not hyperphosphorylation, is critical in Lafora disease. EMBO Mol Med 9(7):906–917. https://doi.org/10.15252/emmm.201707608
6. Wasserman DH (2009) Four grams of glucose. Am J Physiol Endocrinol Metab 296(1): E11–E21. https://doi.org/10.1152/ajpendo.90563.2008
7. Krebs HA, Bennett DAH, DeGasquet P, Gascoyne T, Yoshida T (1963) The effect of diet on the Gluconeogenic capacity of the rat-kidney-cortex slices. Biochem J 86(22):22–27
8. Brandstrup N, Kretchmer N (1965) The metabolism of glycogen in the lungs of the fetal rabbit. Dev Biol 11:202–216
9. Spicer SS (1957) Histological localization of glycogen in the urinary tract and lung. J Histochem Cytochem 6:52–60
10. Evans G (1934) The glycogen content of the rat heart. J Physiol 82(4):468
11. Brown AM, Ransom BR (2007) Astrocyte glycogen and brain energy metabolism. Glia 55(12):1263–1271. https://doi.org/10.1002/glia.20557
12. Fernandes J (1995) The history of the glycogen storage diseases. Eur J Pediatr 154: 423–424
13. Hojlund K, Staehr P, Hansen BF, Green KA, Hardie DG, Richter EA, Beck-Nielsen H, Wojtaszewski JFP (2003) Increased phosphorylation of skeletal muscle glycogen synthase at NH2-terminal sites during physiological hyperinsulinemia in type 2 diabetes. Diabetes 52:1393–1402
14. Roach PJ (2015) Glycogen phosphorylation and Lafora disease. Mol Asp Med 46:78–84. https://doi.org/10.1016/j.mam.2015.08.003
15. Zois CE, Harris AL (2016) Glycogen metabolism has a key role in the cancer microenvironment and provides new targets for cancer therapy. J Mol Med (Berl) 94(2):137–154. https://doi.org/10.1007/s00109-015-1377-9
16. Postler E, Sindern E, Vorgerd M, Schmitz I, Malin J, Müller K (2002) Fatal cardiomyopathy in adult in polyglucosan body disease. Pathologe 23(3):229–234
17. Stojkovic T, Chanut A, Laforêt P, Madelaine A, Petit F, Romero N, Malfatti E (2018) Severe asymmetric muscle weakness revealing glycogenin-1 polyglucosan body myopathy. Muscle Nerve 57(5):E122–E124
18. Lu J-Q, Phan C, Zochodne D, Yan C (2016) Polyglucosan bodies in intramuscular nerves: association with muscle fiber denervation atrophy. J Neurol Sci 360:84–87
19. Valles-Ortega J, Duran J, Garcia-Rocha M, Bosch C, Saez I, Pujadas L, Serafin A, Canas X, Soriano E, Delgado-García JM (2011) Neurodegeneration and functional impairments associated with glycogen synthase accumulation in a mouse model of Lafora disease. EMBO Mol Med 3(11):667–681
20. Sinadinos C, Valles-Ortega J, Boulan L, Solsona E, Tevy MF, Marquez M, Duran J, Lopez-Iglesias C, Calbó J, Blasco E (2014) Neuronal glycogen synthesis contributes to physiological aging. Aging Cell 13(5): 935–945
21. Orhan Akman H, Emmanuele V, Kurt YG, Kurt B, Sheiko T, DiMauro S, Craigen WJ (2015) A novel mouse model that recapitulates adult-onset glycogenosis type 4. Hum Mol Genet 24(23):6801–6810
22. Nitschke S, Petkovic S, Ahonen S, Minassian BA, Nitschke F (2020) Sensitive quantification of alpha-glucans in mouse tissues, cell cultures, and human cerebrospinal fluid. J Biol Chem 295(43):14698–14709. https://doi.org/10.1074/jbc.RA120.015061
23. DePaoli-Roach AA, Contreras CJ, Segvich DM, Heiss C, Ishihara M, Azadi P, Roach PJ (2015) Glycogen phosphomonoester distribution in mouse models of the progressive

myoclonic epilepsy, Lafora disease. J Biol Chem 290(2):841–850. https://doi.org/10.1074/jbc.M114.607796

24. Young LEA, Brizzee CO, Macedo JKA, Murphy RD, Contreras CJ, DePaoli-Roach AA, Roach PJ, Gentry MS, Sun RC (2020) Accurate and sensitive quantitation of glucose and glucose phosphates derived from storage carbohydrates by mass spectrometry. Carbohydr Polym 230:115651. https://doi.org/10.1016/j.carbpol.2019.115651
25. Brust H, Orzechowski S, Fettke J (2020) Starch and glycogen analyses: methods and techniques. Biomolecules 10(7):1020. https://doi.org/10.3390/biom10071020
26. Roth Z, Yehezkel G, Khalaila I (2012) Identification and quantification of protein glycosylation. Int J Carbohydr Chem 2012:1–10. https://doi.org/10.1155/2012/640923
27. Lee YJ, Perdian DC, Song Z, Yeung ES, Nikolau BJ (2012) Use of mass spectrometry for imaging metabolites in plants. Plant J 70(1): 81–95
28. Hansen RL, Lee YJ (2018) High-spatial resolution mass spectrometry imaging: toward single cell metabolomics in plant tissues. Chem Rec 18(1):65–77
29. Camuzeaux S, Timms JF (2016) Disease profiling by MALDI MS analysis of biofluids. In: Advances in MALDI and laser-induced soft ionization mass spectrometry. Springer, Cham, pp 185–196
30. Boggio KJ, Obasuyi E, Sugino K, Nelson SB, Agar NY, Agar JN (2011) Recent advances in single-cell MALDI mass spectrometry imaging and potential clinical impact. Expert Rev Proteomics 8(5):591–604
31. Jackson SN, Wang H-YJ, Woods AS (2005) In situ structural characterization of phosphatidylcholines in brain tissue using MALDI-MS/MS. J Am Soc Mass Spectrom 16(12): 2052–2056
32. Cornett DS, Reyzer ML, Chaurand P, Caprioli RM (2007) MALDI imaging mass spectrometry: molecular snapshots of biochemical systems. Nat Methods 4(10):828–833
33. Lemaire R, Desmons A, Tabet J, Day R, Salzet M, Fournier I (2007) Direct analysis and MALDI imaging of formalin-fixed, paraffin-embedded tissue sections. J Proteome Res 6(4):1295–1305
34. Drake RR, Powers TW, Norris-Caneda K, Mehta AS, Angel PM (2018) In situ imaging of N-Glycans by MALDI imaging mass spectrometry of fresh or formalin-fixed paraffin-embedded tissue. Curr Protoc Protein Sci 94(1):e68. https://doi.org/10.1002/cpps.68
35. Powers TW, Neely BA, Shao Y, Tang H, Troyer DA, Mehta AS, Haab BB, Drake RR (2014) MALDI imaging mass spectrometry profiling of N-glycans in formalin-fixed paraffin embedded clinical tissue blocks and tissue microarrays. PLoS One 9(9):e106255. https://doi.org/10.1371/journal.pone.0106255
36. Drake RR, Powers TW, Jones EE, Bruner E, Mehta AS, Angel PM (2017) MALDI mass spectrometry imaging of N-linked glycans in cancer tissues. Adv Cancer Res 134:85–116. https://doi.org/10.1016/bs.acr.2016.11.009
37. Yokobayashi K, Misaki A, Harada T (1970) Purification and properties of pseudomonas isoamylase. Biochim Biophys Acta 212: 458–469
38. Kokkat TJ, Patel MS, McGarvey D, LiVolsi VA, Baloch ZW (2013) Archived formalin-fixed paraffin-embedded (FFPE) blocks: a valuable underexploited resource for extraction of DNA, RNA, and protein. Biopreserv Biobank 11(2):101–106. https://doi.org/10.1089/bio.2012.0052

Chapter 16

MALDI Mass Spectrometry Imaging of Lipids on Free-Floating Brain Sections and Immunohistochemically Colocalized Markers of Neurodegeneration

Štěpán Strnad, Veronika Strnadová, David Sýkora, Josef Cvačka, Lenka Maletínská, and Vladimír Vrkoslav

Abstract

In mass spectrometry imaging (MSI), the essential steps in sample preparation include collection and storage. The most widely used preservation procedure for MSI consists in freezing samples and storing them at temperatures below −80 °C. On the other hand, the most common method for preserving biological samples in clinical practice is their fixation in paraformaldehyde. The storage of free-floating sections is a particular type of the preservation of paraformaldehyde-fixed tissues that is used in immunohistochemistry. This chapter describes the approach of the multimodal imaging of free-floating brain sections using the MSI of lipids and the immunohistochemistry of neurodegeneration markers.

Key words Colocalization, Free-floating, Immunohistochemistry, MALDI MSI, Multimodal, Tissue section

1 Introduction

Mass spectrometry imaging (MSI) is a technology that enables the mapping of the spatial distribution of compounds within tissues. High-quality results require proper sample handling (namely collection) and storage. The enzymatic activity needs to be stopped after the collection of biological tissue samples to reduce the delocalization and degradation of molecules. A typical preservation method for MSI is the flash freezing of freshly dissected tissues. These can be frozen in liquid nitrogen, liquid isopentane, ethanol, or hexane on dry ice [1–3]. In order to prevent tissue deformation and fracture, the tissues should be wrapped in aluminum foil and frozen on dry ice-chilled isopentane [2]. Properly frozen samples can be stored at −80 °C until sectioning with no significant degradation for at least 1 year [4].

Young-Jin Lee (ed.), *Mass Spectrometry Imaging of Small Molecules: Methods and Protocols*, Methods in Molecular Biology, vol. 2437, https://doi.org/10.1007/978-1-0716-2030-4_16,
© The Author(s), under exclusive license to Springer Science+Business Media, LLC, part of Springer Nature 2022

However, the traditional fixation method used in pathology, histology, and so on is paraformaldehyde (PFA) fixation [5]. The fixation is usually done in 4% paraformaldehyde in phosphate-buffered saline (PBS) for approximately 24 h [2], which causes protein crosslinking. In most cases, the fixed tissues are then paraffin-embedded and stored at room temperature [3]. Because of the paraffin embedding, it is necessary to dewax the samples with organic solvents before analysis. However, this process is incompatible with lipid analysis because organic solvents dissolve lipid molecules [6–9]. Several studies have been published on paraformaldehyde-fixed tissues without paraffin embedding, showing that such tissues are suitable for lipid analysis [6, 10, 11].

Free-floating section storage is a type of sample preparation typically used in immunohistochemistry (IHC). It is referred to as free-floating because the whole staining process is performed with the sections floating in a staining or washing solution [12, 13]. After paraformaldehyde fixation, the tissues are cut and the sections are stored in a cryoprotectant solution. This solution usually consists of a combination of glycerol, ethylene glycol, and PBS. The prepared floating sections are then stored in a freezer at −20 °C.

Our protocol focuses on the preparation and use of free-floating brain sections for multimodal analysis [14]. The procedure makes it possible to utilize the same tissue section for MALDI MSI and immunohistochemical experiments for the localization of lipids with other chemical markers, which allows a better correlation of MSI data with immunohistochemical analysis. Moreover, the MSI analysis of the free-floating sections improves the usability of biological material. Described workflow reduces the time of sample preparation and the number of animals needed for the experiments.

2 Materials

Use analytical-grade reagents and MS grade solvents. Prepare the aqueous solutions using ultrapure water (18 MΩ.cm at 25 °C).

2.1 Paraformaldehyde-Fixed Free-Floating Sections

1. Phosphate-buffered saline (PBS, pH 7.4): Weigh 8.00 g of NaCl (final concentration 137 mM), 0.20 g of KCl (final concentration 2.7 mM), 1.78 g of Na_2HPO_4 $2H_2O$ (final concentration 10 mM), and 0.24 g of KH_2PO_4 (final concentration 1.8 mM). Add approximately 800 mL of water. Mix and adjust pH to 7.4 with HCl (1 M) or NaOH (1 M). Then adjust the volume with water to 1 L. Store at room temperature (*see* **Note 1**).

2. 4% Paraformaldehyde (PFA) in 0.1 M phosphate buffer (PB): Weigh 35.6 g of $Na_2HPO_4 \cdot 2H_2O$ (final concentration 0.1 M) and adjust the volume to 2 L with water. Weigh 7.8 g of NaH_2PO_4 $2H_2O$ (final concentration 0.1 M), add water to 500 mL, and dissolve. Mix 373 mL of 0.1 M NaH_2PO_4 solution and 1627 mL of 0.1 M Na_2HPO_4 solution and adjust pH to 7.4 using 0.1 M Na_2HPO_4 solution. Weigh 14.02 g of NaCl (final concentration 120 mM) and mix with NaH_2PO_4/ Na_2HPO_4 solution to the final volume of 2 L. Transfer 1.8 L of 0.1 M PB pH 7.4 to an appropriate glass beaker and heat it to 58–60 °C while stirring on a magnetic stirrer (*see* **Note 2**). Weigh 80 g of PFA and add it to heated 0.1 M PB. Adjust pH with 0.1 M NaOH drop by drop until the solution clears and adjust volume with 0.1 M PB to the final volume of 2 L (*see* **Note 3**).
3. 30% Sucrose in PBS with 0.1% sodium azide: Weigh 0.10 g of sodium azide, add PBS (pH 7.4) to the volume of 50 mL, and mix. Weigh 30 g of sucrose and add it to the NaN_3 and PBS solution. Mix until dissolved. Adjust to a final volume of 100 mL with PBS (*see* **Note 4**).
4. Freeze spray (*see* **Note 5**).
5. Cryoprotectant solution: Mix 300 mL of 30% ethylene glycol with 250 mL of 25% glycerol in PBS. Adjust to a final volume of 1 L with PBS (pH 7.4) and mix. Store in the fridge.

2.2 Sample Preparation for MSI

1. Acetonitrile (ACN).
2. Acetic acid (AcOH).
3. 10 mg/mL 1,5-diaminonaphthalene (DAN) matrix solution in 70% ACN. Weigh 30 mg of DAN and add 3 mL of 70% ACN. Sonicate in an ultrasonic bath for 5 min. The matrix solution should be prepared fresh before matrix spraying (*see* **Note 6**).
4. ITO glass slides for MALDI imaging (Bruker, Bremen, Germany).
5. Red phosphorus: 10 mg/mL in ACN. Weigh 1 mg of red phosphorus and add 0.1 mL of ACN (*see* **Note 7**).

2.3 Immunohistochemistry

1. 99% Ethanol (EtOH).
2. Antigen-retrieval buffer: 100-fold water-diluted citrate buffer, pH 6.0 (or 0.01 M sodium citrate buffer, pH 6.0).
3. Hydrophobic barrier pen (*see* **Note 8**).
4. Normal goat serum (diluted to 5% solution in PBS-T).
5. Primary antibodies: rabbit anti-GFAP pAb, mouse anti-APP/β-amyloid mAb (diluted 1:400 in PBS-T).
6. Secondary antibodies: goat anti-mouse secondary antibody Alexa Fluor 488, goat anti-rabbit secondary antibody Alexa Fluor 586 (diluted 1:400 in PBS-T).

7. 4,6-Diamidino-2-phenylindole (DAPI): Prepare according to the manufacturer's instructions. Prepare aliquots of stock solution and store at −20 °C for a maximum of 12 months (*see* **Note 9**).
8. 0.2% Triton X-100 in PBS (PBS-T): Add 2 mL of Triton X-100 into 1 L of PBS (*see* Subheading 2.1) and mix well (*see* **Note 10**).
9. A mounting medium with DAPI.
10. Nail polish.

2.4 Instruments and Software

1. Cryostat.
2. iMatrixSpray automatic sprayer (Tardo, Subingen, Switzerland).
3. ultrafleXtreme MALDI TOF/TOF instrument (Bruker, Bremen, Germany).
4. FlexControl and FlexImaging software (Bruker, Bremen, Germany).
5. SCiLS Lab 2016b software (Bruker, Bremen, Germany).
6. Graphical software with overlay function (e.g., commercial Photoshop or opensource Fiji (ImageJ).
7. Fluorescent confocal microscope.
8. Scanner.

3 Methods

3.1 Fixation and Sectioning

1. Fix the dissected brains in 4% PFA in 0.1 M PB at 4 °C for 24 h.
2. Wash the tissues twice in PBS and transfer them to 30% sucrose in PBS with 0.1% sodium azide for 24 h (store in the fridge at 4 °C) (*see* **Note 11**).
3. Set the cryostat temperature to −15 °C.
4. Snap-freeze the samples using freeze spray before sectioning (*see* **Note 5**).
5. Cut the tissue into 30-μm-thick sections.
6. Place the brain sections in the cryoprotectant solution and store them in the freezer (−20 °C).

3.2 MALDI MSI

3.2.1 Sample Preparation for MALDI MSI

1. Wash the brain sections in PBS for 5 min and then 3 times in ultrapure water, each time for 5 min.
2. Mount the brain sections on ITO-coated glass slides using a fine-tip brush and place them in a desiccator for 60 min (*see* **Note 12**).

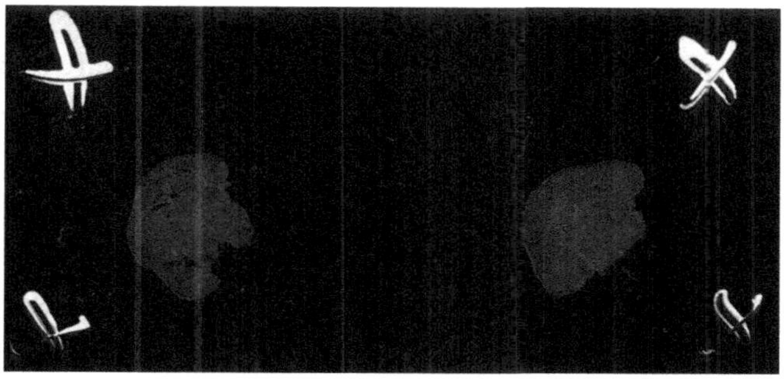

Fig. 1 A high-resolution scan of the slide with teaching points around the samples

3. After drying the tissue, mark at least three teaching points around the tissue using a white correction fluid (*see* Fig. 1).
4. Make a high-resolution scan of the slide and save it as a .jpeg (4800 dpi).

3.2.2 Matrix Deposition for MALDI MSI

1. Prepare a fresh MALDI matrix solution (10 mg/mL DAN).
2. Place the sample in the iMatrixSpray [15] automatic spraying device. Perform spraying in a fume hood. The following procedure has been optimized for lipid analysis [16].
3. Fill the bottle with 3 mL of DAN matrix solution.
4. Adjust nitrogen pressure to 1.2 bar and set the extension glass capillary out of the steel capillary to 0.5 mm.
5. Connect your device (smartphone, laptop, etc.) to iMatrixSpray WiFi and use it to connect to a web interface (type http://10.10.10.10 in your browser) and select the spray parameters.
6. Set following parameters: height 60 mm, speed 180 mm/s, line distance 1 mm, density 2 μm/cm^2, break of 60, cycles 5.
7. Start the spraying sequence. The process automatically starts with a rinsing cycle, delivers the spray solution, and then cleans the system.
8. Spot 0.5 μL of red phosphorus calibrant on the slide (*see* **Note 13**).
9. Place the slide wrapped in aluminum foil in a desiccator for at least 30 min before MSI analysis (*see* **Note 14**).

3.2.3 MALDI MSI Measurement and Analysis

1. Attach the tissue slide with sample to the adapter slide holder and insert it into the MALDI mass spectrometer.
2. Load a suitable positive or negative reflectron MALDI method in flexControl software (*see* **Note 15**).
3. Optimize the instrument parameters (instrument-specific) (*see* **Note 16**).

4. Calibrate the instrument using red phosphorus calibrant.
5. Open flexImaging software and create a new sequence.
6. Enter analysis details: sequence name, data directory, raster width, and autoXecute method (*see* **Note 17**).
7. Import a scan of your sample from Subheading 3.2.1 and set the position of the slide using three teaching points.
8. Choose the area for the imaging experiment with the tool "add measurement region."
9. Start the imaging experiment using the "start autoXecute run" icon in the flexImaging checklist menu (*see* **Note 18**).
10. After finishing the acquisition, data are automatically displayed in flexImaging software.
11. The data can be processed directly in flexImaging or imported to SCiLS Lab software for better visualization and statistical analysis (*see* **Note 19**).

3.3 Immunohistochemistry Staining

1. After MALDI MSI analysis, rinse the samples on MALDI glass slide with 99% EtOH for 120 s and then fix them in 95% EtOH–5% AcOH at −20 °C for 8 min, 70% EtOH at −20 °C for 30 s, and finally in 70% EtOH at room temperature for 30 s (*see* **Note 20**).
2. Wash the samples 3 times in PBS, each time for 5 min. Use an orbital shaker for wash steps.
3. Place the slides in an antigen-retrieval buffer. Incubate the samples in the buffer in a water bath at 90 °C for 30 min and then cool them down to room temperature (*see* **Note 21**).
4. Wash the samples 3 times in PBS, each time for 5 min, then once in PBS-T for 10 min and once in PBS for 5 min.
5. Draw a hydrophobic barrier on the glass slide around each sample with a hydrophobic barrier pen and allow it to dry for a few seconds. Work in a ventilated hood.
6. Block the samples to prevent nonspecific binding of antibodies in 5% normal goat serum at room temperature for 1 h. Pipette the solution carefully on the samples to produce a drop (*see* **Note 22**).
7. Incubate the samples with primary antibodies (rabbit anti-GFAP pAb) against the glial fibrillary acidic protein (GFAP) and Aβ (mouse anti-APP/β-amyloid mAb,) overnight at 4 °C (*see* **Note 23**).
8. Wash the samples twice in PBS-T, each time for 10 min, and once in PBS for 10 min.
9. Incubate the samples with the goat anti-mouse secondary antibody Alexa Fluor 488 and the goat anti-rabbit secondary antibody Alexa Fluor 586 at room temperature in the dark for 1 h (*see* **Note 24**).

10. Wash the samples twice in PBS-T, each time for 10 min, and once in PBS for 10 min.
11. Incubate the samples with 4′,6-diamidino-2-phenylindole (DAPI) for 10 min according to manufacturer's instructions (*see* **Note 25**).
12. Wash the samples twice in PBS-T for 10 min and once in PBS for 10 min.
13. Mount the samples with the mounting medium with DAPI and cover them with a cover glass. Seal the glass with nail polish (*see* **Note 26**).
14. Examine the samples with a fluorescence confocal microscope. The samples can be stored at 4 °C for at least 1 month.

3.4 The Colocalization of Images Acquired from MALDI MSI and Confocal Microscope

1. Open Fiji (ImageJ) software for scientific image analysis or any commercial graphical software (Photoshop etc.) with an image-overlay function. It is used for the spatial overlap between different signals, for example ion distribution map and fluorescent labels.
2. Use the overlay function in Fiji to merge images acquired from MALDI MSI and confocal microscope in the Image section of the menu.
3. Use images of the same scale and significant morphological transitions (tissue edge, cracks) to accurately overlay of the images. An example is shown in Fig. 2 and final image is shown in Fig. 3 (*see* **Note 27**).

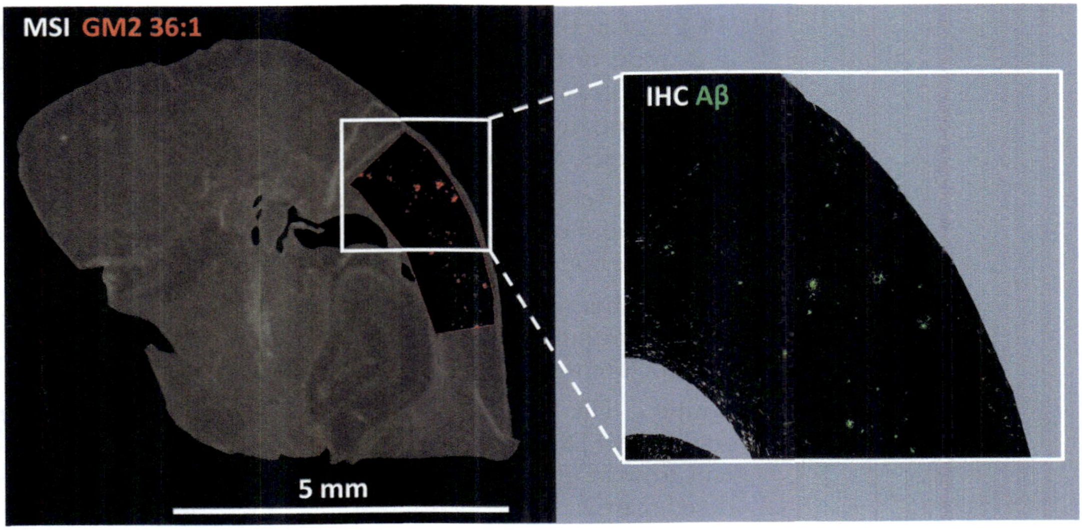

Fig. 2 Example of overlay of the images. Use images of the same scale and significant morphological transitions (tissue edge, cracks) to accurately overlay MALDI MSI and fluorescent images

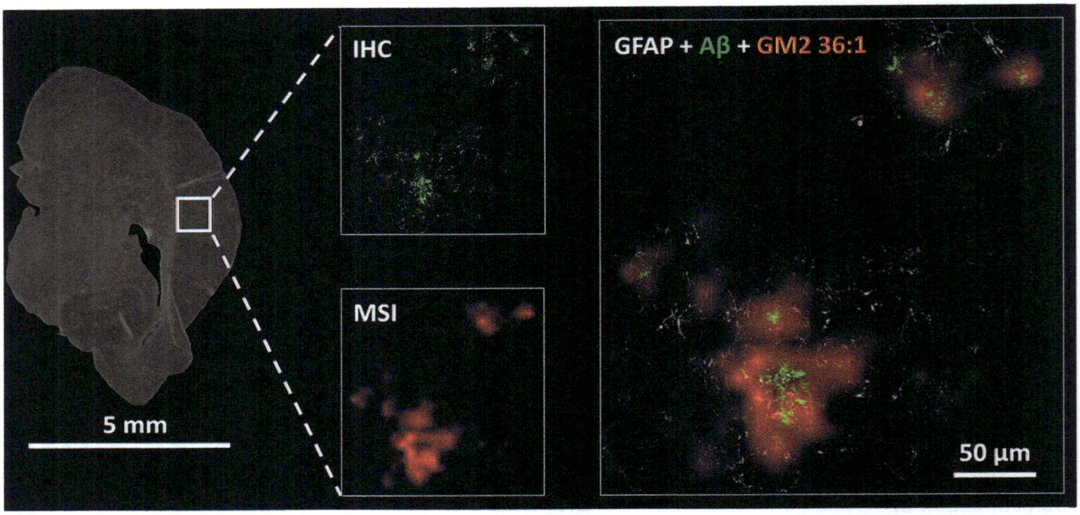

Fig. 3 The colocalization of the lipid changes observed by MALDI MSI and fluorescent staining in the APP/PS1 mouse brain. The ion image of ganglioside (GM2 36:1, *m/z* 1382.7) was obtained in the negative ion mode at a spatial resolution of 15 μm (red). The fluorescent staining of Aβ plaques (green) and the astrocyte marker GFAP (white) were obtained using a fluorescent confocal microscope. [14] - Adapted by permission of The Royal Society of Chemistry

4 Notes

1. PBS can also be made as a 10× stock solution (0.1 M); to prepare 10× PBS, increase the weight of the compounds.
2. Ensure that the solution does not boil, constantly check the temperature.
3. 4% PFA can be aliquoted and stored in the fridge at 4 °C for 1–2 weeks or frozen and stored up to a year at −20 °C. Since PFA is toxic, work in a ventilated hood. Please read the MSDS before working with this chemical. You should wear gloves, safety glasses, and a respirator.
4. Work in a ventilated hood because NaN_3 is very toxic as well. Please read the MSDS before working with this chemical. You should wear gloves, safety glasses, and a respirator.
5. Caution: freeze spray is highly flammable. Avoid the use of the product within the cryostat chamber. Please read the instructions before working with freeze spray.
6. Caution: The DAN matrix is a suspected carcinogen. Please read the MSDS before working with this chemical. It is important to wear gloves and FFP2 masks when handling the powdered matrix.
7. Suspension can be used repeatedly; store it at room temperature.

8. Work in a ventilated hood. Please read the MSDS before working with this chemical.
9. DAPI is a known mutagen and should be handled with care. To prepare its stock solution, dissolve DAPI in water to a final concentration of 1–5 mg/mL. Do not use any buffers.
10. Triton X-100 is a viscous liquid. Use a very slow aspiration and dispensing speed, or use reverse pipetting. Never cut the pipette tip. Please read the MSDS before working with this chemical.
11. Due to sodium azide, the samples can be stored in a fridge at 4 °C for several months.
12. Dried samples can be stored in a freezer at −80 °C in separate boxes for later use.
13. Red phosphorus in ACN is a suspension. Vortex it for 15 s before spotting on the slide. Singly charged phosphorus clusters are formed by laser ionization/desorption and cover the mass range up to *m/z* 3000 in both positive and negative ion modes [17].
14. The DAN matrix is light sensitive and should be stored in the dark.
15. Predefined methods for the negative and positive mode in the low mass range were used.
16. For the ultrafleXtreme MALDI TOF/TOF mass spectrometer equipped with a 1 kHz laser and operated in reflector mode, we set the instrument as follows: the accelerating voltage of the ion source 20 kV in negative polarity and 25 kV in positive polarity, one hundred laser shots per spectrum, laser intensity 75%, global laser attenuator offset 44%, attenuator offset 5%, attenuator range 30%, small spot size and detector gain voltage 2200 V for both polarities.
17. The AutoXecute method can be edited in flexControl. This method includes such features as the optimum laser power and the number of the required shots from Subheading 3.2.3, **step 3**, which involves the optimization of these parameters. For MSI experiments in tab "movement," the random walk should be off.
18. The analysis time depends on the measured area, lateral resolution and method parameters (number of shots, etc.). With the lateral resolution of 15 μm and the parameters specified in **Note 16**, the measurement time was approximately 1.4 h/mm^2. Too long measurement times (more than 9 h) can lead to nonreproducible results due to the loss of the matrix in vacuum [16].

19. For importing data to SCiLS Lab, we set TopHat baseline removal to 150 and normalization to the total ion count (TIC). Receiver operating characteristics (ROC) analysis was used to evaluate the differences between datasets from two groups statistically. The ROC threshold was set to 0.8.
20. Rinse the samples carefully to avoid their detachment from the slider.
21. The water bath is very hot. Be careful when handling the samples.
22. Avoid pipetting an excessive volume. The hydrophobic barrier can hold only a limited volume of the liquid.
23. Mix both primary antibodies together in an appropriate dilution. Avoid pipetting bubbles with a diluted solution of the primary antibody. Place the samples on a straight tray suitable for immunohistochemistry with a cover to retain humidity. Do not shake.
24. Mix both secondary antibodies. Place the samples on a tray suitable for immunohistochemistry with a cover to retain humidity. Do not shake.
25. This step could be skipped if the signal from the mounting medium with DAPI is optimal.
26. Allow the nail polish to dry properly at room temperature on a tray suitable for immunohistochemistry with a dark cover.
27. Figure 3 shows MALDI MSI and fluorescent staining in the APP/PS1 mouse brain. The ion image of ganglioside (GM2 36:1, *m/z* 1382.7) was obtained in the negative ion mode at a spatial resolution of 15 μm (red). The fluorescent staining of Aβ plaques (green) and the astrocyte marker GFAP (white) were obtained using a fluorescent confocal microscope.

Acknowledgments

This work was supported by project No. 20-00546S of the Czech Science Foundation.

References

1. Buchberger AR, DeLaney K, Johnson J, Li L (2018) Mass spectrometry imaging: a review of emerging advancements and future insights. Anal Chem 90(1):240–265
2. Goodwin RJA (2012) Sample preparation for mass spectrometry imaging: small mistakes can lead to big consequences. J Proteome 75(16):4893–4911
3. Norris JL, Caprioli RM (2013) Analysis of tissue specimens by matrix-assisted laser desorption/ionization imaging mass spectrometry in biological and clinical research. Chem Rev 113(4):2309–2342
4. Schwartz SA, Reyzer ML, Caprioli RM (2003) Direct tissue analysis using matrix-assisted laser desorption/ionization mass spectrometry:

practical aspects of sample preparation. J Mass Spectrom 38(7):699–708

5. Werner M, Chott A, Fabiano A, Battifora H (2000) Effect of formalin tissue fixation and processing on immunohistochemistry. Am J Surg Pathol 24(7):1016–1019
6. Pietrowska M, Gawin M, Polańska J, Widłak P (2016) Tissue fixed with formalin and processed without paraffin embedding is suitable for imaging of both peptides and lipids by MALDI-IMS. Proteomics 16 (11-12):1670–1677
7. Casadonte R, Kriegsmann M, Zweynert F, Friedrich K, Bretton G, Otto M, Deininger S-O, Paape R, Belau E, Suckau D, Aust D, Pilarsky C, Kriegsmann J (2014) Imaging mass spectrometry to discriminate breast from pancreatic cancer metastasis in formalin-fixed paraffin-embedded tissues. Proteomics 14 (7-8):956–964
8. Groseclose MR, Massion PP, Chaurand P, Caprioli RM (2008) High-throughput proteomic analysis of formalin-fixed paraffin-embedded tissue microarrays using MALDI imaging mass spectrometry. Proteomics 8(18):3715–3724
9. Lemaire R, Desmons A, Tabet JC, Day R, Salzet M, Fournier I (2007) Direct analysis and MALDI imaging of formalin-fixed, paraffin-embedded tissue sections. J Proteome Res 6(4):1295–1305
10. Carter CL, McLeod CW, Bunch J (2011) Imaging of phospholipids in formalin fixed rat brain sections by matrix assisted laser desorption/ionization mass spectrometry. J Am Soc Mass Spectrom 22(11):1991
11. Harris A, Roseborough A, Mor R, Yeung KKC, Whitehead SN (2020) Ganglioside detection from formalin-fixed human brain tissue utilizing MALDI imaging mass spectrometry. J Am Soc Mass Spectrom 31(3):479–487
12. Potts EM, Coppotelli G, Ross JM (2020) Histological-based Stainings using free-floating tissue sections. J Vis Exp (162)
13. Bachman J (2013) Immunohistochemistry on freely floating fixed tissue sections. Methods Enzymol 533:207–215
14. Strnad Š, Pražienková V, Holubová M, Sýkora D, Cvačka J, Maletínská L, Železná B, Kuneš J, Vrkoslav V (2020) Mass spectrometry imaging of free-floating brain sections detects pathological lipid distribution in a mouse model of Alzheimer's-like pathology. Analyst 145(13):4595–4605
15. Stoeckli M, Staab D, Wetzel M, Brechbuehl M (2014) iMatrixSpray: a free and open source sample preparation device for mass spectrometric imaging. Chimia (Aarau) 68(3):146–149
16. Strnad Š, Pražienková V, Sýkora D, Cvačka J, Maletínská L, Popelová A, Vrkoslav V (2019) The use of 1,5-diaminonaphthalene for matrix-assisted laser desorption/ionization mass spectrometry imaging of brain in neurodegenerative disorders. Talanta 201:364–372
17. Sládková K, Houška J, Havel J (2009) Laser desorption ionization of red phosphorus clusters and their use for mass calibration in time-of-flight mass spectrometry. Rapid Commun Mass Spectrom 23(19):3114–3118

Chapter 17

Silver-Doped Nano-DESI MSI for Increased Specificity and Sensitivity of Alkenes

Johan Lillja and Ingela Lanekoff

Abstract

Nanospray desorption electrospray ionization (nano-DESI) is a technique that can be used for mass spectrometry imaging (MSI) experiments. Due to the separation of sampling and ionization events, the nano-DESI solvent can be doped with additives to alter the specificity and selectivity of the experiment. In this chapter, we describe the addition of silver ions to the nano-DESI solvent for increased specificity and sensitivity of double bond containing analyte molecules for nano-DESI MSI.

Key words Nanospray desorption electrospray ionization, Nano-DESI, Silver adduct, Prostaglandins, Mass spectrometry imaging, MSI, Double bond, π-moiety

1 Introduction

Since the introduction of nano-DESI in 2010 [1] and nano-DESI mass spectrometry imaging (MSI) shortly after [2, 3] the technique has been increasingly used for MSI [4–10] In nano-DESI, molecules from the sample surface are desorbed into a continuously flowing solvent bridge between two capillaries, as previously described in detail [11]. In short, two fused silica capillaries are positioned at an angle to each other in front of the mass spectrometer. The liquid bridge is formed by supplying a solvent through the primary capillary using a syringe pump and aspirating the solvent through the secondary capillary for subsequent electrospray ionization. The solvent through the secondary capillary can be either self-aspirated [1] by close contact to the mass spectrometer inlet or by using a nebulizer [12]. When the liquid bridge makes contact with the sample surface molecules are desorbed, ionized, and detected. By moving the sample systematically under the liquid bridge, molecules from locations all over the sample surface are detected and an ion image can be generated. An important aspect of nano-DESI is that the sampling event is separated from the ionization event. This

Young-Jin Lee (ed.), *Mass Spectrometry Imaging of Small Molecules: Methods and Protocols*, Methods in Molecular Biology, vol. 2437, https://doi.org/10.1007/978-1-0716-2030-4_17,

© The Author(s), under exclusive license to Springer Science+Business Media, LLC, part of Springer Nature 2022

enables doping of the solvent with additives, such as reagents to increase specificity or sensitivity [13–15], or internal standards to account for matrix effects during ionization and allow for direct quantification without any additional sample preparation [4–7, 16].

By doping the nano-DESI solvent with silver ions, the sensitivity and selectivity of alkenes during nano-DESI MSI experiments can be increased [13, 14]. Silver ions complex with the π-moiety of an alkene and form silver adduct ions to be detected in the mass spectrometer [17]. The characteristic isotope pattern of $^{107}Ag/^{109}Ag$ (51/49) is easily recognizable in the mass spectrum, thereby enabling simple identification of silver-containing ions. In addition, molecules that contain an alkene bond moiety can be identified and separated from isobars because the added silver shifts the ion to a higher *m/z* value. Finally, prostaglandins as silver adducts show ~30 times increase in sensitivity compared to deprotonated prostaglandins in negative ion mode, which enables direct imaging of prostaglandins in thin tissue sections (Fig. 1) [13]. To further maximize sensitivity, monoisotopic silver can be used to keep the analyte signal in a single mass channel [14].

2 Materials

2.1 Solvent

To successfully use silver ion doped nano-DESI MSI the desorption of chloride from the tissue section must be minimized to reduce the formation of $AgCl_{(s)}$. Chloride has low solubility in acetonitrile and therefore acetonitrile–methanol (9:1) is a good solvent to choose (*see* **Note 1**). The addition of 0.1% formic acid to the solvent protonates carboxylic acid moieties and thereby maximizes silver cationization at the location of the double bond. A concentration of 10 ppm Ag^+ is sufficient for tissue imaging using nano-DESI MSI.

1. Pure water (18.2 MΩ at 25 °C).
2. $AgNO_3$ (>95% purity).
3. Acetonitrile (HPLC-MS grade).
4. Methanol (HPLC-MS grade).
5. Formic acid (>95% purity).
6. Heatproof vessel (optional).
7. Monoisotopic silver (optional).
8. Sub boiled nitric acid (optional).

2.2 Nano-DESI

1. Fused silica capillaries 150 × 50 μm (OD × ID) (*see* **Note 2**).
2. Ceramic capillary cutter.
3. PEEK fittings and unions for connecting capillaries.

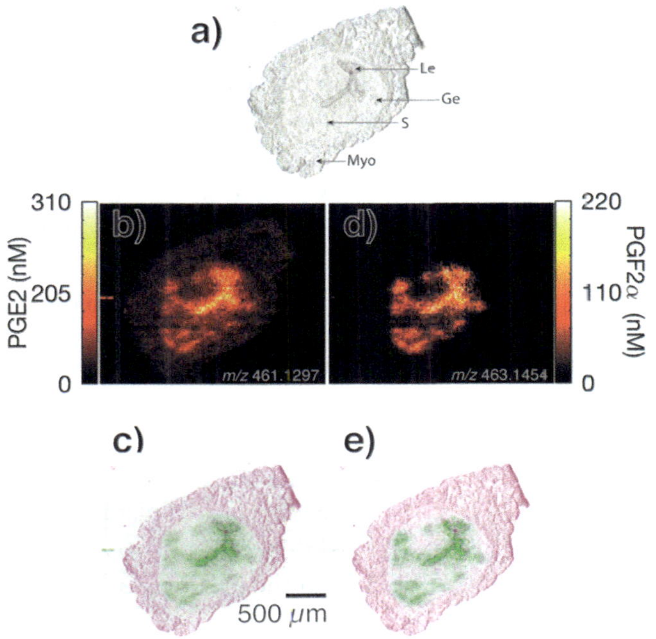

Fig. 1 Quantitative silver-doped nano-DESI MSI of two prostaglandins (PGE_2 and $PGF_{2\alpha}$) in mouse embryo implantation sites using thin capillaries and oversampling to achieve pixel sizes of 10×30 μm, (**a**) Optical image of the tissue (**b**) quantitative nano-DESI image of PGE_2 showing localization of this molecule to the luminal epithelium and detected concentrations up to ~300 nM (**c**) false-color overlay of the PGE_2 distribution onto the optical image. Similarly in (**d**) and (**e**) showing $PGF_{2\alpha}$ also localizing to the luminal epithelium at a detected concentration up to ~200 nM. (Reprinted with permission from K. Duncan et al. [13] Copyright 2018 American Chemical Society)

4. 500 μL Flat head syringe.
5. Syringe pump.
6. Microscope with long working distance.
7. PEEK tubing with an ID that fit the fused silica capillaries (*see* **Note 3**).
8. Two micromanipulators.
9. Modified ion source with voltage cable with alligator clip.
10. Computer controllable XYZ-stage with sample holder.

2.3 Sample Preparation

1. Cryotome.
2. Glass slides (*see* **Note 4**).
3. Liquid nitrogen.

3 Methods

3.1 Preparation of Sample

1. Snap freeze tissue rapidly after removal using liquid nitrogen to quench metabolism.
2. Mount the frozen tissue on the mounting bracket of a cryomicrotome using water (*see* **Note 5**).
3. Cut thin tissue sections and thaw mount them onto the slide (*see* **Notes 6** and 7).
4. Store at −80 °C and rapidly thaw prior to analysis.

3.2 Preparation of Solvent

3.2.1 Preparation of Monoisotopic $AgNO_3$ (Optional)

1. Accurately weigh $^{107}Ag_{(s)}$ or $^{109}Ag_{(s)}$ in a heat proof vessel.
2. Add sub-boiled HNO_3 to the vessel to dissolve the silver (*see* **Note 8**).
3. After the initial reaction with HNO_3 has slowed down put the vessel on a heated plate and gently boil off the excess nitric acid (*see* **Note 8**).
4. Weigh the product to calculate the yield (*see* **Note 9**).

3.2.2 Preparation of Silver-Doped Nano-DESI Solvent

1. Accurately weigh $AgNO_3$ and dissolve it in a known mass of pure water (18.2 MΩ at 25 °C) to a final concentration of ~2000 ppm (*see* **Note 10**).
2. Store the silver solution at 4 °C in a tinted vial or wrap a clear vial in aluminum foil to protect it from sunlight (*see* **Note 11**).
3. Mix a solution containing 9 parts acetonitrile and one part methanol with the addition of 0.1% formic acid and Ag^+ stock solution to a final concentration of 10 ppm Ag^+ (*see* **Note 12**).
4. Store the solvent cold and degas by sonication before analysis.

3.3 Nano-DESI Setup

The nano-DESI setup is depicted in Fig. 2. Capillaries are mounted on micromanipulators and placed on a modified nano-ESI interface above a computer controllable XYZ-stage.

3.3.1 Secondary Capillary

1. Cut an approximately 2 cm-long fused silica capillary using the ceramic cutter tool (*see* **Note 13**).
2. Inspect the cut under a microscope to ensure that no irregularities are present. A good cut is straight and without ragged edges.
3. Cut an approximately one-centimeter-long piece of PEEK tubing and slide the capillary in such that both ends protrude equally long.
4. Glue the PEEK tubing and capillary in place, using a leftover piece of capillary tubing as an applicator (*see* **Note 14**).

Fig. 2 Nanospray desorption electrospray ionization set up in front of a Q Exactive™ instrument (**a**) picture showing micromanipulators fixating the primary and secondary capillaries (A), motorized sample stage (B), digital microscopes to visualize capillary alignment, sampling location, and interactions between the liquid bridge and the surface during acquisition (C). (**b**) Close up picture of the secondary capillary (D) and the primary capillary (E) aligned to the inlet of the MS

3.3.2 Primary Capillary

1. Cut approximately 50 cm of fused silica tubing using the ceramic cutter tool.
2. Inspect the cut under a microscope to ensure that no irregularities are present. A good cut is straight and without ragged edges.
3. Cut a one-centimeter-long piece of PEEK tubing and slide one end of the capillary inside, such that ~2 cm protrudes.
4. Glue the PEEK tubing and capillary in place, using a leftover piece of capillary tubing as an applicator (*see* **Note 14**).
5. Add a micro sleeve at the other end of the capillary and slide a male PEEK fitting over, locking it in place with a PEEK union (*see* **Note 15**).
6. Attach a flat head Hamilton syringe to the setup by firmly pressing it into a male PEEK fitting and screwing it finger tight into the union (*see* **Note 16**).
7. Test the coupling for leaks by gently pushing solvent through the capillary before starting the experiment.

3.3.3 Setting Up the Nano-DESI by the Mass Spectrometer

1. Place the secondary capillary in a micromanipulator, using the PEEK fitted part as an anchoring point, as depicted in Fig. 2a.
2. Align the secondary capillary to the inlet of the mass spectrometer, using the digital microscope for guidance, as shown in Fig. 2b.

3. Place the primary capillary in a second micromanipulator, using the PEEK part as an anchoring point.
4. Align the primary capillary to the other end of the secondary capillary, as shown in Fig. 2b. The primary capillary should be slightly below the secondary capillary, using the digital microscope for guidance (*see* **Note 17**).
5. Supply the solvent using a syringe pump at a flow rate of 0.5 $\mu L min^{-1}$ (*see* **Note 18**).
6. Adjust the position of the capillaries until the liquid bridge between the two capillaries is stable.
7. Finally, attach the voltage cable to the metal part of the syringe using an alligator clip.
8. Adjust the setup until a stable total ion current is obtained.

3.3.4 Running the Imaging Experiment

1. Put the sample slide on a sample holder mounted to a computer controllable stage that can move in XYZ.
2. Move the sample holder using the motorized stage such that the liquid bridge touches the glass surface, ensure that the liquid bridge is steady on the surface and that the TIC is stable.
3. Move the sample under the liquid bridge in subsequent lines while continuously acquiring data. Use a constant velocity (*see* **Note 19**) and step between lines with a defined distance (*see* **Notes 20** and **21**).

3.3.5 Mass Spectrometry Method and Data Analysis

1. Exclude abundant signals from silver-acetonitrile complexes at m/z 147.9316/149.9313 ([ACN + Ag^+]) and 188.9581/190.9579 ([2ACN + Ag]$^+$) by setting the lowest m/z in the scan window to m/z 192 to maximize the intra scan dynamic range.
2. Use monoisotopic silver or a mass spectrometer with sufficient mass resolving power to separate the analytes [M + ^{107}Ag]$^+$ and [M-H_2 + ^{109}Ag]$^+$ ($\Delta m/z$ 0.016 Th), as shown in Fig. 3b.
3. Identify silver ion adducts in the mass spectra by searching for almost equally intense peaks of the natural isotopic abundance of silver (51% ^{107}Ag and 49% ^{109}Ag) with a mass difference of 1.9996 Th, as shown in Fig. 3a.
4. Optionally, use monoisotopic silver to maximize sensitivity by keeping the silver adduct ions in one mass channel instead of splitting the signal into two.

4 Notes

1. Any solvent suitable for electrospray ionization that has a low solubility of chloride can, in theory, be used.

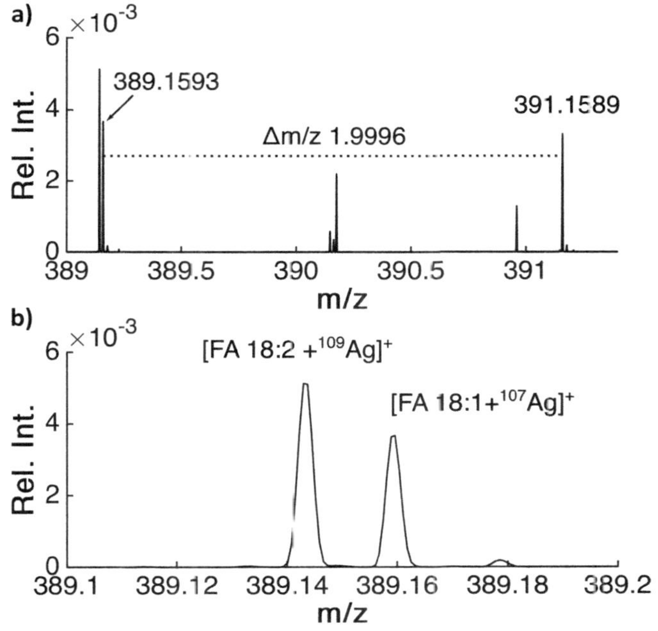

Fig. 3 Mass spectrum obtained from rat brain tissue using a silver-doped nano-DESI solvent (**a**) identification of FA 18:1 using the accurate mass, silver isotopic pattern, and isotopic mass difference (**b**) high-resolution mass spectrum detailing separation of [FA 18:1 + $^{107}Ag]^+$ and [FA 18:2 + $^{109}Ag]^+$ ($\Delta m/z$ 0.016 Th)

2. The dimensions of the capillaries can be changed, smaller capillaries will produce a smaller droplet and therefore higher spatial resolution.
3. The outer diameter of the peek tubing should be chosen such that the micromanipulator can fixate the capillary assembly and the inner diameter such that the capillary fits nicely.
4. Generally, regular microscope glass slides are used, but any material, including plastic and ITO coated slides, will work.
5. Use water as a fixating agent or add OCT only on a small defined part of the tissue to attach it to the mount. If OCT ends up on the tissue section for analysis it suppresses the signal from endogenous molecules and generates contaminations seen as polymer peaks in the mass spectrum.
6. 6 By pressing the slide against a section and using the heat from a finger on the opposite side of the glass slide the section will briefly thaw and adhere to the glass slide.
7. Tissue sections used for imaging are typically 10–12 μm, however, both thinner and thicker sections are possible to use.
8. Work in a fume hood. Boiling nitric acid is potentially hazardous, a safety protocol and appropriate safety measures must be in place before starting the experiment.

9. The yield must be known to add an appropriate amount of $AgNO_3$ to the stock solution.
10. If the concentration of the stock solution is too low there will be too much water added to the nano-DESI solvent, which will solvate chloride. A final concentration of ~2000 ppm is appropriate for the stock solution which adds around 1/200 parts aqueous solvent.
11. Make sure that the liner of the cap is not made of metal to prevent depletion of Ag^+ through redox reactions.
12. To simultaneously account for matrix effects and for quantification an internal standard can be added at this stage.
13. The size of the capillaries can be adjusted to fit the desired pixel size of the ion images. Capillaries with an outer dimension of 150 μm and an inner dimension of 50 μm are regularly used.
14. Cyanoacrylate glue is a good choice but needs to be completely dry before placed in front of the mass spectrometer to avoid detection of polymer peaks originating from the glue.
15. A blank nut can be put at the other side of the union to ensure proper fit to the union.
16. A syringe volume of 500 μL is a good choice for experiments using a flowrate of 0.5 $\mu Lmin^{-1}$.
17. The alignment of the two capillaries is described in detail elsewhere [11].
18. The flow rate is dependent on the dimensions of the capillaries used, for smaller capillaries a lower flow rate should be used.
19. The speed at which the sample is moved determines the size of the pixel in the dimension of sample movement. Typical velocities are in the range of 10–40 μms^{-1} and the sample is generally moved in x-direction under the probe and stepped in y-direction between lines.
20. The lines can overlap to increase the spatial resolution which otherwise is limited on the size of the liquid bridge [18].
21. To simplify data processing, each line should be saved as an individual file. Since the sample is moved with constant velocity the time information in each spectrum is useful for aligning the sampling points in the image.

Acknowledgments

The protocol described in this chapter was developed as part of projects funded by the Swedish Foundation for Strategic research (SSF) and the Swedish Research Council (VR).

References

1. Roach PJ, Laskin J, Laskin A (2010) Nanospray desorption electrospray ionization: an ambient method for liquid-extraction surface sampling in mass spectrometry. Analyst 135: 2233–2236. https://doi.org/10.1039/c0an00312c
2. Laskin J, Heath BS, Roach PJ et al (2012) Tissue imaging using nanospray desorption electrospray ionization mass spectrometry. Anal Chem 84:141–148. https://doi.org/10.1021/ac2021322
3. Lanekoff I, Heath BS, Liyu A et al (2012) Automated platform for high-resolution tissue imaging using nanospray desorption electrospray ionization mass spectrometry. Anal Chem 84:8351–8356. https://doi.org/10.1021/ac301909a
4. Lanekoff I, Thomas M, Carson JP et al (2013) Imaging nicotine in rat brain tissue by use of nanospray desorption electrospray ionization mass spectrometry. Anal Chem 85:882–889. https://doi.org/10.1021/ac302308p
5. Lanekoff I, Thomas M, Laskin J (2014) Shotgun approach for quantitative imaging of phospholipids using nanospray desorption electrospray ionization mass spectrometry. Anal Chem 86:1872–1880. https://doi.org/10.1021/ac403931r
6. Bergman HM, Lundin E, Andersson M, Lanekoff I (2016) Quantitative mass spectrometry imaging of small-molecule neurotransmitters in rat brain tissue sections using nanospray desorption electrospray ionization. Analyst 141:3686–3695. https://doi.org/10.1039/c5an02620b
7. Nguyen SN, Kyle JE, Dautel SE et al (2019) Lipid coverage in nanospray desorption electrospray ionization mass spectrometry imaging of mouse lung tissues. Anal Chem 91: 11629–11635. https://doi.org/10.1021/acs.analchem.9b02045
8. Unsihuay D, Qiu J, Swaroop S et al (2020) Imaging of triglycerides in tissues using nanospray desorption electrospray ionization (Nano-DESI) mass spectrometry. Int J Mass Spectrom 448:116269. https://doi.org/10.1016/j.ijms.2019.116269
9. Mavroudakis L, Stevens SL, Duncan KD et al (2020) CpG preconditioning reduces accumulation of lysophosphatidylcholine in ischemic brain tissue after middle cerebral artery occlusion. Anal Bioanal Chem 413(10):2735–2745. https://doi.org/10.1007/s00216-020-02987-w
10. Zhao J, Xie X, Lin Q et al (2020) Next-generation Paternò-Büchi reagents for lipid analysis by mass spectrometry. Anal Chem 92: 13470–13477. https://doi.org/10.1021/acs.analchem.0c02896
11. Lanekoff I, Laskin J (2014) Mass spectrometry imaging of small molecules. In: He L (ed) Mass spectrometry imaging of small molecules. Humana Press, New York, NY, pp 1–213
12. Duncan KD, Bergman HM, Lanekoff I (2017) A pneumatically assisted nanospray desorption electrospray ionization source for increased solvent versatility and enhanced metabolite detection from tissue. Analyst 142:3424–3431. https://doi.org/10.1039/c7an00901a
13. Duncan KD, Fang R, Yuan J et al (2018) Quantitative mass spectrometry imaging of prostaglandins as silver ion adducts with nanospray desorption electrospray ionization. Anal Chem 90:7246–7252. https://doi.org/10.1021/acs.analchem.8b00350
14. Lillja J, Duncan KD, Lanekoff I (2020) Determination of monounsaturated fatty acid isomers in biological systems by modeling MS 3 product ion patterns. J Am Soc Mass Spectrom 31(12):2479–2487. https://doi.org/10.1021/jasms.0c00194
15. Unsihuay D, Su P, Hu H et al (2021) Imaging and analysis of isomeric unsaturated lipids through online photochemical derivatization of C=C bonds. Angew Chemie Int Ed Engl 60(14):7559–7563. https://doi.org/10.1002/anie.202016734
16. Lanekoff I, Stevens SL, Stenzel-Poore MP, Laskin J (2014) Matrix effects in biological mass spectrometry imaging: identification and compensation. Analyst 139:3528–3532. https://doi.org/10.1039/c4an00504j
17. Frenking G, Fröhlich N (2000) The nature of the bonding in transition-metal compounds. Chem Rev 100:717–774. https://doi.org/10.1021/cr980401l
18. Duncan KD, Lanekoff I (2018) Oversampling to improve spatial resolution for liquid extraction mass spectrometry imaging. Anal Chem 90:2451–2455. https://doi.org/10.1021/acs.analchem.7b04687

Part V

Innovations in Data Analysis

Chapter 18

Extract Metabolomic Information from Mass Spectrometry Images Using Advanced Data Analysis

Xiang Tian, Zhu Zou, and Zhibo Yang

Abstract

Mass spectrometry imaging (MSI) data generally contains large sizes and high-dimensional structures due to their inherent complex chemical and spatial information. A variety of data analysis methods have been developed to comprehensively analyze the MSI experimental results and extract essential information. Here, we describe the protocols of data preprocessing and emerging methods for data analyses, including multivariate analysis, machine learning, and image fusion, that have been applied to the data generated from the Single-probe MSI technique. These strategies and methods can be potentially applied to handling data produced from other MSI techniques.

Key words Single-probe mass spectrometry imaging, Multivariate data analysis, Machine learning, Image fusion

1 Introduction

Mass spectrometry imaging (MSI) techniques are powerful analytical tools for mapping the spatial distribution of molecules (e.g., metabolites, lipids, peptides, proteins, and glycans) on complex biological sample surfaces [1]. To date, a variety of MSI methods have been developed, and they can be divided into two general categories, nonambient and ambient techniques, based on their sampling and ionization environment. Two common nonambient ionization techniques are matrix-assisted laser desorption ionization (MALDI) [2] and secondary ion mass spectrometry (SIMS) [3]. Ambient ionization techniques include desorption electrospray ionization (DESI) [4], nano desorption electrospray ionization (Nano-DESI) [5], liquid micro junction surface sampling probe (LMJ-SSP) [6], laser ablation electrospray ionization (LAESI) [7], and the Single-probe imaging technique [8]. Among these M S

Xiang Tian and Zhu Zou contributed equally to this work.

Young-Jin Lee (ed.), *Mass Spectrometry Imaging of Small Molecules: Methods and Protocols*,
Methods in Molecular Biology, vol. 2437, https://doi.org/10.1007/978-1-0716-2030-4_18,
© The Author(s), under exclusive license to Springer Science+Business Media, LLC, part of Springer Nature 2022

imaging methods, nonambient methods have excellent spatial resolution and sensitivity, whereas the ambient techniques require less or no sample preparation.

MS images contain rich chemical information (e.g., *m/z* values and intensities of large numbers of ions) of spatially resolved individual spots on samples (e.g., tissue slices) [9]. Thus, data analysis methods have become increasingly important to effectively extract essential information from a large amount of MSI data (e.g., several to hundred GB can be typically generated from MALDI MSI experiments [10]). A variety of methods, ranging from direct construction images of selected ions to more advanced strategies (e.g., multivariate analysis, machine learning, and image fusion), have been developed for MSI data analysis. Among them, the most common approach is to directly construct MS images using selected ions to generate their spatial distributions along with their relative intensities (e.g., using heatmap) [4]. However, this method needs prior knowledge to select the target ions, and therefore, it is challenging to generate an adequate number of individual MS images to fully represent the molecular information present in large amounts of MSI data [11]. To efficiently mine information contained in MSI data, a series of multivariate analysis methods have been developed. Among them, a commonly used one is the principal component analysis (PCA) [12]. In PCA results, the majority information of the image can be represented by the score images of the first few components. For example, Dill et al. [13] used PCA to successfully differentiate the bladder cancer tissue from normal tissue in the DESI imaging data. However, PCA-based methods possess multiple drawbacks such as meaningless negative score plots (i.e., the mass intensity cannot be negative), incapability to define the region of interest, and inconsistence with the corresponding loading plot [11, 14]. To overcome these drawbacks, other methods, such as probabilistic latent semantic analysis (pLSA) and nonnegative parallel factors analysis (NN-PARAFAC), can be applied to achieve nonnegative components decomposition; however, the number of components needs to be specified prior to analysis [15]. As another group of Multivariate Analysis methods, clustering analyses (e.g., hierarchical clustering [16] and *k*-means [10]) have been commonly used to segment MSI data according to their profiles. Recently, machine learning methods, including the supervised and unsupervised algorithms, have also been implemented to MSI data analysis due to their advantages of dealing big data. Unsupervised machine learning is used to cluster imaging pixels into different groups according to their similarity without having any prior knowledge [17]. For instance, Makoto et al. [18] utilized unsupervised machine learning to cluster white and gray matters in rat brains. As multimodal analytical techniques have been developed for MSI studies, image fusion tools are accordingly used to combine both MS image and other image data (e.g., florescence image [19], microscopy image [20], electron image [21], and infrared spectroscopic imaging [22]). These multimodal methods take the advantages of different techniques (e.g., rich chemical information from MSI and high spatial-resolution from fluorescence microscopy image), rendering results can hardly be produced from a single method.

In this chapter, we present the step-by-step protocols for data analysis (Fig. 1) utilized to extract metabolomic information from the data acquired using the Single-probe MSI method [19, 23, 24]. Details of this experimental technique have been reported in our previous studies, whereas the design and working mechanisms are briefly described here [8, 25]. The Single-probe is a miniature sampling and ionization device, which can be coupled to MS for different types of experiments, including single cell metabolomics [26–30], extracellular molecules in spheroids [31], and MSI [8, 19, 23, 25]. Although the data analysis methods described here (Fig. 1) are used to analyze the data collected using the Single-probe MSI technique, we expect the general procedures can be potentially applied to handling experimental results acquired from other MSI methods.

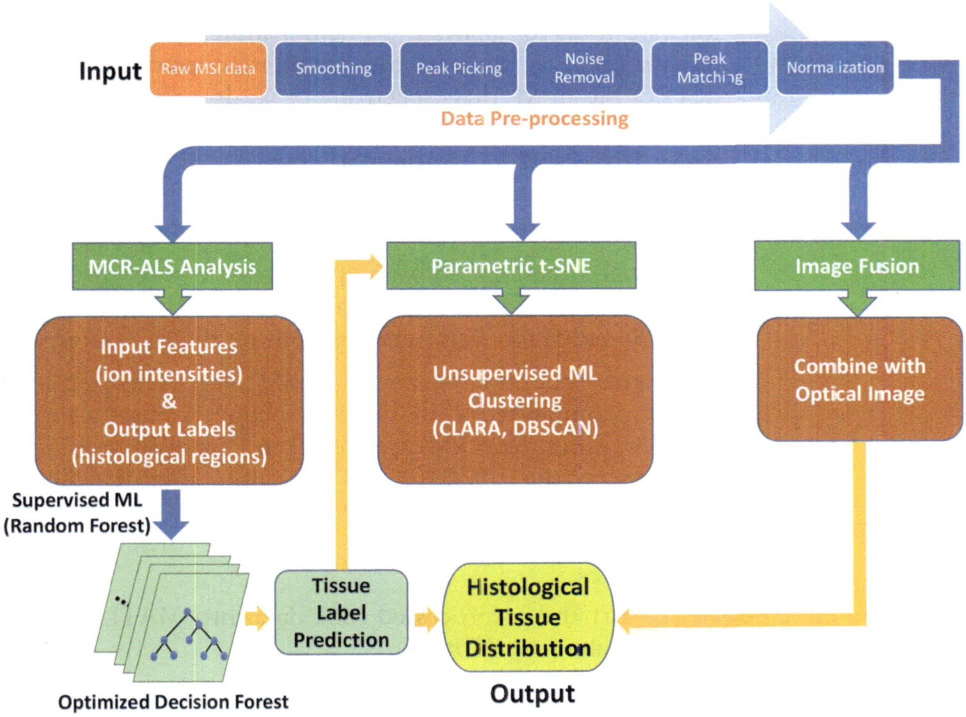

Fig. 1 Flowchart of data preprocessing and data analyses

2 Materials

Several common software packages are needed to perform the data analysis.

1. R (www.r-project.org).
2. MATLAB (MathWorks, Inc., Natick, MA).
3. Molecular Image Fusion (https://fusion.vueinnovations.com/) [20].
4. ProteoWizard Tools (http://proteowizard.sourceforge.net/tools.shtml) [32, 33].
5. imzMLConverter (https://www.cs.bham.ac.uk/~ibs/imzMLConverter/) [34].
6. MCR-ALS GUI 2.0 (https://mcrals.wordpress.com/download/) [35, 36].

3 Method

3.1 Data Preprocessing

The original MSI data may have different formats, depending on the actual mass spectrometers utilized in experiments. Without data pretreatment, the original data may not be directly processed by software tools for analysis. To demonstrate the pretreatment procedures (Fig. 1), we used the .raw files collected from a Thermo LTQ Orbitrap XL mass spectrometer as described in our previous studies [24]. These general protocols and common software tools can be likely used for the pretreatment of MSI data collected using other experimental platforms, whereas the scripts may need to be modified.

1. Install necessary software packages. Ensure that MATLAB (including MATLAB Bioinformatic Toolbox), ProteoWizard, and imzMLConverter are installed on the computer.
2. Convert the .raw files to the .imzML files. This conversion can be accomplished using two steps: convert .raw files to .mzML files (using the mzML converter in ProteoWizards), and then convert the .mzML files to the .imzML files (using imzMLConverter). More detailed user instructions of these software packages can be found on their corresponding websites.
3. Import the preprocessed MSI data into MATLAB to extract *m/z* values (*see* **Note 1**).

   (a) Open MATLAB, and add the imzMLConverter JAR to the Java class path using the following command:

```
javaclasspath('C:\Users\imzML\imzMLConverter.jar');
```

(b) Locate the .imzML file and read them using functions implemented in imzMLConverter. Then, imzML variable should appear in your workspace:

```
filename = 'C:\Users\MS data\mouse.imzML';
imzML = imzMLConverter.ImzMLHandler.parseimzML(filename);
```

The number of rows and columns of your MSI dataset can also be extracted using the implemented function:

```
nColumns = imzML.getWidth();
nRows = imzML.getHeight();
```

(c) Extract *m/z* and intensity values using a loop through all rows and columns (x and y are used here as example) using the following commands:

```
mz_vector = imzML.getSpectrum(x,y).getmzArray();
int_vector = imzML.getSpectrum(x,y).getIntensityArray();
```

4. Mass spectral smoothing: We used Savitzky-Golay filter to perform mass spectra smoothing. Assign values to 'SPAN' and 'DEGREE' parameters, which are two keywords reserved for mssgolay function. For example, SPAN = 7 and DEGREE = 2 have been used in our previous studies. Then execute the following command:

```
smoothed_int = mssgolay(mz_vector, int_vector, 'Span', SPAN,
'Degree', DEGREE);
```

5. Peak picking: Pick peaks from your MSI data using mspeaks function:

```
peak_list = mspeaks(mz_vector, smoothed_int);
```

6. Noise removal: First, set the threshold of signal based on the median absolute deviation (MAD), which can be determined using the following commands:

```
snr_ = 1.4826 * mad(peak_list(:, 2), 1);
SNR_THRES = 1;
```

Any peak with intensity lower than this threshold will be considered as noise. Second, generate the peak list with removed noise using the following commands:

```
mask = peak_list(:, 2) >= SNR_THRES * snr_;peak_list = peak_-
list(mask, :);
```

7. Peak alignment: Because the measured *m/z* value of the same peak may slightly vary from scan to scan, peak alignment is necessary to assign *m/z* values with small difference as the same peak [37, 38]. This can be accomplished using the following command:

```
[CMZ, aligned_peaks] = mspalign(raw_peaks, 'EstimationMethod',
'histogram', ...
'CorrectionMethod', 'nearest-neighbor','Quantile', 0.8);
```

The preprocessed MSI data are then saved in MATLAB workspace as a matrix composed of two dimensions: the aligned peaks and the intensities for corresponding pixels in the MS image. Although the current datasets are ready for analysis, additional steps, such as normalization and logarithmic transformation, are often needed to generate betters results.

8. Intensity normalization and logarithmic transformation: Ion intensity normalization allows for a fair comparison of the relative abundances of metabolites among all scans, whereas logarithmic transformation can improve the illustration of low-intensity ions. Ion intensity normalization to the total ion current (TIC) is commonly used for MS studies (for other normalization methods, please *see* **Note 2**), and this normalization method can be achieved using MATLAB. First, calculate TIC for each pixel by adding up all intensities for aligned peaks:

```
TIC = sum(unnormalized,2);
```

Then, use the calculated TIC to normalize intensities via a loop for all peaks from all pixels:

```
normalized[i,j] = unnormalized[i,j]/TIC[i,1];
```

A logarithmic transformation can be done using the following command:

```
peak_list(:,2)=log(peak_list(:,2));
```

3.2 Multivariate Curve Resolution Alternating Square

Because a certain number of metabolites may possess similar spatial distributions, it is inefficient to manually assign all metabolites into individual groups. Instead, a more practical way is to use multivariate analysis tools to achieve this goal. For example, Multivariate Curve Resolution Alternating Square (MCR-ALS) Toolbox running under MATLAB environment can be used to decompose the preprocessed MSI data matrix into major components and grouped ions. Using MCR-ALS, the MS image data of a mouse kidney slice

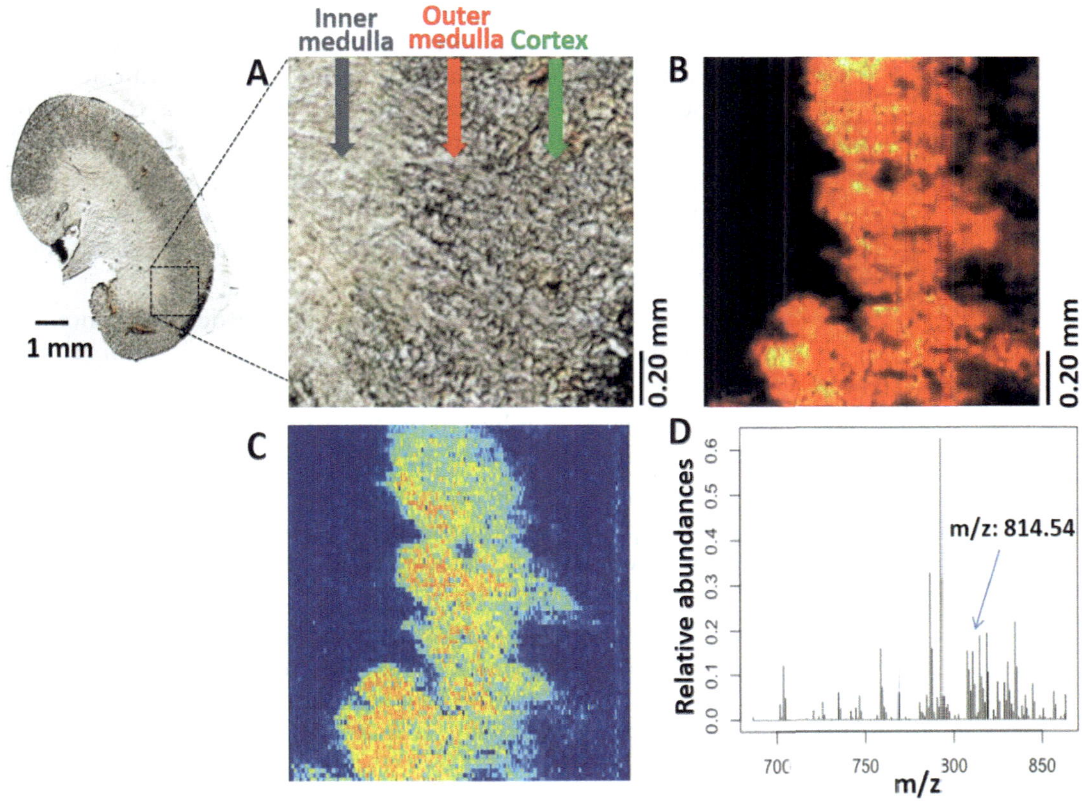

Fig. 2 MSI data analysis using the selected ions or MCR method. (**a**) Optical image of a mouse kidney illustrating three histological regions: inner medulla, outer medulla, and cortex. (**b**) An example of MS image constructed using the selected ion ([PC (38:5) + Na]$^+$; *m/z* 814.5726). (**c**) An example of the spatial distribution pattern obtained from MCR-ALS analysis. (**d**) The mass spectrum of the grouped molecules possessing the similar pattern of spatial distribution shown in **c**. (Reproduced from [24] with the permission from Elsevier)

(Fig. 2a) were decomposed into three major groups representing the inner medulla, outer medulla, and cortex, respectively (Fig. 2c). A representative ion ([PC (38:5) + Na]$^+$) with dominant distribution in the outer medulla is illustrated in Fig. 2b, whereas all ions with high abundances in this region are grouped in Fig. 2d [36, 39].

1. Import the preprocessed datasets into MCR-ALS. Open MCR-ALS GUI 2.0 in the MATLAB bioinformatics Toolbox using the following command:

```
mcr_main
```

A GUI window will appear, and the drop-down menu in the GUI can be used to select and import the preprocessed data matrix. The following procedures can be accomplished using this GUI.

2. Use singular value decomposition (SVD) to analyze and estimate the appropriate number of eigenvalues. Each eigenvalue represents a unique pattern of the spatial distribution of a group of ions and these ions can be assigned to the same group. An eigenvalues representation plot will be produced.
3. Select the transition point. The number of major components is determined at the transition point in the eigenvalues representation plot. Generally, the selected components can explain >90% variance of the matrix.
4. Set the noise level. In the "pure variable detection method" section, concentration was selected in the direction of variable, and the noise level can be set (e.g., 10%).
5. Set the matrix number. In the "selection of dataset" section, set the number of matrix (e.g., 1).
6. Set the number of major components. In the "image distribution maps and spectra section" section, the number of major components can be selected, and the corresponding spectra and image will show up.
7. Export plots. The spatial distribution of each component is exported from the MCR-ALS using the drop-down menu (click the "save" button), and the corresponding ions are plotted as mass spectra using R language.

3.3 Unsupervised Machine Learning

Unsupervised algorithms, such as Cluster Large Applications (CLARA) and Density-Based Spatial Clustering of Applications with Noise (DBSCAN), can be used to analyze the preprocessed MSI data matrix and to cluster pixels on MS images into different groups based on their mass spectra profiles (*m/z* values and intensities), which can be achieved by determining the similarities among all pixels through the calculated Euclidean distance. Using CLARA and DBSCAN (Fig. 3), we performed data clustering of the same experimental results of the mouse kidney (Fig. 2). Six clusters of data were determined by CLARA (Fig. 3a), and each dot represents a pixel on the MS image. To better interpret the results, all pixels were reconstructed into a histological tissue image (Fig. 3b). Similarly, DBSCAN was used to analyze the MSI data (Fig. 3c, d). Similar to MCR-ALS (Fig. 2), both CLARA and DBSCAN exhibit the ability to cluster the MSI data of the mouse kidney into three major components (i.e., inner medulla, outer medulla, and cortex). Because MSI data have high dimensions, dimensionality reduction is needed for most downstream analyses. Here, *t*-SNE was used to perform dimensionality reduction (*see* **Note 3** for other methods).

3.3.1 Cluster Large Applications

1. Open R and load the pretreated data matrix. Find out the exact number of rows 'nrow' and columns 'ncol' after the data matrix is imported in R (*see* **Note 1**).t

```
read.csv ("file location")
```

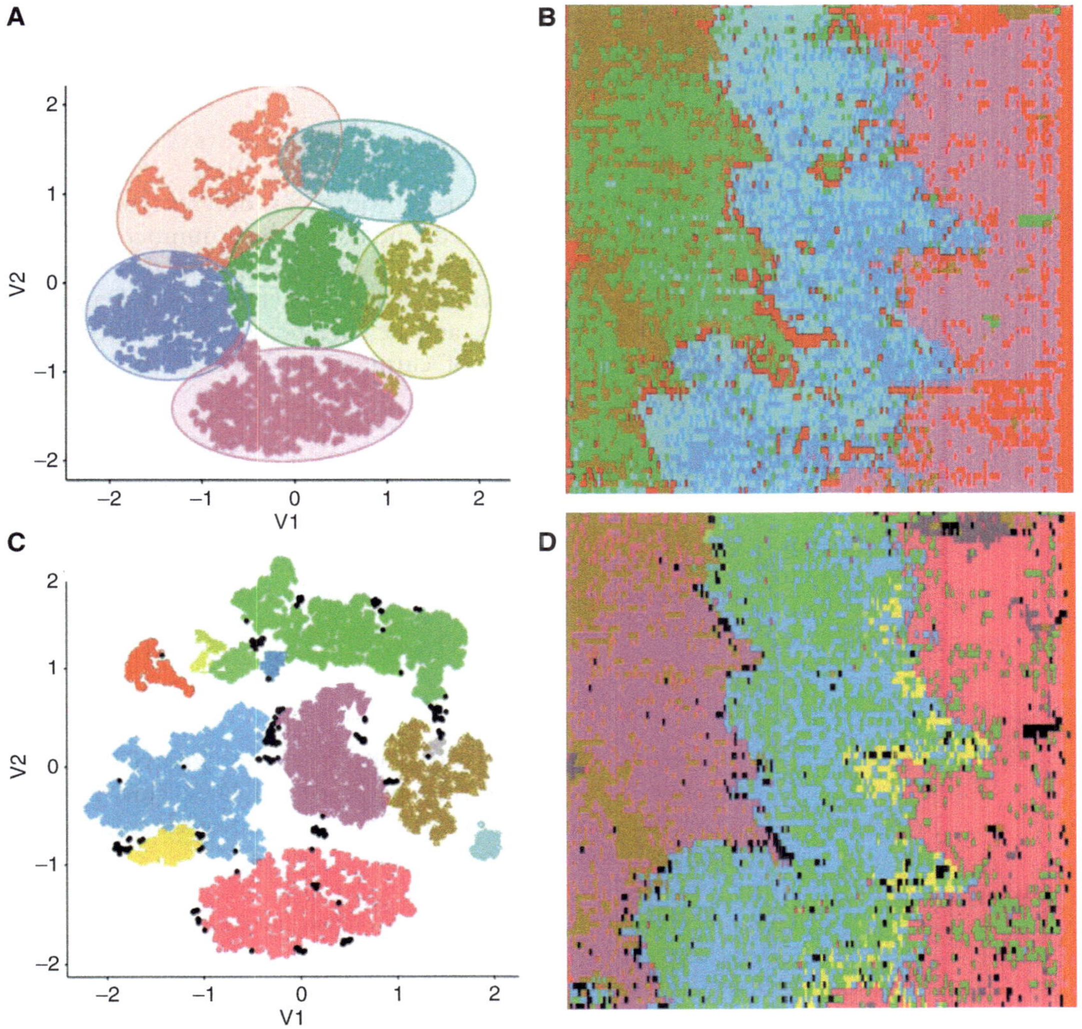

Fig. 3 MSI data analysis using unsupervised ML methods. (**a**) CLARA was used to cluster the MSI data into six major groups upon optimization. (**b**) Histological tissue distribution constructed using CLARA results. (**c**) DBSCAN was applied to cluster the MSI data into seven major groups. (**d**) Histological tissue distribution constructed using DBSCAN results. (Reproduced from [24] with the permission from Elsevier)

2. Use *t*-SNE package in R to reduce the dimensions of MSI datasets. *t*-SNE is widely used for dimensionality reduction due to its advantages of clustering and visualizing highly complex multi-dimensional data [40]. The optimized parameters of *t*-SNE include 'options', statset ('MaxIter', 5000), iterations = 5000, perplexity = 30, Number of Dimensions = 2, Number of PCA components = 50, verbose = 1, learn rate = 500, and exaggeration = 6 (the parameters can be changes for different matrixes). After the dimensionality reduction by t-SNE, the dimension of matrix is reduced to 2.

3. Conduct CLARA analyses. Use "average silhouette width" to find the optimal number of clusters using the following command with *t*-SNE result from last step noted as t_SNE_result:

```
fviz_nbclust (t_SNE_result, clara, method = "silhouette")
```

The average silhouette width for different numbers of clusters can be calculated, and the optimal number of clusters is determined from the calculation producing the minimum average silhouette width. Then, generate CLARA result to cluster t-SNE results, and be sure to use the actual number for optimal number of clusters calculated in the following command:

```
clara_result = clara(t_SNE_result, optimal number of clusters,
samples=50)
```

4. Visualize the clustered data. The clustered data from CLARA can be visualized using function provided in the package 'factoextra' in R:

```
fviz_cluster(clara result)
```

5. Construct the spatial distribution of the clustered data. To exploit the physical meanings of CLARA clusters, their corresponding histological tissue distributions can be plotted using the following command:

```
data = matrix(clarax$clustering, nrow, ncol),
plot_ly(z = data),
```

Replace nrow and ncol with the actual number of rows and columns in your MSI data for plot_ly() function to visualize your result on a two-dimensional (2D) plot.

3.3.2 Density-Based Spatial Clustering of Applications with Noise

DBSCAN, which possesses an excellent capability of picking up system outliers, can be also used as an unsupervised machine learning for data analysis (Fig. 3c, d) [24].

1. Read data matrix and use *t*-SNE package to reduce the dimensions of MSI datasets. This can be accomplished by following the procedures utilized in CLARA analyses.
2. Import "dbscan" and "fpc" packages.
3. Cluster data. Set up "MinPts" and "epsilon." Although there is no golden rule of choosing "MinPts," the square root of sample numbers is commonly used. The value of "epsilon" is determined by the transition point in the kNN distance plot.

For example, "MinPts = 50" and "epsilon = 0.1) were used in our previous studies [24]. DBSCAN can be performed using the following command:

```
dbscan_result=dbscan(t_sne_result, MinPts, epsilon)
```

4. Visualize the clustered data. The clustered data from DBSCAN can be visualized using the following command:

```
fviz_cluster(dbscan_result)
```

5. Construct the spatial distribution of the clustered data. To exploit the physical meanings of DBSCAN clusters, their corresponding histological tissue distributions can be plotted using the following commands:

```
data=matrix(clarax$clustering, nrow, ncol)
plot_ly(z=data)
```

Replace nrow and ncol with the actual number of rows and columns in your MSI data for plot_ly() function to visualize your result on a 2D plot.

3.4 Supervised Machine Learning

Supervised machine learning (ML) has been used in MSI data analysis such as to define the histological subtypes and estimate tumor cell abundances in a gliomas tissue measured by DESI MSI [41]. Multiple supervised ML methods, including support vector machines (SVM) [42] and random forest [43], have been developed for data analysis. In our previous studies, random forest algorithm was selected due to its robustness of classification without over-fitting issues [44], high prediction accuracy, and capability of handling a large number of input variables [45]. As a supervised ML method, prior knowledge is needed for model training. In our studies, we utilized the classification results obtained from MCR analyses to train the random forest model (Fig. 4). Briefly, 3D cultured multicellular spheroid (using *HCT*-116 colorectal cancer cells) were used as tumor models for MSI experiments. MCR was used to decompose MSI data, and pixels from three regions (background, outer region, and inner region) were manually selected to train a random forest model (Fig. 4a). The trained model was then used to label each pixel on the MS image (Fig. 4b).

1. Open R program and load the pretreated data matrix using the following command (*see* **Note 1**):

```
read.csv ("file location");
```

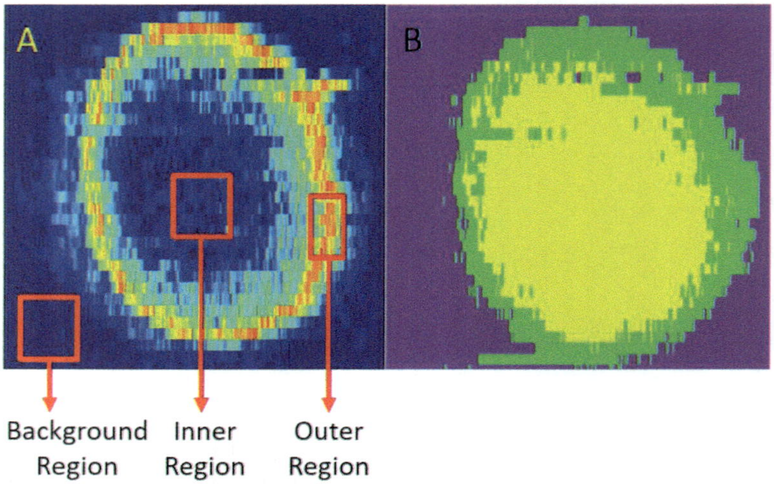

Fig. 4 Using ML methods to analyze MSI data from a spheroid treated by anticancer drug irinotecan (5 μM, 1 h). (**a**) The selection of the training data for supervised ML (random forest) was based on MCR-ALS classification results. (**b**) Color-coded spatial distributions obtained from random forest. (Reproduced from [23] with the permission from ACS)

2. Select the training and testing data. Use MCR-ALS results to assist the data selection for random forest model training. A region of pixels on MS image from the data matrix read in the previous step can be manually assigned as selected data. Then the selected data can be randomly divided into the training data and testing data for the machine learning model [24].
   (a) Calculate the number of pixels in selected data with the following command:

```
data_size=nrow(selected_data)
```

   (b) Randomly divide the selected datasets into the training set and testing set. In our previous study, we selected 90% and 10% for training and testing data, respectively. This ratio can be optimized according to the actual data. To randomly choose 90% of the indexes:

```
indexes=sample(1:data_size, size=0.9*data_size)
```

   (c) Label this 90% portion as training set:

```
training_data=data[indexes,]
```

   (d) Label the pixels not included in this portion as testing set:

```
testing_data=data[-indexes,]
```

3. Train model. The parameters of a random forest model have a significant impact on its performance. Optimized parameters, particularly the number of trees (ntree) and the number of variables randomly sampled at each split (mtry), need to be obtained.

   (a) Build a random forest model using default parameters (without using the optimized ntree and mtry):

```
rf=randomForest(label~., data=training)
```

   A random forest model, rf, is ready for the classification of "label" property.

   (b) Check the prediction accuracy of this random forest model with different number of trees via the following command:

```
plot(rf)
```

   A graph showing "ntree" (x axis) vs the error rate (y axis) is then produced. The optimized "ntree" is the one that produces a low enough error rate but high enough calculation efficiency.

   (c) Optimize "mtry" using tuneRF function, in which the "opt_ntree" should be replaced by the optimized 'ntree':

```
rf2=tuneRF(training_data[,-1], training_data[,1], ntree-
Try=opt_ntree, stepFactor=0.5, trace=T)
```

   A graph showing the "mtry" (x axis) vs the OOB (out-of-bag) error (y axis) is then generated. The optimized "mtry" is the one that produces a low enough OOB error.

   (d) Use the optimized "ntree" and "mtry" to build the optimized random forest model ("opt_model") using the following command:

```
opt_model=randomForest(labels~.,data=training, ntree=opt_n-
tree, mtry=opt_mtry,
type=classification)
```

   Ensure to use the optimized "ntree" and "mtry" to replace "opt_ntree" and "opt_mtry", respectively.

4. Test model.

   (a) Evaluate the performance of the optimized model by checking the prediction accuracy using the testing dataset.

```
prediction_result=predict(opt_model,testing_data)
```

Now, the labels of testing data are predicted along with their correct labels.

(b) Determine if the predicted labels match the correct labels by producing a confusion matrix:

```
t=print(table(testing_data$label,result))
```

In this matrix, the diagonal elements are correct predictions, whereas the off-diagonal elements are incorrect predictions.

(c) Calculate the prediction accuracy using the following command:

```
prediction_accuracy=sum(diag(t))/sum(t)
```

(d) If the prediction accuracies are higher than the preset criteria (e.g., 90%), implement the optimized models to process the rest of data in your MSI datasets with same data structure:

```
prediction_result=predict(opt_model,other_data)
```

(e) As random processes (e.g., the generation of tree, training data, and testing data) are involved, different results can be obtained from each trial. Multiple trials are needed to generate an optimized model with a satisfactory performance.

5. Generate plots. Similar to the procedures to handling the CLARA results, the color-coded labels can be constructed using R language (plot_ly() function in the package "plotly") to easily visualize the spatial distribution of ions in individual categories.

3.5 Image Fusion

Different from the above approaches used to decompose MSI data into subgroups, image fusion is a powerful tool taking the advantages of multimodal analytical techniques (e.g., MS imaging and optical imaging) [20]. MS images provides rich chemical information of samples, but their spatial resolutions are generally limited (e.g., it is challenging to go beyond 1 μm) [1, 46, 47]. In contrast, optical images, such as microscopy photos, can hardly provide versatile chemical information, although they render much higher spatial resolution (e.g., hundreds of nanometers). To fuse both types of images, a model will be built to correlate patterns observed in optical images with those in MS images. The software package, Molecular Image Fusion, was originally designed to fuse optical images with MALDI MS images, in which the coordinates of each

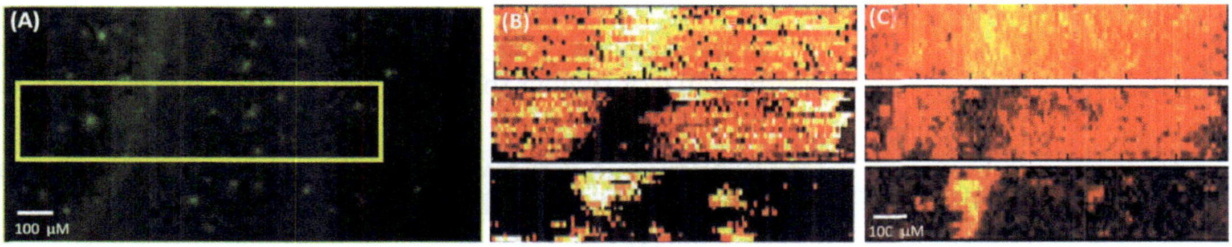

Fig. 5 Fusion of the fluorescence microscopy image and the MS image of a brain slice collected from a mouse with Alzheimer's disease. (**a**) Fluorescence microscopy image of a 5XFAD mouse brain slice stained using thioflavin S. (**b**) Original MS images of metabolites ($[PC(34:1) + H]^+$ (*m/z* 760.5851) (top), $[PC(38:6) + H]^+$ (*m/z* 844.5218) (middle), and $[LPC(18:0) + H]^+$ (*m/z* 524.3693) (bottom)) and (**c**) their fused images. All metabolites were directly identified using MS^2 from the tissue slice, and the results were compared with METLIN. (Reproduced from [19] with the permission from ACS)

pixel are determined. Thus, to accurately correlate coordinates of each pixel to those of optical images, modifications are needed to handle data acquired from our Single-probe MSI method. After training the model with large amounts of data from thousands of pixels, one can combine the chemical information and optical information, leading to enhanced resolution of visualized chemical profiles. In our studies, brain samples were collected from mice with Alzheimer's disease. Fluorescence microscopy image were taken from brain slices (Fig. 5a), and MSI experiments of the neighboring slices were performed (Fig. 5b). Image fusion was then conducted to integrate results from fluorescence microscopy and MS images (Fig. 5c) [19].

1. Extract the .zip file of Molecular Image Fusion into a local folder. The original files include the executable program (i.e., fusiontoolproto.exe), six template files (i.e., mousebrain_msi_info.xml, mousebrain_msi_data.txt, mousebrain_micro_info.xml, mousebrain_micro_data.txt, mousebrain_registration_info.xml, and mousebrain_fusion_params.xml), and a user manual.
2. Prepare the MSI information file using mousebrain_msi_info.xml as a template (*see* **Note 1**).
   (a) Replace <data_label> variable value with a desired name for your MS image.
   (b) Replace <spatial_grid_size> variable values with the number of rows and columns of your MS image's pixels, with a space in between.
   (c) Replace <nr_spatial_grid_elems> variable value with the product of the number of rows and columns of your MS image's pixels.
   (d) Replace <spatial_resolution_um > variable value with the spatial resolution of your MS image.

(e) Replace <nr_obs > variable value with the number of pixels scanned. <nr_obs > variable value should always be equal or smaller than <nr_spatial_grid_elems> variable value.

(f) Replace <nr_vars> variable value with the total number of peaks in your MSI dataset.

3. Prepare the MSI data file using mousebrain_msi_data.txt as a template.

   (a) The first line in this file shows "row,col,peak 1,peak 2,. . ., peak n,". Replace "peak 1," "peak 2," and so on, with the actual *m/z* value of each peak in your MSI dataset.

   (b) Match the total number of peaks with the <nr_vars> variable in your MSI information file.

   (c) For the second line and beyond, each line represents a pixel and should be in form of "row number, column number, intensity 1, intensity 2, . . ., intensity n". "row number" and "column number" need to be represented in integers describing the position of the pixel. Replace "intensity 1," "intensity 2," . . ., "intensity n" with the ion intensity from the pixel for peak 1, peak 2, etc. This file can be automatically generated via an in-house MATLAB script.

4. Prepare the optical image information file using mousebrain_-micro_info.xml as a template.

   (a) Replace <data_label> variable value with a desired name for your optical image.

   (b) Replace <spatial_grid_size> variable values with the number of rows and columns of your optical image's pixels, with a space in between.

   (c) Replace <nr_spatial_grid_elems> variable value with the product of the number of rows and columns of your optical image's pixels.

   (d) Replace <spatial_resolution_um > variable value with the spatial resolution of your optical image.

   (e) Replace <nr_obs > variable value with the number of pixels scanned. <nr_obs > variable value should always be equal or smaller than <nr_spatial_grid_elems> variable value.

   (f) Replace <nr_vars> variable value with the total number of channels (e.g., the default value is 3 for a regular RGB image) of your optical image.

5. Prepare the optical image data file using mousebrain_micro_-data.txt as a template. This file can be automatically generated via an in-house MATLAB script. The first line of this file is

"row,col,Red (R),Green (G),Blue (B)". For the second line and beyond, each line represents a pixel and should be in form of "row number, column number, R-value, G-value, B-value".

   (a) Replace "row number" and "column number" by integers describing the position of the pixel.
   (b) Replace "R-value," "G-value," and "B-value" with the RGB values (to one decimal place) at the specific pixel.

6. Prepare the registration information file using mousebrain_registration_info.xml as a template (*see* **Note 4**). Registration information describes how MSI space is transformed into optical imaging space. The transformation between these two 2D planes can be mathematically described using an affine transformation matrix, which can be found using three specific locations on each of two images and calculations from these six (x,y) pairs on two planes. The matrix can be generated by "cp2tform" function in MATLAB. Replace the <affine_transformation_matrix> variable value with the affine transformation matrix that describes how your MSI space is transformed into your optical imaging space.
7. Prepare the fusion parameter file using mousebrain_fusion_params.xml as a template. Detailed description for each parameter is included in the template file, and parameters can be changed if needed.
8. Place all six prepared files in the same folder containing the executable program, fusiontoolproto.exe.
9. Open a command prompt window (cmd) and change the directory to that containing fusiontoolproto.exe along with six prepared files.
10. In command prompt window, type in "fusiontoolproto file1 file2 file3 file4 file5 file6". "file1" through "file 6" need to be replaced with the names of the six files in the sequence as described above. Then press "enter" for the program to run with provided information.

4 Notes

1. Make sure all the formats, spellings, and sequences are correct when executing the software.
2. In data preprocessing step, normalization can also be performed using other internal standard compounds. For example, standard lipids could be used in the MSI lipidomic studies.
3. In addition to *t*-SNE, other methods, such as UMAP (Uniform Manifold Approximation and Projection) and PCA, can also be used to achieve dimensionality reduction.

4. For the image fusion methods, the registration information, which is used to locate the fuse area of optical image, is critical. Ensure to select the corresponding area to generate the affine transformation matrix.

Acknowledgments

This research project is partially supported by grants from National Institutes of Health (R01GM116116), National Science Foundation (OCE-1634630), and Research Council of the University of Oklahoma Norman Campus.

References

1. Buchberger AR, DeLaney K, Johnson J, Li LJ (2018) Mass spectrometry imaging: a review of emerging advancements and future insights. Anal Chem 90(1):240–265. https://doi.org/10.1021/acs.analchem.7b04733
2. Angel PM, Caprioli RM (2013) Matrix-assisted laser desorption ionization imaging mass spectrometry: in situ molecular mapping. Biochemistry 52(22):3818–3828. https://doi.org/10.1021/bi301519p
3. Benninghoven A (1973) Surface investigation of solids by the statical method of secondary ion mass spectroscopy (SIMS). Surf Sci 35:427–457. https://doi.org/10.1016/0039-6028(73)90232-X
4. Takáts Z, Wiseman JM, Gologan B, Cooks RG (2004) Mass spectrometry sampling under ambient conditions with desorption electrospray ionization. Science 306(5695):471. https://doi.org/10.1126/science.1104404
5. Roach PJ, Laskin J, Laskin A (2010) Nanospray desorption electrospray ionization: an ambient method for liquid-extraction surface sampling in mass spectrometry. Analyst 135(9):2233–2236. https://doi.org/10.1039/c0an00312c
6. Van Berkel GJ, Kertesz V, King RC (2009) High-throughput mode liquid microjunction surface sampling probe. Anal Chem 81(16):7096–7101. https://doi.org/10.1021/ac901098d
7. Nemes P, Vertes A (2007) Laser ablation electrospray ionization for atmospheric pressure, in vivo, and imaging mass spectrometry. Anal Chem 79(21):8098–8106. https://doi.org/10.1021/ac071181r
8. Rao W, Pan N, Yang Z (2015) High resolution tissue imaging using the single-probe mass spectrometry under ambient conditions. J Am Soc Mass Spectrom 26(6):986–993. https://doi.org/10.1007/s13361-015-1091-4
9. Amstalden van Hove ER, Smith DF, Heeren RM (2010) A concise review of mass spectrometry imaging. J Chromatogr A 1217(25):3946–3954. https://doi.org/10.1016/j.chroma.2010.01.033
10. McCombie G, Staab D, Stoeckli M, Knochenmuss R (2005) Spatial and spectral correlations in MALDI mass spectrometry images by clustering and multivariate analysis. Anal Chem 77(19):6118–6124. https://doi.org/10.1021/ac051081q
11. Alexandrov T, Becker M, Deininger SO, Ernst G, Wehder L, Grasmair M, von Eggeling F, Thiele H, Maass P (2010) Spatial segmentation of imaging mass spectrometry data with edge-preserving image denoising and clustering. J Proteome Res 9(12):6535–6546. https://doi.org/10.1021/pr100734z
12. Klerk LA, Broersen A, Fletcher IW, van Liere R, Heeren RMA (2007) Extended data analysis strategies for high resolution imaging MS: new methods to deal with extremely large image hyperspectral datasets. Int J Mass Spectrom 260(2–3):222–236. https://doi.org/10.1016/j.ijms.2006.11.014
13. Dill AL, Eberlin LS, Costa AB, Zheng C, Ifa DR, Cheng LA, Masterson TA, Koch MO, Vitek O, Cooks RG (2011) Multivariate statistical identification of human bladder carcinomas using ambient ionization imaging mass spectrometry. Chem Eur J 17(10):2897–2902. https://doi.org/10.1002/chem.201001692
14. Alexandrov T (2012) MALDI imaging mass spectrometry: statistical data analysis and current computational challenges. BMC

Bioinformatics 13 Suppl 16:S11. https://doi.org/10.1186/1471-2105-13-S16-S11

15. Hanselmann M, Kirchner M, Renard BY, Amstalden ER, Glunde K, Heeren RM, Hamprecht FA (2008) Concise representation of mass spectrometry images by probabilistic latent semantic analysis. Anal Chem 80(24):9649–9658
16. Deininger S-O, Ebert MP, Fütterer A, Gerhard M, Röcken C (2008) MALDI imaging combined with hierarchical clustering as a new tool for the interpretation of complex human cancers. J Proteome Res 7(12):5230–5236. https://doi.org/10.1021/pr8005777
17. Alexandrov T, Kobarg JH (2011) Efficient spatial segmentation of large imaging mass spectrometry datasets with spatially aware clustering. Bioinformatics 27(13):i230–i238. https://doi.org/10.1093/bioinformatics/btr246
18. Nampei M, Horikawa M, Ishizu K, Yamazaki F, Yamada H, Kahyo T, Setou M (2019) Unsupervised machine learning using an imaging mass spectrometry dataset automatically reassembles grey and white matter. Sci Rep 9(1):13213. https://doi.org/10.1038/s41598-019-49819-1
19. Tian X, Xie BE, Zou Z, Jiao Y, Lin LE, Chen CL, Hsu CC, Peng JM, Yang ZB (2019) Multimodal imaging of amyloid plaques: fusion of the single-probe mass spectrometry image and fluorescence microscopy image. Anal Chem 91(20):12882–12889. https://doi.org/10.1021/acs.analchem.9b02792
20. Van de Plas R, Yang JH, Spraggins J, Caprioli RM (2015) Image fusion of mass spectrometry and microscopy: a multimodality paradigm for molecular tissue mapping. Nat Methods 12(4):366–U138. https://doi.org/10.1038/Nmeth.3296
21. Vollnhals F, Audinot J-N, Wirtz T, Mercier-Bonin M, Fourquaux I, Schroeppel B, Kraushaar U, Lev-Ram V, Ellisman MH, Eswara S (2017) Correlative microscopy combining secondary ion mass spectrometry and electron microscopy: comparison of intensity–hue–saturation and Laplacian pyramid methods for image fusion. Anal Chem 89(20):10702–10710. https://doi.org/10.1021/acs.analchem.7b01256
22. Neumann EK, Comi TJ, Spegazzini N, Mitchell JW, Rubakhin SS, Gillette MU, Bhargava R, Sweedler JV (2018) Multimodal chemical analysis of the brain by high mass resolution mass spectrometry and infrared spectroscopic imaging. Anal Chem 90(19):11572–11580. https://doi.org/10.1021/acs.analchem.8b02913
23. Tian X, Zhang G, Zou Z, Yang Z (2019) Anticancer drug affects metabolomic profiles in multicellular spheroids: studies using mass spectrometry imaging combined with machine learning. Anal Chem 91(9):5802–5809. https://doi.org/10.1021/acs.analchem.9b00026
24. Tian X, Zhang G, Shao Y, Yang Z (2018) Towards enhanced metabolomic data analysis of mass spectrometry image: multivariate curve resolution and machine learning. Anal Chim Acta 1037:211–219. https://doi.org/10.1016/j.aca.2018.02.031
25. Rao W, Pan N, Tian X, Yang ZB (2016) High-resolution ambient MS imaging of negative ions in positive ion mode: using dicationic reagents with the single-probe. J Am Soc Mass Spectrom 27(1):124–134. https://doi.org/10.1007/s13361-015-1287-7
26. Sun M, Yang Z (2019) Metabolomic studies of live single cancer stem cells using mass spectrometry. Anal Chem 91(3):2384–2391. https://doi.org/10.1021/acs.analchem.8b05166
27. Standke SJ, Colby DH, Bensen RC, Burgett AWG, Yang ZB (2019) Integrated cell manipulation platform coupled with the single-probe for mass spectrometry analysis of drugs and metabolites in single suspension cells. J Vis Exp 148:e59875. https://doi.org/10.3791/59875
28. Rao W, Pan N, Yang Z (2016) Applications of the single-probe: mass spectrometry imaging and single cell analysis under ambient conditions. J Vis Exp 112:e53911. https://doi.org/10.3791/53911
29. Pan N, Rao W, Standke SJ, Yang Z (2016) Using dicationic ion-pairing compounds to enhance the single cell mass spectrometry analysis using the single-probe: a microscale sampling and ionization device. Anal Chem 88(13):6812–6819. https://doi.org/10.1021/acs.analchem.6b01284
30. Pan N, Rao W, Kothapalli NR, Liu RM, Burgett AWG, Yang Z (2014) The single-probe: a miniaturized multifunctional device for single cell mass spectrometry analysis. Anal Chem 86(19):9376–9380. https://doi.org/10.1021/Ac5029038
31. Sun M, Tian X, Yang Z (2017) Microscale mass spectrometry analysis of extracellular metabolites in live multicellular tumor spheroids. Anal Chem 89(17):9069–9076. https://doi.org/10.1021/acs.analchem.7b01746

32. Chambers MC, Maclean B, Burke R, Amodei D, Ruderman DL, Neumann S, Gatto L, Fischer B, Pratt B, Egertson J, Hoff K, Kessner D, Tasman N, Shulman N, Frewen B, Baker TA, Brusniak MY, Paulse C, Creasy D, Flashner L, Kani K, Moulding C, Seymour SL, Nuwaysir LM, Lefebvre B, Kuhlmann F, Roark J, Rainer P, Detlev S, Hemenway T, Huhmer A, Langridge J, Connolly B, Chadick T, Holly K, Eckels J, Deutsch EW, Moritz RL, Katz JE, Agus DB, MacCoss M, Tabb DL, Mallick P (2012) A cross-platform toolkit for mass spectrometry and proteomics. Nat Biotechnol 30(10):918–920. https://doi.org/10.1038/nbt.2377
33. Kessner D, Chambers M, Burke R, Agusand D, Mallick P (2008) ProteoWizard: open source software for rapid proteomics tools development. Bioinformatics 24(21):2534–2536. https://doi.org/10.1093/bioinformatics/btn323
34. Race AM, Styles IB, Bunch J (2012) Inclusive sharing of mass spectrometry imaging data requires a converter for all. J Proteome 75(16):5111–5112. https://doi.org/10.1016/j.jprot.2012.05.035
35. Klinkert I, Chughtai K, Ellis SR, Heeren RMA (2014) Methods for full resolution data exploration and visualization for large 2D and 3D mass spectrometry imaging datasets. Int J Mass Spectrom 362:40–47. https://doi.org/10.1016/j.ijms.2013.12.012
36. Jaumot J, Tauler R (2015) Potential use of multivariate curve resolution for the analysis of mass spectrometry images. Analyst 140(3):837–846. https://doi.org/10.1039/c4an00801d
37. Nakayama K, Inoue T, Sekiya S, Terada N, Miyazaki Y, Goto T, Kajihara S, Kawabata SI, Iwamoto S, Ikawa K, Oosaga J, Tsuji H, Tanaka K, Ogawa O (2014) The C-terminal fragment of prostate-specific antigen, a 2331 Da peptide, as a new urinary pathognomonic biomarker candidate for diagnosing prostate cancer. PLoS One 9(9):e107234. https://doi.org/10.1371/journal.pone.0107234
38. Inglese P, McKenzie JS, Mroz A, Kinross J, Veselkov K, Holmes E, Takats Z, Nicholson JK, Glen RC (2017) Deep learning and 3D-DESI imaging reveal the hidden metabolic heterogeneity of cancer. Chem Sci 8(5):3500–3511. https://doi.org/10.1039/c6sc03738k
39. Jaumot J, de Juan A, Tauler R (2015) MCR-ALS GUI 2.0: new features and applications. Chemometr Intell Lab Syst 140:1–12. https://doi.org/10.1016/j.chemolab.2014.10.003
40. van der Maaten L, Hinton G (2008) Visualizing data using t-SNE. J Mach Learn Res 9: 2579–2605
41. Eberlin LS, Norton I, Dill AL, Golby AJ, Ligon KL, Santagata S, Cooks RG, Agar NY (2012) Classifying human brain tumors by lipid imaging with mass spectrometry. Cancer Res 72(3):645–654. https://doi.org/10.1158/0008-5472.CAN-11-2465
42. Schölkopf B, Smola AJ (2002) Learning with kernels: support vector machines, regularization, optimization, and beyond. Adaptive computation and machine learning. MIT Press, Cambridge, MA
43. Breiman L (2001) Random forests. Mach Learn 45(1):5–32. https://doi.org/10.1023/A:1010933404324
44. Svetnik V, Liaw A, Tong C, Culberson JC, Sheridan RP, Feuston BP (2003) Random forest: a classification and regression tool for compound classification and QSAR modeling. J Chem Inf Comput Sci 43(6):1947–1958. https://doi.org/10.1021/ci034160g
45. Hanselmann M, Kothe U, Kirchner M, Renard BY, Amstalden ER, Glunde K, Heeren RMA, Hamprecht FA (2009) Toward digital staining using imaging mass spectrometry and random forests. J Proteome Res 8(7):3558–3567. https://doi.org/10.1021/pr900253y
46. Rocha B, Ruiz-Romero C, Blanco FJ (2017) Mass spectrometry imaging: a novel technology in rheumatology. Nat Rev Rheumatol 13(1):52–63. https://doi.org/10.1038/nrrheum.2016.184
47. Bouschen W, Schulz O, Eikel D, Spengler B (2010) Matrix vapor deposition/recrystallization and dedicated spray preparation for high-resolution scanning microprobe matrix-assisted laser desorption/ionization imaging mass spectrometry (SMALDI-MS) of tissue and single cells. Rapid Commun Mass Spectrom 24(3):355–364. https://doi.org/10.1002/rcm.4401

INDEX

Young-Jin Lee (ed.), *Mass Spectrometry Imaging of Small Molecules: Methods and Protocols*, Methods in Molecular Biology, vol. 2437, https://doi.org/10.1007/978-1-0716-2030-4,

© The Author(s), under exclusive license to Springer Science+Business Media, LLC, part of Springer Nature 2022

R

S

T

Z

9781071620298